Mastering Elliott Wave

Version 2.0

Written, Edited and Illustrated
by Glenn Neely

Additional Editing, Production and Computerized Rendering
by Eric Hall

Cover Design and Production
by Evans Design Associates

Published by Windsor Books for the Elliott Wave Institute

Mastering Elliott Wave

Version 2.0
Copyright © 1990 by Glenn Neely

This is a revised edition of a work originally published
under the title of "Elliott Waves In Motion"
Copyright © 1988 by Glenn Neely

For information contact:
Windsor Books
P.O. Box 280
Brightwaters, N.Y., 11718

Manufactured in the United States of America

ISBN 0-930233-44-1

- - - - -

For information on related products and services, write or call:

Elliott Wave Institute

1278 Glenneyre - Suite 283
Laguna Beach, California (CA) 92651
Office: (714) 497-0949 • TeleFax: (714) 497-0983

Disclaimer

Acknowledgments

This book has been nearly a decade in the making. During that period of time, a multitude of friends, clients and associates have provided me with support, helped me solidify my ideas regarding the Wave Theory and have indirectly assisted me in creating the knowledge which is presented in "Mastering Elliott Wave." Some listed may not have realized or, through time, may have forgotten their impact on my work. Here, I would like to remember them for their contributions.

First, the current and past "Elliott Wave Telephone Course" clients too numerous to all be mentioned, but who were instrumental in the culmination of this treatise. A few should be singled out due to their support beyond normal expectations: John Lux (my very first Course client -- you probably thought I forgot about you); Jim Cottone, who gave me the encouragement and assistance to write my first book (no longer available); George Limberg (a great student); John Porter (thanks for all the help during the early newsletter days); Jo Roberts (for all the valuable statistics she provided); Percy Stokes (for his generous support); Jay Walman (thanks for letting me on your show and for being a friend); Robert Blum (a true believer); Bob Stark (who has been my friend longer than I can remember, and has always been there when I needed him); Mark Brantley (one of the most devoted and helpful students I have ever worked with); Mike Walker (who was not only a client, but a close, personal friend from my home town, Lafayette, La. and who I will never forget - thanks for all the free computer time, charts and support); Walter Koval (who, for personal reasons, gives me the greatest sense of accomplishment for having taught); Donald Anderson (who currently holds the record for the highest number of retakes of my Telephone course [six-times]) and finally and especially Jack Sumich (much more than a course client, his belief and support have been incalculable).

Next, the close associates I would like to thank are: David Reif (who provided me with some of the most valuable knowledge I have ever learned about market behavior); Gerald Friedman (it has been a pleasure working with you); Jeannine Henry (hope our relationship continues for a long time); Jim Barth (the best broker I have ever known); Jeff Bower of Financial News Network for allowing me to be on his program and for being the first one to establish my "Guru" status; Richard Mogey of the "Foundation for the Study of Cycles" for maintaining the integrity of that organization and for allowing me to be one of their occasional guest speakers; Larry Jacobs for inviting me to participate in his annual Conferences; Irwin Mintz for his interest in expanding the Institute's horizons; Phyllis Kahn (of Gann World) for being such a caring and generous friend and a true lady; Perrin Gower for his promotional "plugs," ardent letter writing and stimulating conversation; Robert Debnam, my best contact with Europe; Walter Murphy (Merrill Lynch's resident Elliott Wave analyst); Michael Jachtschitz (for his valuable assistance and effort above and beyond the call of duty); Evans Design Associates for creating the powerful graphics on the front cover and finally, Steve Schmidt of Windsor Books for his interest in publishing "Mastering Elliott Wave" and for letting me do it my way.

At this point I would like to thank a special group of friends who were invaluable in finalizing the "infamous" rewrite of Chapter 3 (which took eight months), I call them my "beta" testers: Gary Long (a long-time devotee of the Elliott Wave Institute, his ideas for Chapter 3 were pivotal to its design), John Kozma (the most enthusiastic tester of Chapter 3 who had some great ideas of his own), Richard Schmoeller (thanks for all the time you spent error checking my rough drafts), John Lomas (thanks for your ideas), Bill Wilson (the most energetic Telephone Course client I have ever taught, thanks for your input) and to Bill Abrams (that towering icon of the east coast who benevolently holds the "Big Apple" in the palm of his hand).

This next person has been such a believer and promoter of my approach to Elliott Wave analysis you would think he ran a P.R. firm and was charging me for the exposure. He has done so much to help me over the years and has never asked for anything in return (but my friendship) that I feel it only appropriate to devote an entire paragraph to giving him a long over-due heart-felt "**thank you**." This wonderful fella's name is Tim West. Thanks, Tim, for your assistance at my talks in New York City and Chicago in 1989, thanks for proof-reading so many of my manuscripts over the last three years and for giving me your opinions on promotional literature **and** thanks for all your terrific ideas regarding Chapter 3. Tim West is everything you could expect from a friend and more. Even though we live on opposite coasts and I virtually never see him in person, I feel Tim possesses all the qualities most people could only hope to have. Tim, your a great friend *(I hope this makes up for any occasion in the past [and future] when I have not remembered to thank you for your help)*.

Furthermore, the following personal friends who have never stopped believing in me: Jeannie Bishop (who has meant more to me than she will ever know); all the Aucoins (where can I begin -- Mrs. Aucoin - you're an inspiration to us all, Carla - I'll never forget you, you are the best friend I have ever had, Kevin - the most creative person I know, Kim and Doug, Keith and Mr. Aucoin); Eric Hall (without whom this book could not have been done); Billy Leblanc - thanks for being a friend; Claude - a terrific and long-term friend, Kip-thanks for helping me get started; Travis, Chuck, Hanna and Wright -- the people to know in New Orleans; Myrtle Blanchett, a realtor and an inspiration - she wants things her way and everyone steps aside to let her have it; Renee' Havig, the most feminine and well adjusted Chiropractor on the west coast and to my wonderful brothers and sister (Bobby, Brent, Karen and Mark).

Finally, a standing ovation to my mother, Doris L. Neely, who provided the most ideal upbringing imaginable, who has been an incredible pillar of strength (morally, financially, and psychologically) all these years, and who has unconsciously implanted the drive for perfection and the confidence that enabled me to create this book and who I miss very much *(additionally, who is the best proof-reader in the state of Louisiana, helping to make "Mastering Elliott Wave" as error-free as possible --- any errors still remaining are exclusively of my own making)*.

Glenn Neely - 1989
(revised April 1990)

Table of Contents

General Concepts

Preliminary Analysis

Intermediary Observations

Central Considerations

Chapters 6 through 12 are almost exclusively the

Neely Extensions
of Elliott Wave Theory

6

11

Advanced Progress Label Application

Appendix

Forecast to the Year 2060

Dedication

This book is dedicated to Richard J. Teweles, Charles V. Harlow and Herbert L. Stone. None of these people have I met or talked with, but they are the authors of the book "The Commodity Futures Game." It was while reading that book that I first set my eyes on the phrase "Elliott Wave Theory." For me, that is where it all began.

T his book was written with the intent of significantly advancing the viability of Elliott Wave Theory by introducing a scientific, objective approach to market analysis. This innovative approach, known as the **Neely Method** of Elliott Wave analysis, is the result of a decade of trading, teaching and extensive research by the author. In reading this book, you will quickly realize it is like no other you have ever found on market analysis. Presented herein is the <u>first</u> step-by-step approach to Elliott Wave analysis ever devised. No aspect of the Theory is left for you to ponder. Every detail is thoroughly explained and, in the words of Bruce Babcock (Commodity Traders Consumer Reports), "every sentence is devoted to something important."

The concept of the Wave Theory was originally presented to the investment world by R.N. Elliott in the early 1930's. Going well beyond the original discoveries of Mr. Elliott, this book reveals new techniques, rules and market price patterns never before disclosed to the public. These new discoveries will dramatically increase your market forecasting accuracy and your confidence in trading.

"Mastering Elliott Wave" deals in great detail on the subject and application of the **Neely Method**. In support of such detail, the accurate prediction and profitable trading of a market on a continuous basis *requires* a complex understanding of economic and financial price activity. Another reason for this extensive detail is to ensure the book serves as a valuable guide and reference for years to come.

If you desire the ability to forecast and profit from the complex undulations of economic activity, this book will help you achieve that goal like no other. In addition, every effort has been made to create in "Mastering Elliott Wave" a truly interactive learning environment which will allow you to follow the book's instructions while simultaneously analyzing real-time market action. You will definitely find the **Neely Method** to be challenging and I am confident you will find reading this book to be time well spent. Now, turn the page and prepare yourself for a new adventure; your entrance into the world of high-tech market analysis.

Glenn Neely - December 1987
(revised April 1990)

Elementary Discussions

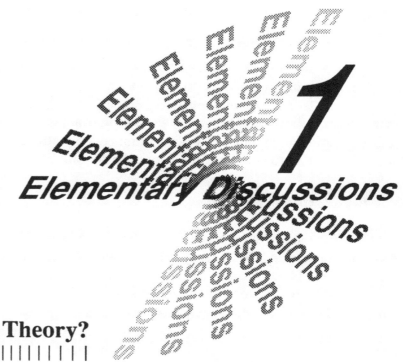

What is the Elliott Wave Theory?

From an Elliott Wave perspective, the plotted price activity of a market is the graphical representation of **crowd** psychology. The Wave Theory describes how local plotted data relates to surrounding data, how data should behave under a multitude of circumstances, when and how psychological trends begin and end, how one psychological environment mandates the unfolding of another and what general shape the price action should exhibit upon completion. In other words, the Elliott Wave Theory organizes the seemingly random flow of market price action into identifiable, predictable patterns based on the natural progression of crowd psychology.

Individuals behave in certain ways depending on their current opinions and feelings. When elated, that individuals actions will invariably be distinctly different from the same individuals actions when depressed. Just the way one person can feel excited or concerned about his or her personal future, the same feelings can permeate and be manifested by society. When a majority of people have similar feelings (favorable, unfavorable, indifferent) about their, or society's, prospects for the future, their cumulative action unfolds in predictable patterns. In addition, the same way an individual can change his or her mind over time (occasionally in an instant), so can the crowd. The Crash of 1987 was a good example of an almost universal, "instant" change in the crowd's opinion of the stock market and the future of the United States economy.

Once a certain psychological mood has run its course, people tend to get bored with an idea or concept (sometimes financially ruined by them), so they grab onto new ideas. It is this changing from one mood to the next that terminates one psychological trend (price pattern) and begins another.

Natural, Non-Periodic Phenomenon

The amount of time consumed by a specific pan-societal mood appears to have no <u>absolute</u> minimum or maximum time requirement, but there are general limits. By analyzing previously completed patterns and applying certain principles of time behavior, it is possible to approximate the *best* time periods for completion of a formation. These techniques, many of which I have developed over the last ten years, will be discussed in later chapters.

The "structure" of market movement should almost always supersede time observations. Some people consider this an annoying aspect of the Theory since they want to know "*when*" to make a trade, *in advance!* Elliott virtually proves that absolute anticipation of market movement is impossible. The greatest degree of predictability occurs immediately after a move has terminated. In other words, you wait for a pattern to complete, then you take action.

A Unique Analytical Tool

Unlike most forms of technical analysis, the Wave Theory was developed strictly from a price behavior standpoint. The Theory helps the analyst realize that whatever the market does, it does for a reason. There is no "chance" in market activity, no aberrations and no unclassifiable price behavior.

A large portion of market technicians spend their time trying to *manipulate* price data to find that "magical" key-indicator. How can a transformation of price data be better than the original data? In other words, price should be considered the ultimate indicator. It is the only indicator of market action that can be absolutely depended on and the only one that *directly* affects your bottom line. It does not make any difference what someone thinks or what an indicator signals, the bottom line is, "what did the market do?" If prices are going up (and you are long), you make money. If prices are going down (and you are long) you lose money, all other indicators notwithstanding.

Some of the remarkable aspects of the Elliott Wave Theory are:

---

1. **Its adaptability to new technological developments and unexpected fundamental news;**
2. **Its thoroughness in describing ALL possible market behavior; and**
3. **Its progressive and dynamic characteristics.**

The Theory inherently indicates man, and his markets, continually charts new territory and types of behavior. It stipulates that at no point in history is a market's action or psychological environment identical to any other period. Similarities are allowed, but not exact duplication. This can be a problem for traders, and especially "system" developers, who endeavor to formulate strategies based on **historic** price action and behavior. Unlike most systems and forms of analysis, the Wave Theory warns the analyst to look for change and warns him when and where a market **will not** behave as it has in the past.

The "*Elliott*" point of view, that "history *does not* repeat exactly," is the perfect explanation of why most mechanical systems (and other forms of analysis based on the expectation of exact repetition) fail. This is particularly evident when a market changes from a bullish to bearish environment, or vice versa.

The computer age has dramatically (and probably permanently) altered certain aspects of the trading environment and thus market behavior. This is an example of why working with historical data way back to "who knows when" will not necessarily make a mechanical system function any better. Ever expanding technology continuously transforms the decision element of the market, *people*. The way people react to, use and learn from new technology cannot be decided by historical study. A form of analysis which explains, categorizes and structures *progress*, not *repetition*, is needed. That is what the Elliott Wave Theory is capable of doing.

The 1987 crash was a perfect illustration of how a study of history could not have prepared one for what happened October 19, 1987. Even the largest declining day during the crash of 1929 was only half as detrimental (percentage-wise) as the one experienced in 1987. Anyone using the 1929 crash as a barometer may have decided to buy the market after it had fallen 10% or so the day of October 19, 1987, expecting no day to be worse than that experienced in 1929. Of course, anyone taking such an approach would have been in for a very unpleasant surprise.

During the 1987 advance, I remember continually hearing people talk about (and seeing ads espouse) the extreme similarities between 1982-1987 and 1920-1929. They assumed everything would continue on course *exactly* (same time and percentage advance), with the 80's bull market repeating the nine year boom as if it were another 1920-1929. Unfortunately, those who believed such an easy comparison of history was their road to riches may now be doing something besides investing in the stock market.

Why Learn Elliott Wave Theory?

To the beginning student, the Theory's complexity, multiple pattern variations, "alternate count" possibilities and apparent subjectivity seem almost insurmountable and appear to nullify the Theory's validity. This is due to the extreme demand of specifics required of the Elliott Wave analyst. All general pattern categories, and each of its variations, requires different relationships, channeling, price behavior and technical characteristics. This creates the illusion that the Theory is purely subjective and can be molded to fit anyone's opinion. While there may have been some justification for this stance in the past, before the publication of this book, this simply is not true anymore. The **Pre-Constructive and Post-Constructive Rules of Logic** (Chapters 3 & 6, respectively), in collaboration with numerous, completely new techniques developed by the author over many years, will greatly enhance your ability to arrive at a single wave pattern conclusion. <u>Caveat</u>: These techniques, and the specifics involved in applying them, are so numerous and varied that their proper application can take years of practice and real-time trading. As a result, until you thoroughly understand the concepts presented in this book, your conclusions may frequently be incorrect.

Multiple Benefits

The long-term benefits of "Mastering Elliott Wave" are numerous. As a business person or investor, you can frequently anticipate major changes in the economic environment allowing you to sidestep disaster or even profit from the winds of change. Proper understanding of crowd psychology (the fundamental basis of the Theory) can help you avoid financially dangerous business dealings at the end of an economic cycle.

Multi-area application

The Wave Theory can be applied to virtually all areas of mass human endeavor: the stock market, commodities, real estate, production of goods and services, etc.; all that is required is accurate, consistent data. The benefits of the Theory's multi-area application are obvious.

Numerous Techniques Unnecessary

When incorporating Elliott's original rules and observations with the **Neely Method** and the new techniques revealed in this book, you have at your disposal a virtually all encompassing description of market behavior and how to decipher it. This allows you to know, under most circumstances, a great deal about the current position of a market without any information other than price action. You will not need to keep track of numerous time consuming (and sometimes subjective) indicators.

Remember, no matter how good a mechanical system or indicator works during a particular period in history, the minute the pattern (which was unfolding during the formation of the system) completes, that system or indicator will usually not work anymore. The environment in which it was designed to perform so perfectly no longer exists and (according to the Wave Theory) will never exist again.

Helpful in Indicator Clarification

If you decide to use other indicators (outside of Elliott) to "clarify" wave counts, the Wave Theory is actually instructive in qualifying *where* and *when* a particular indicator does and does not work.
Example:
Readings of investor sentiment normally indicate a significant market turning point **only** when in overbought or oversold territory (usually considered to be around 75% and 35%, respectively). A comprehensive understanding of Elliott patterns allowed me to recognize in 1986 and 1987 that sentiment could oscillate around neutral readings even if a major market turning point was forming. This was a logical deduction based on an understanding of Horizontal Triangles and Terminal patterns (ex. Gold Jan.-June 1986 and S & P 500 Jan. - Sept. 1987). When most were confused by the lack of public excitement generated during the 1987 stock market advance, I constantly warned clients that was a sign a Terminal Impulse wave (diagonal triangle) started in the last quarter of 1986 and would portend a **crash**, in three months or less, back to 1900 on the DOW and 230 on the S&P 500 Cash Index.

Signals Infrequent, but Reliable

Only after an identifiable pattern is complete is it safe or desirable to enter a market. This helps avoid over-trading and prevents entering the market when there is little potential. On the other hand, it promotes trading when the probabilities are greatly in your favor and risk is at a minimum. The Theory also allows for the placement of very objective stops, enabling you to know *when in time* **and** *where in price* your interpretation is wrong. What else could a trader ask for?

Why the Controversy?

Complexity

The Wave Theory is probably the most <u>complex</u> and all encompassing form of market analysis ever devised. It requires a lot of time and practice to master. As a result, most people do not have the time nor inclination to properly learn or apply it. The Theory takes more time to perfect than most people are willing to commit. In addition, the idea of an analytical technique allowing someone to call market turning points to the day, even the hour, is considered by most to be ridiculous or at least suspect. When a majority of investors have trouble being on the right side of a market, attempting to pick the highest or lowest tick of a trend is not even considered a serious possibility. A Theory which professes to produce such astounding results will always create controversy and be treated with skepticism by the uninitiated.

Public Mind Set

Profitable application of the Wave Theory requires a mentality divergent from the norm. To accurately call market turning points demands you buck the crowd precisely at the time the majority is most convinced an event will occur (you may possibly even be ridiculed for your beliefs, one of the most reliable signs your analysis is correct). Extreme self-confidence is a must; control of greed, strict money management (which includes risk control), an open mind and the ability to change your entire perspective (from bullish to bearish or vice versa) in the blink of an eye are all necessary attributes and abilities of the successful analyst and/or trader. Most of these characteristics are not all present in one person, making it difficult to be a prognosticator **and** profitable investor. Many people can predict future market action when no money of their own is at risk. When trading becomes personal, everything changes.

Years to Master

Even if you had all of the qualities mentioned above, it could take years of practice to confidently and accurately apply the Theory to real-time. Since most people **will not** spend the time necessary to properly learn the Theory, they misapply it and lose money in the market. This creates disgruntled "ex-students" who, of course, blame their misfortune on the Theory, thus continuing the controversy.

Application Requires Time

Even if you have the mentality and the years of experience and knowledge, you still must spend time analyzing everyday or so to stay in practice and keep up with the current position of a market. If you are following numerous markets, this alone can take several or more hours everyday.

On many occasions, I have spent an entire weekend looking at one market, sometimes just one chart, trying to figure a way to fit all the important elements of a good count into the price patterns. Most people want making money in the market (and all other endeavors) to be *easy*. They work all day at another job, go out to eat, go to the movies, then right before going to bed in the evening, spend 5 or 10 minutes on their charts to "decide" what the market is going to do the next day.

This is just one of many reasons most people never make money trading. If you are in the above mentioned category, you are competing against a handful of devoted, full-time professionals who derive their sustenance from profitable trading or investing. The chances of a part-timer consistently winning against this group are small. If you are not able to spend the time required to follow markets properly, it is recommended you find a professional who can advise you on your investments.

Endless Array of Specifics

To apply Elliott Wave properly under all circumstances requires the application of an almost endless array of specific criteria fitting particular situations. Relationships, channeling, the significance of certain turning points to the over all picture, Structure (Impulsive or Corrective and what type of Impulsion and Correction) all need to be considered when piecing together a pattern's count.

Memorization

From the statements on the previous pages, it should be obvious that a large amount of information needs to be memorized in order to hasten the application of important Elliott Wave and Neely techniques to a price series. This memorization is one of the major obstacles most students encounter to successful application of the Wave Theory to real-time action.

Frequent Indeterminacy

The price or time termination area of a trend is not predictable, with a high degree of confidence, until an Elliott Wave pattern is on the verge of completion. Sometimes, confidence is not obtainable until <u>after</u> a pattern has completed. This fact keeps most of the public suspicious about the Theory. Why? If you ask an Elliottician for his prediction of future market activity **before** a pattern is near completion, the probabilities are high his analysis will be wrong, at least in detail. As the market approaches the end of a move, the pattern in progress begins to become clear. As it moves away from the identified turning point of a previous pattern and moves toward the middle of a new pattern, the number of possible patterns in progress again increases. This is the reason experienced Elliott analysts can disagree on the position of a market during certain time periods; each is choosing what he or she considers to be the best scenario of the interpretations possible.

When experts in the same field continuously disagree on the same subject, that is the very foundation of controversy. In a latter chapter, a whole new concept developed by the author, will be introduced to handle these areas of "uncertainty" (see Expansion of Possibilities - Page 12-43).

Difficulty

The extreme complexity, subtlety, and demanding nature of the Wave Theory prevents most people from pursuing this area of analysis. Also, most of the reasons listed throughout Chapter 1 prevent the Theory from being completely computerized. Application of the Theory sometimes demands forms of abstract thought, not a specialty of the computer realm. For example, application of all the rules under certain circumstances may leave no *obvious* possibilities. These situations require personal brainstorming sessions to "unlock" the count or, sometimes, merely patience to wait for a count to reveal itself.

Mastery of this form of analysis will be applicable for the rest of your life. There is no need to concern yourself with the possibility of heavy public participation, most will apply it wrong. As Elliott has become increasingly popular, the wave patterns in most markets, to compensate, have begun to develop on an ever more complex level preventing amateur Elliotticians from correctly deciphering market behavior. Again, this helps to perpetuate the controversy in favor of the Theory's inadequacy or, at the very least, its supposed subjectivity.

Why This Book May Create More Controversy

The degree of detail covered in this book will <u>significantly</u> increase the number of important rules and guidelines that have to be memorized to properly apply the Theory. Most people feel there are <u>already</u> too many rules associated with Elliott Wave.

Many experienced practitioners will disagree with some of the findings in this book proclaiming "my wave behavior criteria is <u>too</u> specific or demanding to be applied to real-time market action." This is absolutely not true. It is understandable, though, how these feelings might have come about in others. Many practitioners have, for years, used data to plot their charts which is inaccurate or improperly plotted *forcing* them into invalid conclusions. This obviously would affect their perception of the Wave Theory. When data is inconsistent (which it will be when collected and plotted incorrectly), it is <u>impossible</u> to analyze a market's action with exactitude. [For a more detailed description of what constitutes good data and how to plot it, see "**How Do You Plot Data To Analyze Waves**," in Chapter 2.]

What Makes Elliott Wave Theory Unique?

Complete Perspective

Unlike most all mechanical systems and many analytical techniques, the Elliott Wave Theory allows the analyst to study any time frame: hourly, daily, weekly, monthly, yearly, etc. Even more interesting, the analyst can study all time frames simultaneously, allowing a decision to be made on which time frame currently provides the greatest trading environment and how all the long and short-term movements interplay with one another.

Quantification of Mass Psychology

Even if a system or technique tells you what to do in the market, rarely will it give you a feeling for the economic outlook of a society or of the psychology overshadowing all market action. Due to the Theory's mathematical quantification of mass psychology, it allows you to experience the economic "boom and bust" phenomenon with greater interest and understanding.

Detailed Categorization

Almost in a class by itself, Elliott allows for the categorization of moves which cover just a few seconds in length up to hundreds of years. Every move on a price chart has its effect on the larger patterns that are evolving and each can be labeled based on its specific impact. If you know what type of pattern is unfolding, you have a better idea of what to expect from that move in the area of speed, complexity, breadth, volume, etc. (many of these areas will be covered later).

Grand Simplification

The whole Elliott Wave analysis process involves the deciphering of small price patterns, combining those with numerous others into larger price patterns and then reducing those complex patterns back into simpler patterns again. This is what makes the Theory so unique; development is the same on all scales.

Clear Delineation of Price Behavior

No matter how small or large a trend turns out to be, you can clearly label the movement as an upward correction in a downtrend, a downward correction in an uptrend, etc.

How Should You Learn the Theory?

Reading this book is a good start! "Mastering Elliott Wave" (the book and the process) requires a lot of real-time practice. Initially, your charts should be constructed on a short-term basis until you grasp the basic concepts. Memorization of numerous rules is required along with an understanding of general shapes and channeling. It is recommended you follow no more than one market until you have memorized the basic rules and can quickly apply them to a market as it unfolds in real-time.

Why Another Book on Elliott Wave was Necessary

Additional Techniques Required

The purpose of this book is to provide new, introductory methods to assure objective application of the Theory and, in later chapters, discuss more subtle qualifying techniques. These will be required by the experienced analyst to enhance the wave tracking process. These techniques can be employed even when the market is developing on a very complicated basis, eliminating the need to resort to other types of less reliable indicators or techniques.

Specific Procedures Described

Despite the preponderance of information now available on Elliott, most people interested in the subject do not know where to begin when trying to apply it to real-time. The approach in this book is to logically progress from the simplest concepts (many of which were developed for the benefit of my course clients), to the more complex as each stage of wave analysis is understood. In addition, exact start-up procedures and data-plotting methods will be explained in detail.

Realistic Diagrams Previously Non-Existent

Even in Elliott's original works, diagrams of market behavior were very poor and unrepresentative of real-time market action. In my early days of fighting with the Theory, those diagrams were a strong influence on my thinking. The idealized patterns presented *never* reflected actual market behavior. Any student familiar with Elliott's original works will greatly appreciate the attention to "real-life" detail present in the charts of this book. In the early sections of the book, the "typical" diagrams are used to instill the general shape of each Elliott pattern, but they are quickly discarded later for more accurate representations.

Disclosure of Advanced Concepts

Even though general descriptions of Elliott's tenets can be found in all writings on the subject, very little is available on advanced techniques and applications. It is in that area that this book really excels.

The difficulty most students have in the early stages of learning Elliott Wave is mostly due to the standard method of presentation. A lot of importance is placed on what I call "Progress Labels" (1,2,3,4,5,a,b,c) too early in most publications. That area of knowledge is of little importance in the early learning stages; it just tends to confuse if discussed too soon. Also, far too much significance is placed on **Degree** in Elliott's work and its spinoffs. If there is any aspect of the Wave Theory that is subjective, Degree is it. For the above reasons, both concepts are presented after more basic ideas are covered.

New Terminology

The difficulty of discussing the Wave Theory over the phone to clients and in book form required the creation of new terminology in order to convey precise meanings. Some coined words and phrases were also needed to describe new discoveries and techniques covered in this book. In addition, a few old terms have been renamed for clarity. Many "Standard and Non-Standard" Elliott patterns have been broken down into more specific sub-categories. This was done to aid the application of "pattern specific" rules and guidelines which will enhance your forecasting accuracy and pattern interpretation.

New Discoveries, *The Neely Extensions*

Initially, I intended to include a separate section on <u>all</u> the new discoveries (channeling, Fibonacci, price, time, structure, patterns, degree, relativity, momentum) I had made over the years. This proved too difficult. For the completeness of each section, the presentation of only *old* information *first* and then *new* information later would have required introducing all the same categories twice. For reasons of continuity and space preservation, each new technique, concept or discovery has been introduced at its proper place with the assumption that the reader will clearly detect the presence of "never before presented" information. The value of these **Extensions** to Elliott's work will be proven to anyone who applies them properly. For those new techniques, concepts and discoveries which <u>could not</u> be included in the normal flow of the text, there are several sections toward the end of the book which present information that has either never been presented before (*i.e., Rule of Similarity and Balance, The Neely Extensions, Complexity Rule, Power Ratings, Logic Rules, Emulation, Reverse Logic Rule, Missing Waves, etc.*) or have only been <u>implied</u> but never explained in detail (*i.e., Structure Series, the dominance of Structure over Progress Labels, Compaction, Progress Label Application, the importance of Channeling, the differences between Extensions and Subdivisions, etc.*).

How I Discovered these New Concepts and Techniques

The Telephone Course

When I started my studies of the Wave Theory, many things appeared to be missing from available literature on the subject. Questions like; "where do you start a chart, how do you organize charts, how do you plot price action, how do you analyze and treat a near vertical advance or decline on a price chart that exhibits no up and down price action?" came to mind. This, along with the frustration I experienced when wrong about the direction of a market's move, kept me searching for answers. Refusing to believe a market could outwit me, I believed with enough time and work the Theory could be reduced to a scientific approach where all market action could be explained and most future action could be anticipated. My purpose was to eliminate any subjectivity in the analysis process, therefore allowing rational, logical trading decisions to prevail.

In an attempt to explain market action which did not fit into the Wave Theory as R.N. Elliott described it, I set out to quantify every unquantified facet. As part of this study, eight years of notes were kept on all manner of market behavior in an attempt to classify and regiment market action above and beyond Elliott's original discoveries.

Most of what I have discovered has come from teaching a "one of a kind" **Telephone Course** on the Theory since 1983. Due to this detailed teaching process, new ways of presenting the Theory to beginners, important wave behavior discoveries (sometimes very subtle and specific), Channeling techniques, Fibonacci relationships and charting approaches emerged. Through constant application and research, I believe I have quantified *almost* every, up till now, unquantified aspect of the Theory.

Long Hours of Work

Since I generally work seven days a week, ten to fifteen hours a day and have been studying this subject since 1980, I estimate I have logged more than 30,000 hours in the study of Elliott Wave Theory. This alone was a good breeding ground for new ideas.

Where is the Theory Applicable?

The Wave Theory manifests itself in markets that have broad public appeal and involvement. Markets that are more susceptible to weather and other natural events are less reliable followers of the Wave Theory since the weather is not dependent on human thoughts or actions. Many individual stock issues are not broadly based enough to set into motion this natural law of mass human psychology and so may not yield consistent results. On the other hand, markets such as **gold**, broadly based **stock averages** or **real estate** (even though reliable data is harder to obtain for real estate), work well since little can happen to change the value of the commodity or index except human action. Virtually any mass human endeavor will exhibit predictable progress as long as there is consistent, reliable data available.

How Should You Work with and Perceive Elliott?

Meticulously

All market action needs to be accounted for and classified to produce bankable interpretations. Counts should be built from the ground up on a "molecular" basis. Do not start the analysis process by trying to interpret a long-term price chart first. The long-term count is the by-product of all the short-term studies. Keep your short-term analysis accurate and the long-term possibilities will usually become evident as they unfold.

Once charts are organized on a daily, weekly and monthly basis, the analyst should patiently work with the short-term data until a clearly definable pattern appears. Later it will be discussed what to do with the completed pattern and what part of the whole the pattern becomes.

The Elliott Wave Theory is the graphic representation of a natural law which is continuously at work in all areas of mass human activity. As an analyst working with the Theory, do not try to create count scenarios or force the market into a pattern which would favor your current opinion of where the market should be headed. Let each pattern evolve, then objectively deduce how that pattern, based on the highest probabilities, fits into a grander scheme.

Elliott frequently allows you the pleasure (intellectually and financially) of calling major turning points to the day or even the hour. At these junctures you will be a "lone voice in the woods." If you have the courage to act on your convictions, the *Elliott Wave Theory* will produce results for you.

Open-Mindedly

To properly apply the Theory requires an open-mind to all possibilities. Never begin the analysis of a market with a preconceived scenario and then attempt to mold an Elliott count which would verify your opinions. Your conclusions should be the by-product of a carefully constructed wave count. Also, try to avoid all optimistic or pessimistic impulses. When all techniques are followed correctly, you do not need to make a guess, the completed wave count will virtually *tell you* what price level the market should reach and in what time frame that should happen.

What's Next?

The next chapter of this book, *General Concepts*, is designed to answer basic questions you may have about the Wave Theory, but - until now - have gone unanswered. The third chapter, *Preliminary Analysis*, describes how to prepare and maintain chart data, how to identify waves, properly observe wave interaction and much more. Chapter 3 contains the most complete and methodical explanation of price analysis available. These techniques will allow the novice to accurately decipher even the most complex market conditions. The fourth chapter, *Intermediary Observations,* discusses how to combine individual waves into acceptable groups to form "standard" Elliott formations. Chapter 5, *Central Considerations*, advances further into the construction of specific Elliott patterns with numerous rules explained to eliminate unsatisfactory wave development. Chapter 6 introduces the *Post-Constructive Rules of Logic* discovered by the author. These rules will describe how to confirm the authenticity of patterns you have identified. Chapter 7, *Conclusions*, will help you solidify your wave patterns, simplify their Structure and prepare them for future use. Chapters 9-12 contain a vast array of mostly new pattern formulation, testing and verification techniques.

General Concepts

What is a "Wave"?

It seems quite remarkable that the basis of the Wave Theory, "waves," has never - to my knowledge - been defined in any literature ever available on the subject. To make sense and be of use, a wave must be described in specific terms with absolute limits. A more general definition will be offered later, after you have been introduced to some additional concepts.

The obvious place to start with the definition of a "wave" is in its simplest form. Since waves refer to the movement of a market, and markets are measured in price, the definition will depend on references to price action. Clearly, the simplest price movement on a chart would be comprised of a straight line of any length moving in any direction - except parallel to the y-axis on an x-y plane (see Figure 2-1 on the next page).

Figure 2-1

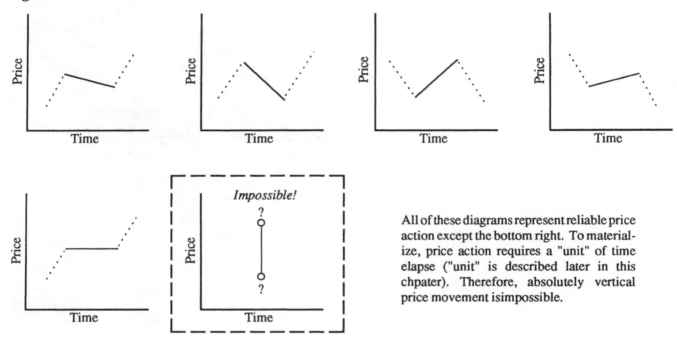

All of these diagrams represent reliable price action except the bottom right. To materialize, price action requires a "unit" of time elapse ("unit" is described later in this chpater). Therefore, absolutely vertical price movement isimpossible.

To describe this type of action (which is paramount to our discussions throughout this book) it was necessary to coin a new term: *Monowave* (see Fig. 2-2). **A Monowave is the movement of a market starting *from* a change in price direction *until the next* change in price direction occurs and is the simplest type of "wave."** In Figure 2-3a, the starting point "m" is where the price changed direction *from* its previous course. The point labeled "n" is where *the next* change in price action occurred. Between any "m" and "n" you usually witness a straight line. Except when the Rule of Neutrality is applied (explained in detail in Chapter 3), even if price action does slow down temporarily and then speeds up again, (see Figure 2-3b) the entire advance or decline should be considered one "wave" (a monowave) until price actually changes direction; when it does, the "Wave" is over.

Monowaves are the micro components (the building blocks) of *all* wave patterns. Learning how to decipher monowaves is your first step toward understanding Elliott Wave Theory. Remember, all market patterns and trends (no matter how large), begin with a single move. Unfortunately, analyzing every move a market makes is tedious, but usually necessary to correctly interpret the larger picture accurately on a consistent basis.

Figure 2-2

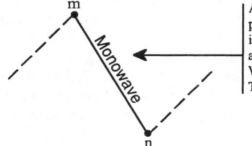

A *monowave* can be of any price <u>and</u> time duration. As long as price continually moves up or down (with no intervening action in the opposite direction), the movement should be considered a monowave. The most important task to learn as an Elliott Wave analyst is to correctly interpet monowaves. The entire Theory is based on the proper identification of monowaves.

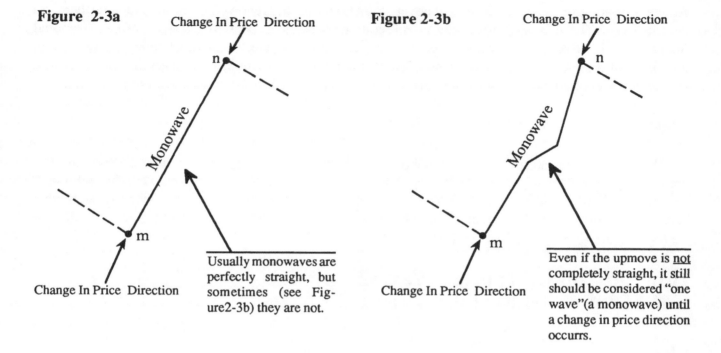

Figure 2-3a

Change In Price Direction

Monowave

n

m

Change In Price Direction

Usually monowaves are perfectly straight, but sometimes (see Figure2-3b) they are not.

Figure 2-3b

Change In Price Direction

Monowave

n

m

Change In Price Direction

Even if the upmove is not completely straight, it still should be considered "one wave"(a monowave) until a change in price direction occurrs.

Why Do Waves Occur?

"Waves" are the result of imbalances which occur between the number of buy and sell orders on an exchange floor or in the open market, for whatever reason. When demand for a product, in relation to its supply, is increasing (or in the case with commodities, the total number of "buy" orders exceeds the total number of "sell" orders) the price goes up; this event could be called an "up-wave." When demand for a product, relative to its supply, is decreasing (in commodities, the sell orders overpower the buy orders), the price falls creating a "down-wave." Each time one force overpowers another, even for the shortest period of time, a change in price direction occurs which starts a new wave. These supply and demand forces are continuously falling into and out of balance on scales of varying magnitude (Degree).

Why are Waves Important?

Finances

Market movement is composed of waves which continually combine to form larger waves with larger reactions. Accurate analysis of a market's wave patterns can give you an overview of current economic conditions. As a businessman, investor or speculator, an understanding of these conditions could directly affect your financial status and standard of living by allowing you to profit from circumstances few are expecting.

Psychology

When a market moves in a particular direction for a long period of time, it tends to attract the media who then alerts the public (the crowd). The longer a market moves in one direction, the easier it becomes for the "crowd" to believe the trend will never stop or believe the trend will continue much farther than current levels. These feelings affect how people conduct business, especially if it is in an area of personal interest. It will affect how they invest, bank, entertain, consume, etc. How much many people spend, and on what they spend it, affects a whole list of other people and industries. Even if a person is not involved in speculating or investing, he can be affected by the events in the stock and commodity markets.

Patterns

Correct combinations of individual monowaves create larger, recognizable patterns. Elliott discovered that certain types of action can be depended upon when specific patterns occur. Through much practice and hard work, it is possible to reach a point where you can predict, occasionally with great accuracy, the price and time extent of a market move. Of course, the regular display and application of such skill can produce great financial rewards.

How Do You Categorize Waves?

Classes

All market movement under Elliott falls into two, logical **Classes**:

Impulsions (trending & terminal patterns): These are the patterns which occur *in the direction of the trend*. When analyzing short-term market movement, Impulsions <u>can</u> be monowaves. If the Impulsion is more complex than a monowave, it will be composed of five (5) segments.

Corrections (non-trending patterns): These are the patterns which occur *against the trend*. Corrections can also be monowaves but, if more complex, will generally appear as sideways consolidations on a price chart and are *usually* composed of three (3) segments.

These concepts will be covered in increasingly greater detail as you progress throughout this book.

Degree

"*Degree*" is a broad, rather nebulous term detailing the hierarchical stratification of waves based on price and time interrelationships. If you closely study a price chart of any market, it should be obvious some waves consume a lot of time, some do not. Some waves consume a lot of price, some do not. A wave which consumes a far greater amount of time and price than another should always be considered a larger degree move. More specific definitions of *Degree* are gradually introduced over the next several Chapters to ease you into the concept.

With waves of different sizes simultaneously present on a chart, *Degree*, to be employable, must be thought of in relative "levels" of <u>price complexity/time duration</u>. *Degree* can **never** be described or applied in absolute terms. In other words, just because a move takes a week, a month, a year, $1, $10, $100 does not make it a particular *Degree*. All that can be derived from a pattern which takes a month or $100 is that a pattern of the same degree will probably take a comparable amount of time and/or price. The specific limits of *Degree* commonality will be partially discussed in *Intermediary Observations* (see "Rule of Similarity and Balance," Chapter 4) along with specific quantification rules.

Once you chooses a *Degree* label and symbol for a particular move *Degree* takes on a more concrete meaning. Why? A frame of reference is created allowing comparisons to be made among all other moves. This frame of reference will provide you with enough information to begin calling certain moves by certain names. The names will imply how a move relates to another, not how big or small it is inherently. Due to the esoteric nature of the *Degree* concept, it is advised if you are a beginning student to not spend much time or pay much attention to this area of the Wave Theory until a greater understanding of more important basics is achieved.

How Do You Label Waves?

Once individual monowaves have been identified, they will have to be *classified*. This involves the application of ***structure*** labels.

Structure Labels

As mentioned in the last section, **all** market action is placed into one of two categories. Out of necessity, each category **must** be given a symbol for use in marking price action. This will enable you to quickly note into which "Class" a particular move falls.

Deciding on the symbols needed to represent wave Structure was not difficult. As indicated in the last section, Impulsive patterns should always be comprised of - or thought to be comprised of - five (5) individual segments. Corrections are usually comprised of three (3) individual segments. Therefore, the symbol to represent Impulsive action is the number five (5) preceded by a colon (i.e., ":5"). To represent corrective action, the symbol ":3" is used. The colon before each number is there to prevent any direct connection to additional forms of labeling which you will implement later in the analysis process.

Structure labels are critical on all levels of application to the Wave Theory. They ***must*** be applied to waves of all shapes, sizes and complexities. Structure labels inform you whether a movement is with or against the trend "of one larger degree." Proper attention to structure labels will aid you in answering that ubiquitous market question, "which way is the market going?"

Organization

After having designated numerous monowaves on a chart with a ":3" or a ":5," you will be able to follow the market's development into more advanced stages. To achieve this, the specific serialization of those ":3's" and ":5's" into groups will be necessary. The groups created from the serialization process will be given different names (another form of categorization beyond Structure labeling) which not only indicate their CLASS (Impulsive or Corrective), but their general appearance (shape).

Progress Labels

As you become more proficient at the basic labeling concepts presented in this book, more advanced forms of labeling will eventually need to be mastered. **Progress Labels** are a dramatic step for the beginner. The complexity of this subject is such that a whole Chapter is devoted to it later. A brief introduction follows to familiarize you with the concept, but complete coverage is scheduled for later.

Progress Labels are an essential and critical part of testing a group of waves for proper development. They direct the analyst "down a narrow path," guiding him through certain conditions and observations which are relevant to the current environment. Progress Labels provide order and limits to a market's movement, allowing the skilled technician to finalize his opinions and prepare strategies to take advantage of a move which "should" take place in the future.

Unlike Structure labels (:3's and :5's), which represent the inherent number of segments in a pattern, Progress Labels identify the position of each segment within a standardized pattern. The five segments in an Impulse pattern are labeled as such: "1,2,3,4,5." Even though there are generally only three segments to a correction, exceptions do occur. For this reason, more than three labels are needed, at times, to describe all types of Corrective behavior. The various segments of Corrective market action are given the first three or more of the following letters: "a,b,c,d,e,x."

When referring to market action with the use of a Progress label, use the word "wave" in front of it. For example, if you are discussing a move which meets all the criteria and inherent characteristics of Progress label "1," you should refer to that move as "wave 1." This may seem mundane to experienced Elliotticians, but it may not be obvious to the beginner, so is included for completeness.

In addition, to make things perfectly clear, each Progress label is placed next to the termination of the market action it represents. The labels are placed in order from left to right as each segment unfolds. In other words, wave-2 cannot be marked until wave-1 is completed.

Finally, keep in mind that Progress labels should not be implemented until you become familiar with numerous preliminary and secondary procedures covered in this book. For this reason, the discussion and application of Progress Labels is reserved for Chapter 5.

What Data Should be Used to Analyze Waves?

Closing Price Data

The use of closing prices on a monthly, weekly, daily, or hourly basis is the most unreliable way to track development of a market from an Elliott Wave standpoint. The closing price of a market has little to do with the action which took place intra-monthly, weekly, daily, or hourly. It will not give you any idea of how high or low the market traded during a particular period (Figures 2-4a & 2-4b illustrate why this is the case), thus making it virtually impossible to apply the wide range of "pattern specific" criteria presented in this book. For example, a market could rally 10 points during the first hour of the day, close near the low of that hour, drop 20 points the next hour, recover and close where it closed the previous hour. Under those circumstances, the plotting of closing data would not accurately represent reality. Therefore, closing figures - over any time frame - should **never** be used unless nothing else is available. If you want to employ hourly data, the correct way of collecting data is revealed on page 2-9 (see **Cash Data**, #2). Proper plotting of data, collected over any time frame, is discussed on page 2-11.

Bar Charts

Bar charts reveal what a market has done in the past by showing the trading range which existed over a standardized portion of time plotted on a single, vertical line (Figure 2- 4c). The problem with bar charts, from an Elliott perspective, is they do not possess the necessary qualities to allow for the analysis of waves in absolute terms. If you were to draw a line across only the highs or lows of some price action, you would have a band of price action similar to Figure 2-4d. This produces the right idea, but is still not the best representation of the **entire** day's price action. The motion is important, but to correctly interpret market action requires data be of a singular nature. The price action on a bar chart consists of dual price factors occurring simultaneously, the high and low. Only one figure per time period should

Figure 2-4a

Intra-Day activity separated by the hour

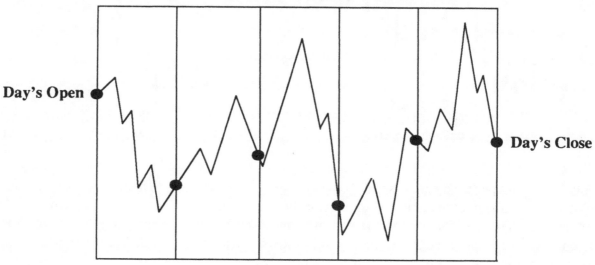

Day's Open

Day's Close

Each heavy **dot** is the final price each hour on the hour (Figure 2-4a). The exact same chart and **dots** are reproduced below, minus the intra-hour action (Figure 2-4b). Each hourly closing was connected with a straight line. The result is not even representative of the above chart. Since market patterns do not develop on a specific time schedule, it is quite illogical to think the closing of each hour (one second out of 3600) is going to occur right at the precise moment a pattern finishes, right at the high or low. Just based on probabilities, the hourly close will virtually <u>never</u> occur at the proper moment to register the highest and lowest price of the day. In future chapters, when relationships are used to verify wave patterns, it is critical the actual highs and lows are recorded (or a reliable representation of those highs and lows). It should be clear from this example that hourly closings are not the way to plot price action.

Figure 2-4b

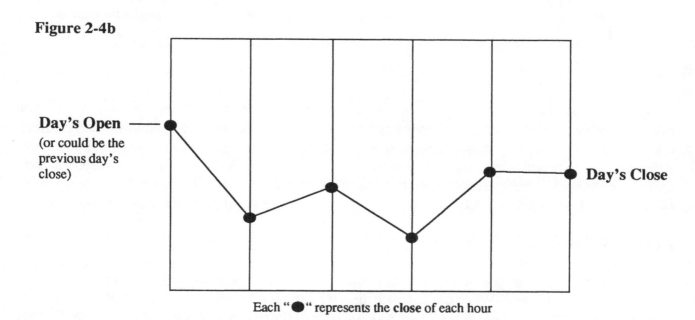

Day's Open —
(or could be the previous day's close)

Day's Close

Each " ● " represents the **close** of each hour

Figure 2-4c

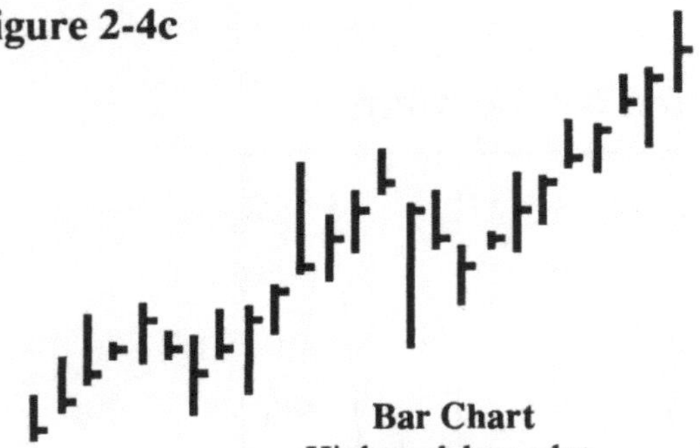

Figure 2-4d

Bar Chart
High and low plot-
ted simultaneously

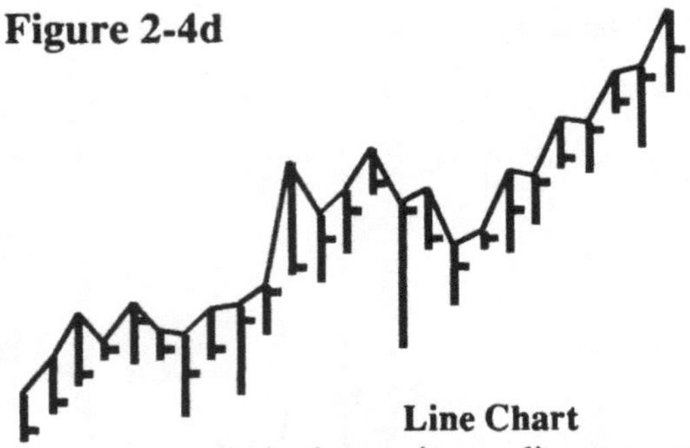

Line Chart
A single, continuous line connects the
highs only of the Bar chart on the left.

Only one figure per time period should be employed so it is possible to decide how each figure changes (if at all) the current expectations and possibilities of the market. This does not mean only one figure per day should be used for your analysis, simply that a single figure <u>per time unit</u> (whatever that time unit is - one day, week, hour, etc.) is optimal. This allows you to arrive at a definitive conclusion on current market conditions and future expectations. A price band (the bar line on a bar chart) does not enable you to make an absolute, immediate decision on current and future market possibilities (see Figure 2-5).

Figure 2-5

Depending on whether the high or low on this day came first, it could change the possibilities from bullish to bearish. If the low came first, a new high here would indicate a fifth wave extension is in progress. If the high came first the previous up move (x-y) might be corrective (3 parts) and the down move could be the start of a new decline. This example illustrates how important is it to know whether the high or the low came first, reinforcing the concept that data needs to be analyzed on a singular basis to have "real-time" value to the trader.

For this example, let us assume the market experiences a violent day. When looking at a chart you notice the price action for one day is above <u>and</u> below the previous day's high and low. This creates a dilemma. Proper analysis requires you know which of the two (the high or the low) came first to make a decision on whether price action is continuing as it should or deviating from the assumed path, indicating changes in opinion are necessary. Therefore, bar charts should be avoided when employing Elliott Wave Theory. Reasons for this will become clearer as you progress through the book.

Futures Charts

Futures charts present a unique problem to the application of the Wave Theory. They contain an element of "deterioration." When dealing with distant Futures contracts in most commodities, the Cash market trades well below the Futures price. As expiration of the Futures contract draws near, the Futures price will approach the cash and, for most markets, will be equal at expiration. Occasionally, supply and demand factors in some of the agricultural commodities can create severe imbalances which prevent the Futures from equaling the cash at expiration. As carrying charges and other costs are subtracted from the Futures contract on a daily basis, a continuous and gradual price erosion occurs during the life of the contract.

Over a period of time, the inherent price distortion in a Futures market can wreak havoc with long-term wave analysis. In addition, some Futures markets (due to limited public participation) are susceptible to intentional or accidental manipulation by "strong hands" (well capitalized individuals or groups). Again, this can create great difficulty in deciphering an Elliott Wave pattern.

Cash Data (the proper data for the creation of wave charts)

Manipulation of a true cash market is far more difficult than a Futures market and requires much more capital (due to a lack of leverage) and time to initiate. Therefore, the return on investment is less when attempting to manipulate a Cash market. Additionally, a Cash position is not as liquid as a Futures position. The result, manipulation of a Cash market is seldom attempted or successful.

The Wave Theory requires a large degree of public involvement (which generally precludes or greatly hinders any manipulation potential) to manifest itself. Since Cash markets always involve more participation by the public (through direct buying and consumption, as is the case with commodities) than Futures markets, wave formations always tend to be more standardized and predictable when cash data is used. In constructing your data series, use cash data whenever possible.

Now, we need to deal with another matter which has to be broken up into three different discussions. As mentioned earlier, proper analysis requires only one price data point occur per predetermined time period. How is this data obtained or decided upon? The next three numerical divisions discuss that problem:

1. For most Cash markets, there is no decision to be made; only a single price quote per day is available. When only one data point per day is available, you simply plot that data over a predetermined, consistent horizontal distance ("base" time unit) which represents one full trading day. Each new data point will be plotted an equal distance from the next (this always applies no matter what time frame is covered on the chart).

2. It becomes more difficult to deal with a Cash market which has data available continuously all day (such as the S&P 500, NYSE, MAXI, Value Line, etc.). First, you need to decide which minimum time frame you plan on following. The shorter the time frame, the more sophisticated the equipment and/or software necessary to track the data and the more charts which will need plotting and deciphering. Following shorter and shorter time frames for trading requires you have a progressively greater understanding of the Wave Theory and a "warehouse" of memorized information at your disposal. The shorter the time frame followed, the greater the likelihood one of the many important, but subtle factors necessary for proper Wave analysis will be overlooked in the rush and excitement of the trading day. We could call this the "Elliott time crunch;" everyone is subject to it, some more than others. Guard yourself against the danger of trying to trade with the Theory over shorter periods of time than can possibly be handled. If you do not, adverse and unexpected market action will cure you of the habit.

The greatest benefit of following a market on a short-term time frame (if you are holding positions overnight) is it increases your ability to determine when a pattern has completed.* Personally, I usually employ the intraday market action as part of my charting arsenal **only when** I am looking for a low risk entry and price patterns indicate the end of a campaign is approaching.

As an analyst, trader or investor, you will need to decide which time frame is best suited to your needs. If you decide one figure a day is all you would like to deal with, you are also deciding the type of move you are interested in is one that lasts several weeks to a few months. You will not be able to day-trade following only one figure per day, obviously. To arrive at one figure per day for a *continuously traded* cash market, add the extreme (real) high and low together and divide by two (2). Plot as described in "How Do You Plot Data," which immediately follows. If you want a little more daily detail by using two figures a day, there are two choices. You can divide the day in two equal parts and take the average of the high and low for each session **or** you can use the high and low for the entire day and plot them in the order they occurred (this last approach is the best). That is, if the high of the day came first, reserve it for the first time slot of the two slot day and plot the low second (or vice versa).

There is a small problem with plotting real-high/low readings that are available for cash market indices based on underlying stocks. Not all stocks open at the same time. The practice of most exchanges is to calculate the opening from the few stocks which are open early in the morning with the previous days closes of all stocks not open during early trading. This forces the cash market to always open approximately where it closed the previous day even if the Futures price "gaps" at the opening. By using averages, you basically eliminate any significant distortion created by this form of tabulating. Without any adjustment, correctly using a cash Index's "real" high/low plot often exhibits horizontal price movement (during the opening period) even when the Futures market opens hundreds of points higher or lower. The only way to avoid this situation is to drop the first 10 to 15 minutes of the trading day from the recorded data, then calculate the high and low for the day. This way you are usually assured most stocks are open and the openings (not the previous day's closes) are part of the current reading.

3. The last situation of concern when plotting data is in reference to 24-hour markets; markets not just traded here in the U.S., but those which continue to trade around the clock on virtually every exchange in the world. Currency markets are a perfect example of a situation. How do you properly keep track of such markets?

Over a period of time, I experimented with plotting the cash markets of Gold, Silver, Australian $/U.S.$ and the Swiss Franc/Australian $ in different ways. When these markets were plotted using data available only during the time they traded in one particular country (the U.S., London, or Australia), the wave patterns were decipherable and followed typical Elliott patterns. Another approach was to equally space the price action of a 24-hour market among the number of markets in which the product is traded and plot that action in a continuous string. Another way is to separate the world into time zones of equal amounts and place the price action of the important world market that falls into that time zone on the chart

---

*Determining when a pattern has completed, to most beginners, is perceived as picking top or bottom tick in a market and is the assumed key to making money in the markets. Always trading with the intent of only getting in at top or bottom tick is a *very* dangerous habit and is not the way large amounts of money are made in this game. Profits can usually be made faster and safer when entering right before a market's acceleration phase. This usually takes place well after the high or low in the market has occurred. In addition, it is generally very important to let the market confirm (even if only in the smallest way) a change in trend has occurred before entering a position. By demanding confirmation before entering a market, you can almost always eliminate the possibility of getting caught on the wrong side of a "gap" opening. Also, if your attention is strictly focused on short-term or overnight trading, large patterns which have been forming for months or years can unwind overnight, creating panic "gap" openings. (A "gap" opening occurs when the market opens substantially higher or lower than the previous day's close and usually indicates the opening is outside of the previous day's trading range.) The Silver market experienced one of these openings two days in a row in mid-April 1987. The "gap" was so large on both days that the market opened "up-the-limit" preventing anyone from exiting the market who was short from the previous day. Wrongly positioned in just one such situation can eliminate an entire year's worth of trading profits (assuming you had a profit to lose).

in a continuous chain of events. The wave patterns under both of the last two approaches tended not to conform to normal Elliott patterns and were quite often indecipherable. The only approach I have not experimented with, that may work, is keeping track of an entire high/low range for all the world markets per every 24-hour time period. The easiest way to do that would be to use the close of the market in your country as the close of each 24-hour period. The minute your country's market closes, the next trading day begins. Just like the other markets which have a high/low range, instead of just one price per day, you could average the range of the 24-hour time period **or** plot the high/low in order (in this situation that might be difficult information to obtain).

Conclusion: From the above experiments, it was proven that each market should be followed in one country **only**! Most people involved in a market get in and out of their positions using the same exchange; this creates multiple closed-loop trading environments for each exchange (and country). In essence, each market, despite its similar name to another market traded in another country, is trading on its own internal perception of the technical and fundamental factors which affect that market in the country of origin. This indicates that as each trading day passes, each country is basically reliving the same market conditions experienced by all the others during their trading day (with minor variations, of course). After a period of time has passed, the technical and fundamental conditions will exert themselves in such a way that each market is poised to make a significant move in unison with all the others. Under these conditions, it takes only one market to "start the ball rolling" and the rest follow suit as they open.

To get to the root of the problem, this is how 24-hour markets should be handled. Keep cash charts on the market you follow, but only during the trading time of your particular country. If it is a very thinly traded vehicle in your country, you may want to follow the cash price in a country where the currency or commodity is more heavily traded and follow the Futures contract in that country.

For protection against a significant move which starts in another country, it is essential you have a brokerage firm which provides an overnight trading desk. This will allow you to place orders to protect against gap openings which may occur in the Futures market of your country. How does it work? The overnight desk keeps tabs on trading as it circles the globe. When a market starts to move in another country, any movement which hits your stop (with the overnight desk) will be activated immediately. You will be out well before the move has had a chance to pick up steam and really cause damage by the time the U.S. markets open. This basically eliminates the possibility of a market opening at a price much worse than where you wanted to be stopped out.

In closing, because of the circumstances surrounding 24-hour markets, if you do not have a broker who can provide you with a night desk to place overnight stops, it is strongly recommended you only day-trade (no position held overnight) OR only intermediate to long-term trade the 24-hour markets. Taking a position with the intent of only holding it overnight OR just for a few days (short-term) can be very risky without the aid of an overnight desk.

How Do You Plot Data?

Once you have decided on the market you wish to follow and the data desired, it is necessary to choose the time frame which will be employed on the chart. A properly plotted data series should be constructed like Figure 2-6 (next page). This diagram shows one figure (represented by each dot) being placed in the center of the time slot which the figure covers. After all figures have been plotted, use a straight line to connect each dot. Monowaves will immediately begin to appear (see Figure 2-7, next page).

Figure 2-6

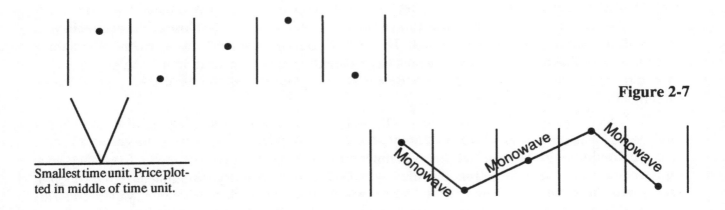

Smallest time unit. Price plotted in middle of time unit.

Figure 2-7

How Many Charts are Necessary?

Normally, a minimum of three charts is necessary on each market followed; *daily, weekly and monthly.* On some of my favorite markets, I keep as many as twenty different charts of varying shapes, sizes and scales, including logarithmic plots of long-term data.

How Complicated Do Waves Get?

Until now, all discussions have been exclusively aimed at monowaves. Monowaves, you remember, are the simplest wave forms possible. More complicated patterns do of course occur. After you understand how to analyze the monowave stage of market development (discussed in Chapter 3), combining monowaves into groups (the purpose of Chapter 4) is the next step in the analysis process.

When combining three or five **monowaves**, you create what I have termed a **Polywave**, one level above monowave development. When you combine three or five polywaves, you generally create a **Multiwave**. Three or five multiwaves generally can be combined to create a **Macrowave**. This is the highest level of development I have named. Anything above that level will continue to be referred to as a **Macrowave**.

Remember, all smaller waves become part of larger patterns. This process continues indefinitely. Therefore, as an answer to the question created by the heading of this section, *there is no limit to how large, time consuming or complicated an Elliott Wave pattern can get.* Despite this fact, all market action remains explainable within the limits of the Theory.

With most theories and techniques, it would be safe to say "as the time period covered increases, the action in a market gets more difficult to decipher." On the contrary, due to increases in the certainty of wave Structure for each segment of a large move, Elliott Wave can actually make long-term prediction easier than short-term prediction.

How is This Knowledge Used to Analyze?

The most important aspect of identifying Impulsive or Corrective monowaves is that it allows you the ability to deduce the trend of a particular market. With a comprehensive understanding of all the rules and factors which influence wave patterns, it is possible to predict the distance a move will travel and approximately how long it will take. You can also accurately describe the psychological environment which should unfold during the development of the predicted pattern. Even more useful and impressive, an accurate view of current action can frequently allow you to predict market action on a day-to-day basis with precision accuracy. Finally, through the process of Compaction (explained in Chapter 7), you can derive larger and larger trends to be taken advantage of.

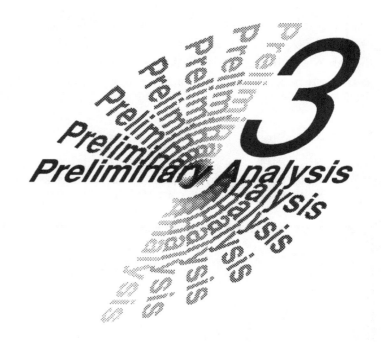

Whether you are a novice or an advanced student of the Elliott Wave Theory, the meticulous study of simple wave patterns is essential to accurate, long-term analysis. Monowaves, being the simplest price action possible on a chart, are the obvious place for you to begin your observations. Unless you learn, early on, how to *indirectly* categorize monowaves into one of the two classes (Impulsive or Corrective), making progress with Elliott is **extremely** difficult. Let's start out by creating the proper charts. Afterward, we can proceed with the observations and quantitative mathematics necessary to decide whether a monowave is *representing* an Impulsion or *representing* a Correction.

Chart Preparation and Data Management

On your initial exposure to this material, you may simply want to follow the illustrations in the book. After reviewing this section, it is recommended you construct your own real time charts as directed.

The first step in making a chart is to pick a market you wish to analyze and then decide on a starting point for your observations. Without a specific starting point, none of the techniques to be described in this chapter will make any sense. Here is what you do: plot a year of data consisting solely of the highest and lowest price of each <u>month</u> in their order of occurrence *(using the techniques described earlier under "What Data Should Be Used To Analyze Waves," page 2-6 and "How Do You Plot Data," page 2-11)*, then look over that wave chart and locate the monthly monowave which is closest to the center of the entire price range (high to low). *[You want a starting point which is not of historical significance. Why? Major turning points are more difficult to work with under Elliott since they are generally associated with abnormal behavior. For our purposes we need to make things as simple as possible.]* Next, ascertain a date which falls around the price center of that middle, monthly monowave.

Once a date has been established, begin constructing your first daily wave chart (in the proper manner) using approximately 60 cash data points covering about 8 inches of horizontal space on a sheet of graph paper. On this scale, monowaves should be relatively easy to identify. After constructing your first chart, pick the earliest important high or low that occurred (see Figure 3-1). From that point you want to draw a <u>second chart</u> that doubles the detail of the first one *(Figure 3-2a starts from the important low marked in Figure 3-1; its time scale has been enlarged to allow for close scrutiny of monowaves).* In other words, the same 8 inches of horizontal space used for the first chart will, on the second chart, cover only half as much time as the first and employ approximately 30 data points.

Figure 3-1

In this example (Figure 3-1), an important low has been made around the center of the time period. From that point a new chart was started (Figure 3-2a) that displays about half the time covered by the first chart.

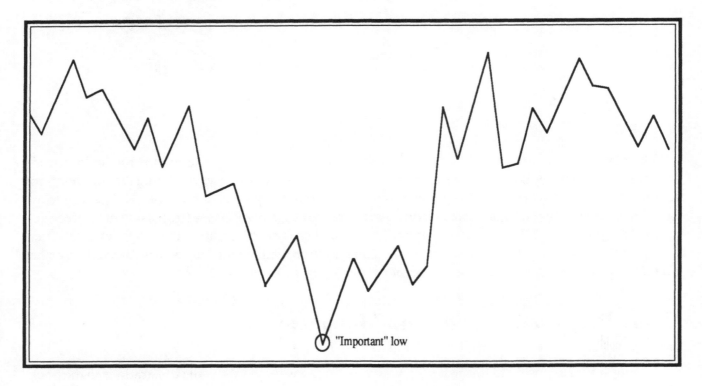

"Important" low

Identification of Monowaves

After the proper charts have been completed, identifying each monowave becomes the next step in the analytical process. Using Figure 3-2a for demonstration purposes, start from the lowest and earliest point (designated "Important Low"). Follow the price action upward, data point by data point, until one data point is lower than the previous, **no matter how minor the amount.** Once that change in price direction occurs, you have just identified your first monowave. Put a **DOT** at the end of that initial monowave (see Figure 3-2a).

Follow the down move from the **dotted** monowave's high (data point by data point). The first data point that is <u>higher</u> than the previous indicates the market completed its second monowave. Put a **dot** next to the second monowave's low (see Figure 3-2a). Continue this identification process until you have **dotted** the end of numerous monowaves. In Figure 3-2b, all monowaves have been identified with their own dot (the <u>end</u> of the last monowave is not **dotted** since a change in price direction has not occurred).

Figure 3-2a

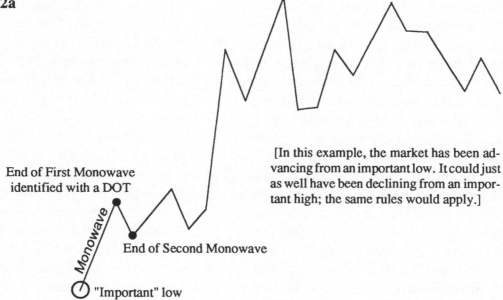

End of First Monowave
identified with a DOT

Monowave

End of Second Monowave

[In this example, the market has been advancing from an important low. It could just as well have been declining from an important high; the same rules would apply.]

"Important" low

Figure 3-2b

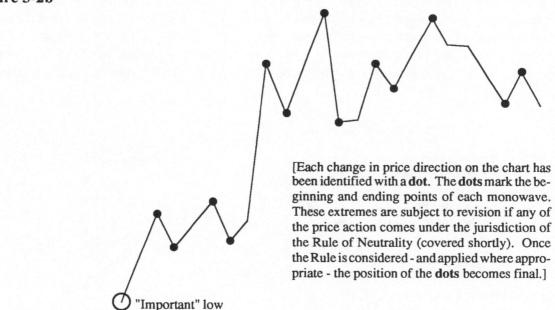

[Each change in price direction on the chart has been identified with a **dot**. The **dots** mark the beginning and ending points of each monowave. These extremes are subject to revision if any of the price action comes under the jurisdiction of the Rule of Neutrality (covered shortly). Once the Rule is considered - and applied where appropriate - the position of the **dots** becomes final.]

"Important" low

Rule of Proportion (proper scaling of charts)

When you pick a particular price and time scale to plot a market, you are also simultaneously deciding which Elliott Wave patterns will be visible and which will not. In other words, each pattern develops based on its own, unique price/time scale. To detect and analyze a particular Elliott Wave pattern, your chart must possess the correct proportions. The proper proportionality of market action is predicated on the pattern developing. Adherence to the Rule of Proportion is essential to the accurate application of the Rule of Neutrality (covered after this section) and to the standardized look of price patterns.

The reason no single **time scale** will work for all analytical purposes is due to the flexible nature of time. Just as Einstein's Theory of Relativity revealed, time is not absolute, but variable. In Einstein's Theory, time was dependent on the velocity of the observer. Under Elliott Wave Theory (as it relates to market action), time is dependent on mass psychology. Due to psychological influences, time expands and contracts as the financial and economic hopes and fears of the masses are recorded through buying and selling on an exchange floor. The reason no single **price scale** will work for all purposes is due to the dynamic and fractal basis of price action under the Wave Theory. Price patterns develop simultaneously on all dimensions large and small.

Before we can discuss the proportions of an ideal Elliott Wave chart, we must define the two ways price action progresses through time; **Directionally** and **Non-Directionally** (this is not to be confused with Impulsive and Corrective activity). Just like all Elliott patterns, if Directional or Non-Directional action begins at a low, it will complete at a high and vice versa. <u>Directional</u> action is always composed of a *series of monowaves* which, on the average and as a whole, create an **increase or decrease** in a market's value (see Figure 3-3 below). As a general rule, the first monowave of a Directional period is retraced no more than 61.8%. Directional action usually terminates when a monowave, traveling in the direction of the Central Oscillation line, is retraced more than 100%. <u>Non-Directional</u> action is composed of a *series of monowaves* which, on the average and as a whole, create **stagnation** of a market's value (see Figure 3-4). The first monowave of a Non-Directional period is always retraced more than 61.8%. Furthermore, all phases (with the occasional exception of one phase) of Non-Directional action <u>must</u> be retraced at least 61.8%. Non-Directional action usually terminates when price action moves beyond the **range** of the entire Non-Directional period by a factor of 161.8% or more (see Figure 3-4).

To properly study a market using the **Neely Method** of Elliott Wave analysis *(discussed in this chapter and throughout the book)*, you must strive to achieve a <u>price-to-time</u> ratio which will produce an angle of ascent or descent predicated on the type of market action unfolding. Irrespective of the price or time consumed by a pattern, this angle should be implemented on charts of **immediate** analytical importance.

Figure 3-3

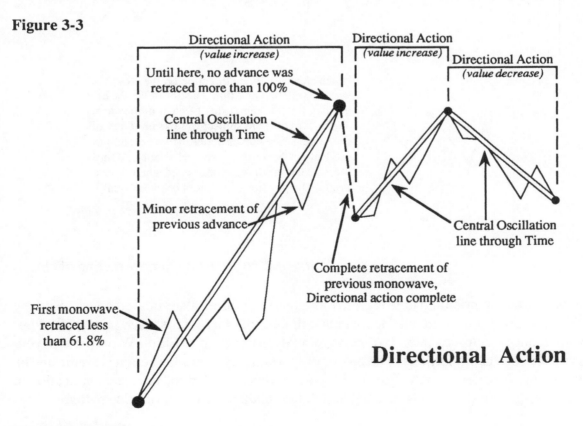

Directional Action

Figure 3-4

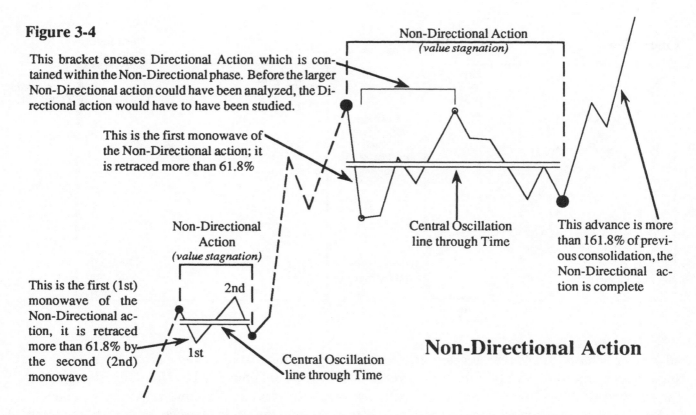

Non-Directional Action
(value stagnation)

This bracket encases Directional Action which is contained within the Non-Directional phase. Before the larger Non-Directional action could have been analyzed, the Directional action would have to have been studied.

This is the first monowave of the Non-Directional action; it is retraced more than 61.8%

Non-Directional Action *(value stagnation)*

2nd

This is the first (1st) monowave of the Non-Directional action, it is retraced more than 61.8% by the second (2nd) monowave

1st

Central Oscillation line through Time

Central Oscillation line through Time

This advance is more than 161.8% of previous consolidation, the Non-Directional action is complete

Non-Directional Action

What is this specific angle of progression needed to view the Elliott Wave phenomenon? When dealing with price action which is **Directional**, plot the data so it travels from the bottom left to top right or top left to bottom right corner of a perfect square. In other words, the data should oscillate around an angle which is approximately 45 degrees from start to finish (see Figure 3-5). This emphasizes the important fact that **the market** dictates how charts should be constructed, thus eliminating the need to subjectively deduce a charts' plotting parameters. This is not meant to suggest that every time the market changes slope or direction, you throw away your current chart and start a new one. Continue to update

Figure 3-5

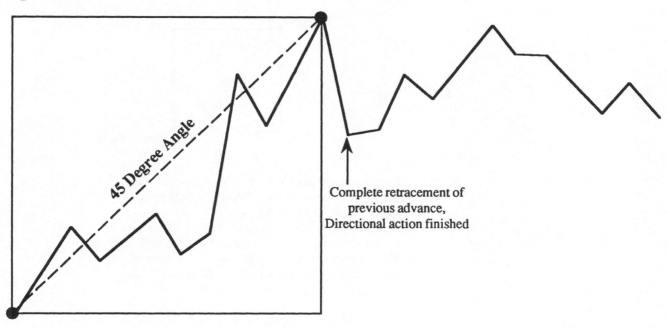

45 Degree Angle

Complete retracement of previous advance, Directional action finished

Figure 3-6

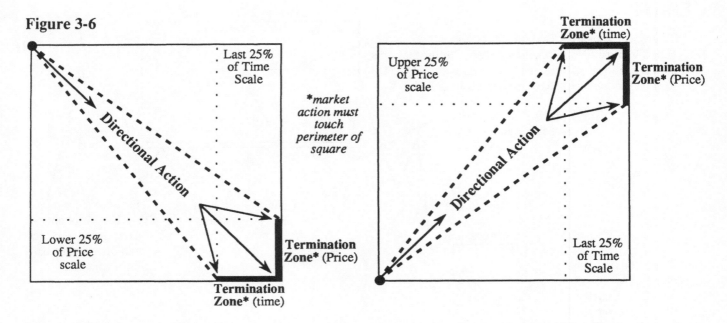

all previous charts, but begin a shorter-term chart at the point the market began a dramatically different angle of progression. An example of this would be a shift from Directional to Non-Directional action (or vice versa). Make use of your longer-term charts to help you identify these important changes.

Realize, when dealing with real-time data, it **is not** essential the angle of ascent or descent is <u>exactly</u> 45 degrees; some variance from ideal is allowed. Figure 3-6 illustrates the price and time **range of error** which is allowed when you attempt to proportion a market's price action. As you can see from the chart, a deviation of no more than 25% (vertically or horizontally) of the length of one side of the square should be your charting goal. An example of a downward sloping Directional period can be found in Figure 3-7. Notice the termination point is slightly to the right of ideal (late), but well within limits of acceptability. Figure 3-8 illustrates a Directional period which terminated a little to the left of ideal (early).

Figure 3-7

Late Termination (acceptable deviation)

The Directional price action within the Square progresses as it should, from the <u>top left to bottom right corner</u>. The pattern terminated a little late, but is close enough to the bottom corner to say that the **Rule of Proportion** has been obeyed. Now, the Elliott Wave pattern within the square can be analyzed with the knowledge that all important Elliott and Neely Rules can be applied with confidence.

Figure 3-8

Early Termination (*acceptable* deviation)

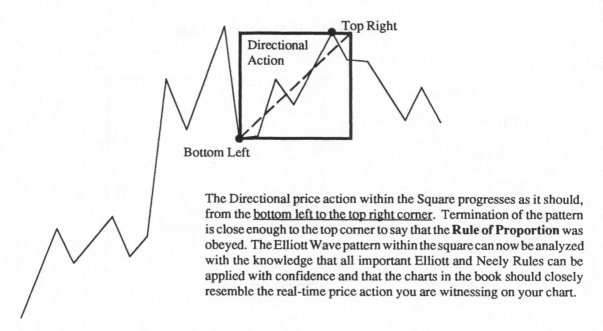

The Directional price action within the Square progresses as it should, from the <u>bottom left to the top right corner</u>. Termination of the pattern is close enough to the top corner to say that the **Rule of Proportion** was obeyed. The Elliott Wave pattern within the square can now be analyzed with the knowledge that all important Elliott and Neely Rules can be applied with confidence and that the charts in the book should closely resemble the real-time price action you are witnessing on your chart.

Figure 3-9 illustrates the **incorrect** application of the Rule of Proportion. Even though the market advances into the top-right corner of the square and the action oscillated around a 45 degree advancing angle, the Directional period ended too early (the Directional action in Figure 3-9 ended at the point identified with an asterisk). To adjust, Figure 3-9 should be redrawn to look like Figure 3-5 (page 3-5).

Figure 3-9

Improper Termination (*unacceptable* deviation)

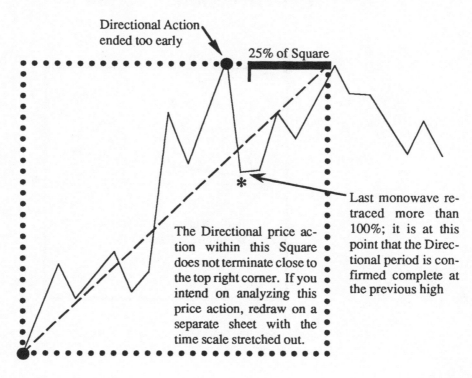

Directional Action ended too early

25% of Square

The Directional price action within this Square does not terminate close to the top right corner. If you intend on analyzing this price action, redraw on a separate sheet with the time scale stretched out.

Last monowave retraced more than 100%; it is at this point that the Directional period is confirmed complete at the previous high

Figure 3-10

Non-Directional Action

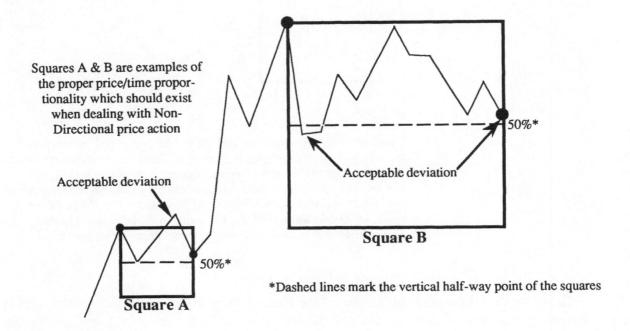

Squares A & B are examples of the proper price/time proportionality which should exist when dealing with Non-Directional price action

Acceptable deviation

Acceptable deviation

50%*

Square B

Square A

50%*

*Dashed lines mark the vertical half-way point of the squares

Figure 3-11

Non-Directional Action

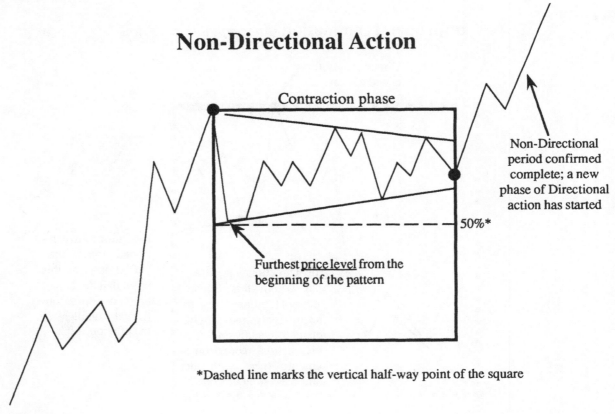

Contraction phase

Non-Directional period confirmed complete; a new phase of Directional action has started

50%*

Furthest price level from the beginning of the pattern

*Dashed line marks the vertical half-way point of the square

When analyzing price action which is **Non-Directional**, you want the up and down periods to cover approximately <u>half</u> of a perfect square (see Figure 3-10). Notice how the Non-Directional action in Figure 3-10 (**Square A**) terminates its horizontal period around the vertical 50% retracement point of the square. Even when a Non-Directional period consumes a large amount of time and contains many monowaves, the goal is the same (see **Square B**). Sometimes this goal may be impossible to obtain if the pattern continuously contracts over a period of time. For those situations, make sure the furthest price level from the beginning of the pattern falls around the 50% point (see Figure 3-11).

To synopsize, when it is time to construct a **new chart**, the Rule of Proportion directs you to design the price and time scale to fit the requirements of the pattern forming. This will help your charts look like those in this book, thus allowing for the most direct comparison and accurate analysis of the market action. It will also prepare your chart's data for the possible application of the Rule of Neutrality.

Rule of Neutrality

After **marking** every change in price direction on your chart, you will find that most monowaves travel at a *diagonal* to the price/time axis. Occasionally, you will find a monowave <u>containing</u> price action which is progressing more horizontally than vertically. It is that type of behavior which comes under the jurisdiction of the Rule of Neutrality. The Rule of Neutrality explains how to deal with these problem monowaves and the **dots** which define their borders.

There are several areas of sideward price action which could come into question if we restudied Figure 3-2b, page 3-3 (reproduced below as Figure 3-12). Those areas have been designated in Figure 3-12 with empty circles. We will spend some time later analyzing those areas. First, a detailed discussion of the Rule of Neutrality - and how you should apply it to real-time - is necessary.

Figure 3-12

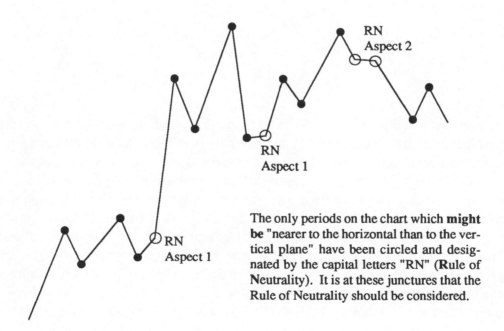

The only periods on the chart which **might be** "nearer to the horizontal than to the vertical plane" have been circled and designated by the capital letters "RN" (Rule of Neutrality). It is at these junctures that the Rule of Neutrality should be considered.

Horizontal price action can separate two waves that are traveling in opposite (**counter**) **directions** (see Figure 3-13a) or two waves that are moving in the **same direction** (see Figure 3-13b). To spot these two types of behavior and to make the Rule of Neutrality useful and applicable, the term "horizontal" must be defined. Creating **perfectly** horizontal price action on a chart requires the same data point be plotted twice in succession. Perfectly horizontal price action is not the only action which can qualify for consideration under the Rule of Neutrality. The term "horizontal" describes price action which operates within limits on either side of perfection; these specific limits are covered below.

Figure 3-13a

Figure 3-13b

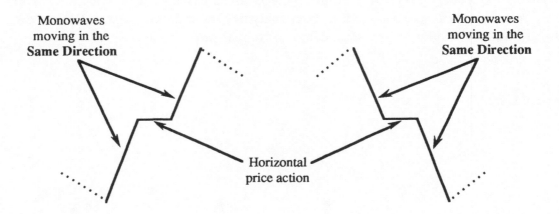

If you discover price action which appears to be moving more horizontal than vertical, apply the following techniques. Draw a vertical (90°) and horizontal (0°) axis on your chart beginning from the start of the suspected price action. Next, if the monowave is moving downward, draw a line from the *origin* of the plane through the lower right quadrant (see left-hand side of Figures 3-14a and 3-14b); if the monowave is moving upward, draw a line from the *origin* of the plane through the upper right quadrant (see right-hand side of Figure 3-14a and of 3-14b). This will create a 45 degree angle to the time (0°) axis. The result is the separation of the quadrant into two equal parts. If price is moving **downward and is <u>on or below</u>** the 45 (top left, Figure 3-14a and 3-14b) <u>or</u> is moving **upward and is <u>on or above</u>** the 45 (top right, Figure 3-14a and 3-14b), do not consider the Rule of Neutrality. If market action is moving **downward and is above** the 45 (bottom left, Figure 3-14a or 3-14b) <u>or</u> is moving **upward and is below** the 45 (bottom right, Figure 3-14a or 3-14b), the Rule should at least be considered. The closer price action moves to the horizontal axis, the more likely the Rule will take effect.

Figure 3-14a

Counter Directional Monowaves

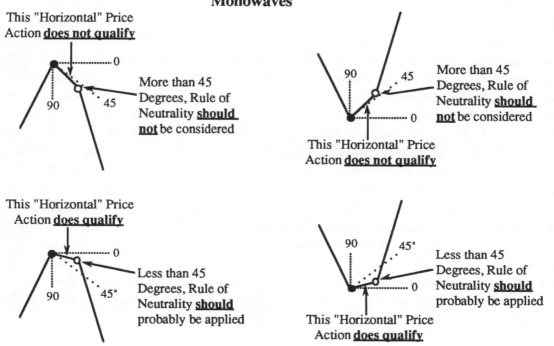

This "Horizontal" Price Action **does not qualify**

More than 45 Degrees, Rule of Neutrality **should not** be considered

This "Horizontal" Price Action **does not qualify**

More than 45 Degrees, Rule of Neutrality **should not** be considered

This "Horizontal" Price Action **does qualify**

Less than 45 Degrees, Rule of Neutrality **should** probably be applied

Less than 45 Degrees, Rule of Neutrality **should** probably be applied

This "Horizontal" Price Action **does qualify**

Figure 3-14b

Same Direction Monowaves

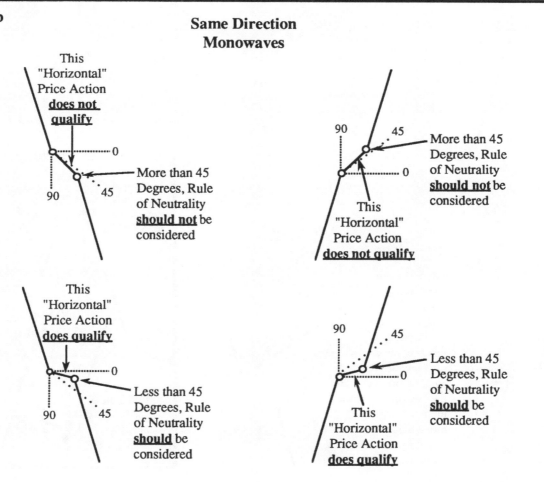

This "Horizontal" Price Action **does not qualify**

More than 45 Degrees, Rule of Neutrality **should not** be considered

More than 45 Degrees, Rule of Neutrality **should not** be considered

This "Horizontal" Price Action **does not qualify**

This "Horizontal" Price Action **does qualify**

Less than 45 Degrees, Rule of Neutrality **should** be considered

Less than 45 Degrees, Rule of Neutrality **should** be considered

This "Horizontal" Price Action **does qualify**

If price action indicates you should consider the Rule of Neutrality, it becomes necessary to observe pre- and post-horizontal action to decide which **Aspect** of the Rule applies to the current situation. If the horizontal price action separates two <u>counter-directional</u> monowaves, Aspect 1 of the Rule applies (see Figure 3-15a). If the horizontal period separates two monowaves which are moving in the <u>same direction</u>, Aspect 2 applies (see Figure 3-16). <u>NOTE</u>: *An easy way to decide which Aspect of the Rule applies in any situation is to count the number of empty circles required to encase the horizontal price action. If only* **one** *circle is placed on the price action, Aspect-1* **(one)** *applies. If* **two** *empty circles must be drawn, then Aspect-2* **(two)** *is in effect (restudy Figure 3-12 [page 3-9] while considering this concept).*

Aspect-1 allows for the termination of monowaves to occur above lows and below highs (Figure 3-15a). Aspect-2, applicable only after the fact, allows the <u>analyst</u> to divide a single advancing or declining phase (under the proper conditions) into three smaller segments (Figure 3-16). **WARNING**: Aspect-1 should not be applied if the monowave, whose dot is being considered for transference, is retraced less than 61.8% and then the market exceeds the end of said monowave (see Figure 3-15b).

In simpler terms, Aspect-1 states that "when dealing with horizontal (or near-horizontal) price action which *separates* two waves moving in **opposite** directions, resolve the first monowave to the far right, at the termination of the horizontal price action." Aspect-2 states that "when horizontal (or near horizontal) price action is found separating two waves moving in the **same** direction, the horizontal action can be ignored, leaving just one larger monowave to contend with, or it may be employed to create three smaller monowaves.

Figure 3-15a

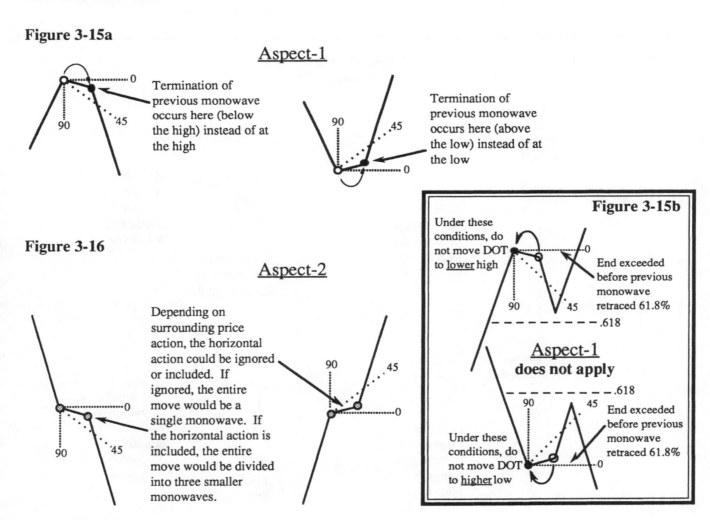

Aspect-1

Termination of previous monowave occurs here (below the high) instead of at the high

Termination of previous monowave occurs here (above the low) instead of at the low

Figure 3-16

Aspect-2

Depending on surrounding price action, the horizontal action could be ignored or included. If ignored, the entire move would be a single monowave. If the horizontal action is included, the entire move would be divided into three smaller monowaves.

Figure 3-15b

Under these conditions, do not move DOT to <u>lower</u> high

End exceeded before previous monowave retraced 61.8%

.618

Aspect-1 **does not apply**

.618

Under these conditions, do not move DOT to <u>higher</u> low

End exceeded before previous monowave retraced 61.8%

Aspect-2 of the Rule of Neutrality does not apply if the "horizontal" price action is actually counter-directional to the two <u>same-direction</u> waves it separates. The horizontal price action must drift in the same general direction as the two waves it separates. **For example**, *if the market begins moving upward then hesitates for a time period, but no data point is lower than the last, and then the market begins to advance again, Aspect-2 of the Rule may be used. If at any time a single data point is lower than the last, Aspect-2 CANNOT be applied.* <u>WARNING</u>: Anytime horizontal (or near-horizontal) price action consumes more than one time unit **and** separates two <u>same-direction</u> waves, Aspect-2 <u>must</u> be applied to break the horizontal period into three segments.

Application of the Rule of Neutrality can, at times, be conditional. When a market contains horizontal price action of only one time unit, the decision to apply Aspect-2 frequently rests with the relation of past and future market action to the current. For example, if alternation of Intricacy is enhanced **or** Levels of Complexity between patterns is improved **or** "missing" waves are eliminated by employing Aspect-2, then it should be used. If use of the Rule actually creates these or other problems with standard pattern formation, it should be ignored. If the termination of a monowave would occur beyond 38.2% from the end of said monowave, do not apply Aspect-1 (see explanation in Figure 3-17, below).

Until you have read most of this book and have come to understand the complex concepts discussed in the previous paragraph which would prevent the use of Aspect-1 or Aspect-2 of the Rule of Neutrality, the best approach is to employ the Rule in a simplistic manner. If your chart is properly scaled according to the Rule of Proportion (see page 3-3) and price action progresses at an angle less than 45 degrees (upward or downward) from the horizontal **and** the Rule can be applied to the current situation, use it. Ignore the Rule when "horizontal" action progresses at an angle greater than 45 degrees.

The previous Figure 3-12 (page 3-9) contained several "horizontal" periods which may require consideration under the Rule of Neutrality. Employing the concepts revealed in this section to those periods, the horizontal action has been identified, discussed and finalized in Figure 3-17.

Figure 3-17

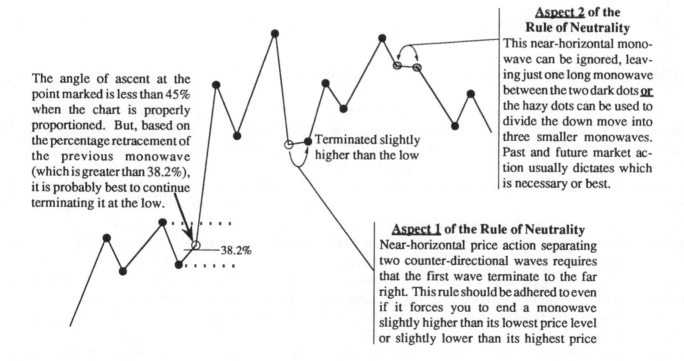

The angle of ascent at the point marked is less than 45% when the chart is properly proportioned. But, based on the percentage retracement of the previous monowave (which is greater than 38.2%), it is probably best to continue terminating it at the low.

Terminated slightly higher than the low

Aspect 2 of the Rule of Neutrality
This near-horizontal mono-wave can be ignored, leaving just one long monowave between the two dark dots <u>or</u> the hazy dots can be used to divide the down move into three smaller monowaves. Past and future market action usually dictates which is necessary or best.

Aspect 1 of the Rule of Neutrality
Near-horizontal price action separating two counter-directional waves requires that the first wave terminate to the far right. This rule should be adhered to even if it forces you to end a monowave slightly higher than its lowest price level or slightly lower than its highest price

Chronology

The numbering of monowaves is useful when discussing wave analysis with others over the phone or through the mail as I do for my Telephone course and in my correspondent publication, "WaveWatch," respectively. The numbering of monowaves is not necessary if you have no plans of discussing your wave interpretations with other people. When doing so, all monowaves which do not contain verified Progress labels should be referred to by their Chronological number (i.e., Chronological 1, Chronological 2, etc., or abbreviate them as Chrono-1, Chrono-2). To properly label each monowave, begin numbering with the first monowave on the far left of your chart (see Figure 3-18). Notice the neutral period (Chrono-13) has also been numbered prior to the decision to apply Aspect-2.

Figure 3-18

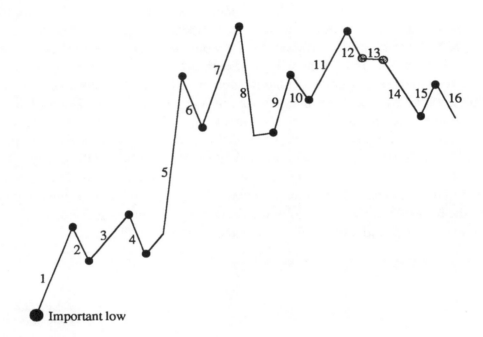

Rules of Observation (determining relative position of monowaves)

All wave movement, no matter how large or small, must be classified into one of two categories: *Impulsive* (:5) or *Corrective* (:3). The term "Impulsive" defines action which moves in the direction of the trend. "Corrective" represents price action which travels against the trend. For reasons which will become clear later, the number ":5" is short for Impulsive action and ":3" is short for Corrective action. These numerical representations (or symbols) have been given the name *Structure Labels* and are useful when beginning real-time chart analysis. [For a general description of Structure Labels, refer back to page 2-5.] Some new analytical terms, techniques and Rules have been developed by the author to assist you in the accurate placement of Structure Labels to real-time action even before you have a comprehensive understanding of the intricacies and complexities inherent in Elliott Wave Theory. These techniques will soon be applied to each of the monowaves recently identified in Figure 3-18. If you are following along with your own chart, apply the same techniques in the order indicated.

Since monowaves have no indicative features, their *Structure* (:3 or :5) must be detected <u>indirectly</u>. This is accomplished through the observation of preceding and proceeding market action. Obviously, before you can actually begin the placement of Structure labels on price action, you must learn how to associate preceding and proceeding market action to the monowave currently under analysis. The presentation of these concepts will require what I call "relative perspective" diagrams.

Under all circumstances, the indirect analysis of monowaves depends on the relation of the "**current**" monowave to surrounding monowaves. Whenever studying a chart, the monowave under analysis will be considered monowave number one ("m1," see Figure 3-19). The monowaves immediately <u>after</u> m1

Figure 3-19

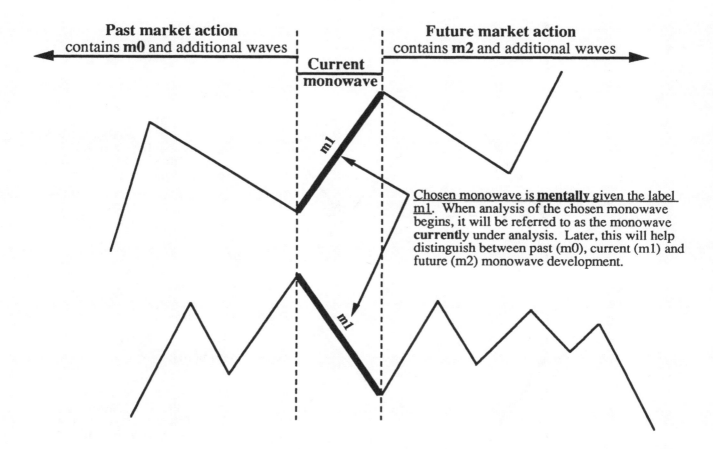

Past market action
contains **m0** and additional waves

Future market action
contains **m2** and additional waves

Current
monowave

m1

m1

<u>Chosen monowave is **mentally** given the label</u>
<u>m1</u>. When analysis of the chosen monowave begins, it will be referred to as the monowave **current**ly under analysis. Later, this will help distinguish between past (m0), current (m1) and future (m2) monowave development.

will consist of m2 and additional monowaves. The monowaves which occurred immediately <u>before</u> m1 consist of m0 and additional waves. Figure 3-19 illustrates how the choosing of any monowave on a chart **mentally** creates the *relative* label "m1" for the chosen monowave.

Figure 3-20a presents the observations required to discover where m2 completed when m2 is a monowave. Figure 3-20a also demonstrates the importance of breaking either the high or low of m1 before the finalization of m2 can take place. Figure 3-20b presents a similar process which is used in uncovering the termination of m0. These techniques will be used in the next section, "***Retracement Rules,***" to calculate the percentage relation of m1 to m2, m0 to m1, m0 to m2, etc.

In Figures 3-20a & 3-20b, m0 and m2 were monowaves. Unlike m1, which <u>must</u> be a monowave (or a compacted* Elliott wave pattern), m0 and m2 can be composed of one (or any <u>odd</u> number) of monowaves, compacted or uncompacted. The composition of m0 and m2 is dependent upon the amount of market action that takes place **within** the confines of the high and low of m1. Using an upward slanting

*Compaction is covered in detail in Chapter 7

Figure 3-20a

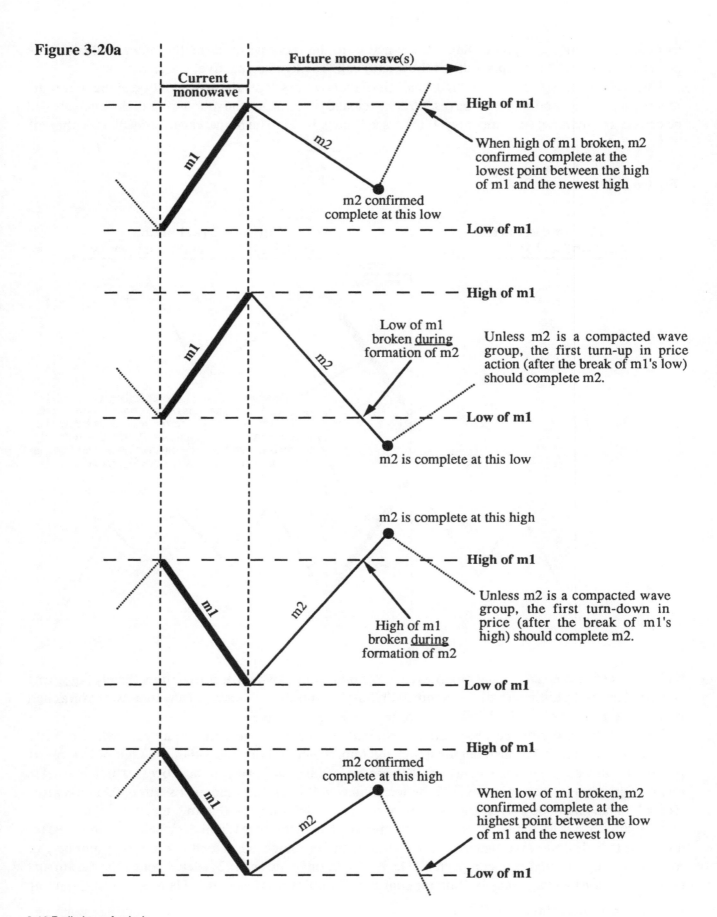

Current monowave

Future monowave(s)

High of m1

m1

m2

When high of m1 broken, m2 confirmed complete at the lowest point between the high of m1 and the newest high

m2 confirmed complete at this low

Low of m1

High of m1

m1

m2

Low of m1 broken during formation of m2

Unless m2 is a compacted wave group, the first turn-up in price action (after the break of m1's low) should complete m2.

Low of m1

m2 is complete at this low

m2 is complete at this high

High of m1

m1

m2

High of m1 broken during formation of m2

Unless m2 is a compacted wave group, the first turn-down in price (after the break of m1's high) should complete m2.

Low of m1

High of m1

m2 confirmed complete at this high

m1

m2

When low of m1 broken, m2 confirmed complete at the highest point between the low of m1 and the newest low

Low of m1

Figure 3-20b

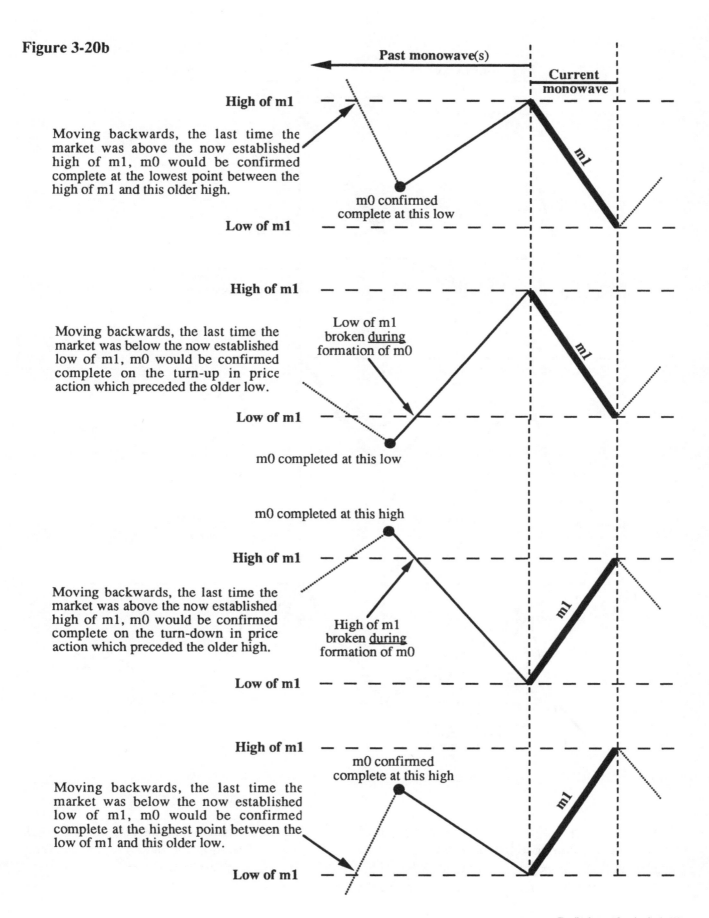

Past monowave(s)

Current monowave

High of m1

Moving backwards, the last time the market was above the now established high of m1, m0 would be confirmed complete at the lowest point between the high of m1 and this older high.

m0 confirmed complete at this low

Low of m1

m1

High of m1

Moving backwards, the last time the market was below the now established low of m1, m0 would be confirmed complete on the turn-up in price action which preceded the older low.

Low of m1 broken <u>during</u> formation of m0

Low of m1

m1

m0 completed at this low

m0 completed at this high

High of m1

Moving backwards, the last time the market was above the now established high of m1, m0 would be confirmed complete on the turn-down in price action which preceded the older high.

High of m1 broken <u>during</u> formation of m0

Low of m1

m1

High of m1

m0 confirmed complete at this high

Moving backwards, the last time the market was below the now established low of m1, m0 would be confirmed complete at the highest point between the low of m1 and this older low.

Low of m1

m1

Figure 3-21a

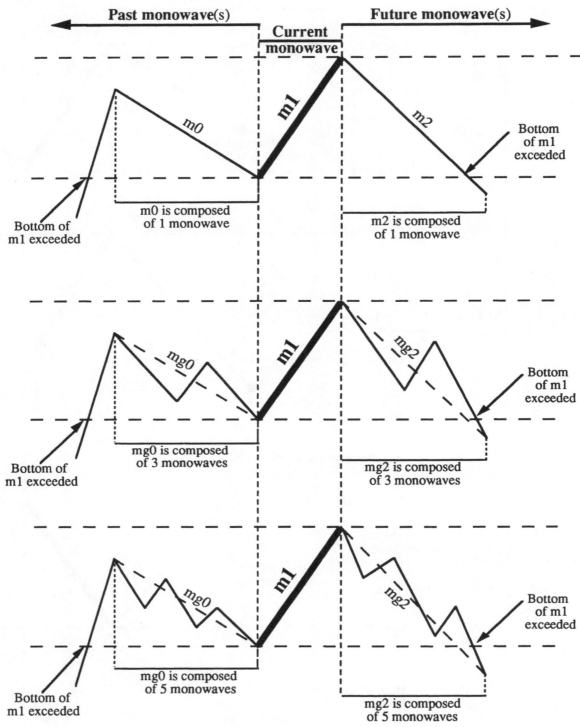

m1 for this example, the right-hand side of Figure 3-21a illustrates what market action <u>after</u> m1 may look like when **m2** is composed of one or more monowaves. The left-hand side of the same Figure illustrates what market action <u>before</u> m1 may look like when **m0** is composed of one or more monowaves.

If more than one monowave occurs <u>before</u> the high or low of m1 is exceeded, m0 and m2 can be

Figure 3-21b

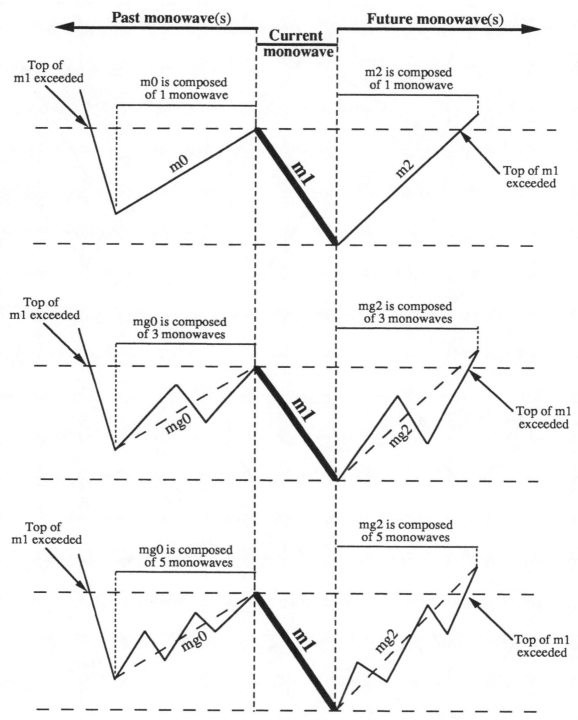

considered (if you prefer) "**m**onowave **g**roups (mg)" and designated mg0 and mg2 respectively (study Figures 3-21a & b). When m1 is a monowave, mg0 and mg2 will <u>usually</u> consist of <u>no more</u> than five monowaves, but exceptions can and do occur. To provide you with both perspectives, Figure 3-21a depicts "m1" as a rising monowave while Figure 3-21b depicts "m1" as a declining monowave.

Figure 3-22a

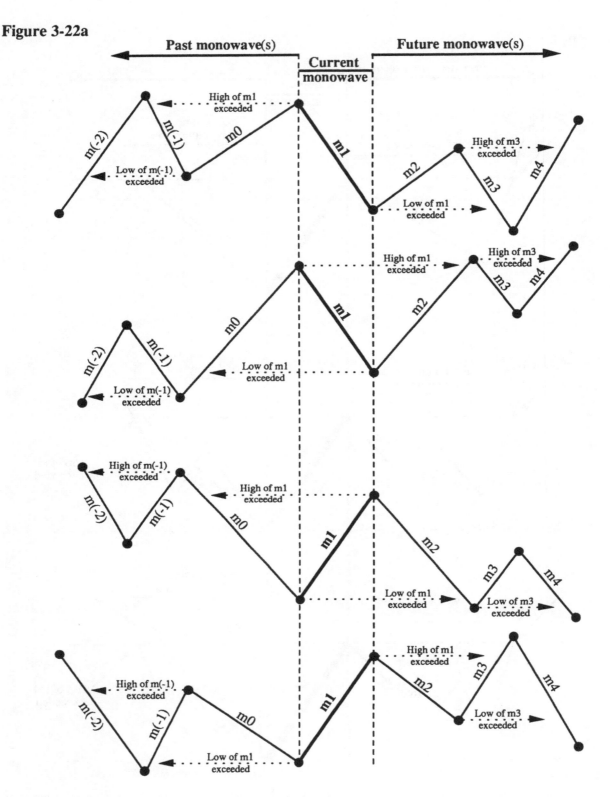

Occasionally, when implementing the Retracement Rules (which follow this section) to determine the internal Structure of a monowave, the study of monowaves other than m0 or m2 is necessary. Figure 22a, above, demonstrates the **Rules of Observation** as they apply to monowaves which occur <u>before m0</u> and <u>after m2</u>. Careful study of the chart will demonstrate how to determine each monowaves' starting and stopping points based on the high or low of the monowave before or after the current one.

Figure 3-22b

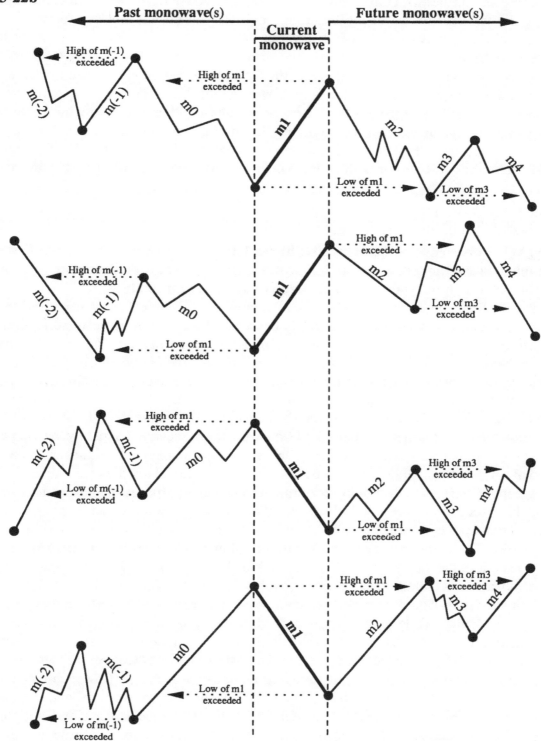

Sometimes, breaking the high or low of m1 (or any other monowave) requires the development of more than one monowave. Figure 3-22b illustrates how m(-2), m(-1), m0, m2, m3 or m4 may be composed of one or more monowaves. It is important to remember that the termination of all monowaves **to the right of m1** must break a <u>previous</u> high or low; the termination of all monowaves **to the left of m1** must be figured backwards in time based on when they "break" a <u>latter</u> high or low.

Before going any further with this chapter, it is imperative you understand the procedures necessary in obtaining the terminal points of m(-2) through m4. If you do not feel comfortable with these procedures, continue to study the Rules of Observation until you do; no additional work can be done until this process is understood.

After you have become proficient at applying the Rules of Observation to actual market action, the next analytical step involves the application of the ***Retracement Rules***. This will require calculating the relationship of <u>m2 to m1</u> and <u>m0 to m1</u> (on a percentage basis), then deciding which preestablished **relational range** the calculated ratio falls within. The m2/m1 ratio describes which *Rule* Identifier applies to "m1." Then, the m0/m1 ratio indicates which alphabetic *Condition* of each Rule applies. Keep in mind, the Retracement Rules work the same whether m1 is slanting upward or downward.

Retracement Rules

WARNING: This section <u>is not</u> intended to be read straight through. If this is your first reading, skim through to grasp the general ideas, but do not spend much time on the details. The next time you read this section, have your own chart handy and ready for analysis.

Once you have finalized the terminal points of all monowaves on your chart, you are ready to identify the specific **Retracement Rule** which accurately describes the behavior of surrounding monowaves. *[NOTE: Your chart should never be so complicated that this process involves the identification of more than 20 monowaves at any one time. For the best results (and the maintenance of your sanity), it is recommended you apply the Rules in Chapter 3 to just a few monowaves of real-time action each day.]*

To begin this process, pick the earliest monowave on your chart which you wish to analyze. That monowave, as described earlier, should *mentally* be considered m1 (monowave-1). Applying the **Rules of Observation** discussed earlier (see page 3-14), measure the retracement of m1 by m2 - in percentage terms - using the terminal points (**dots**) of each. These measurements can be made accurately and swiftly with the aid of a <u>proportional divider</u> set to a 61.8% ratio. Proportional dividers are available from most engineering supply stores and are highly recommended for the purposes of this section. If you are performing this procedure more precisely (and tediously) with the aid of an electronic calculator, measure the vertical price coverage of m2 and divide it by the vertical price coverage of m1. Multiply the result by 100. The number remaining is the relationship of m2 to m1 in percentage terms.

Refer to the table at the top of the next page (titled RULE IDENTIFICATION); in that listing, locate the relationship just calculated. That listing provides you with a **Rule** Identifier which represents particular characteristics of current market behavior. Next, move forward to the heading which matches your findings (i.e., **Rule 1, Rule 2**, etc.). Further clarification of the market environment is obtained by calculating the m0/m1 ratio. Measure the price coverage of m0, divide it by m1's price coverage and multiply the result by 100. Locate that percentage relationship under the "Conditions" section following each Rule diagram and translate the m0/m1 ratio into a **Condition** Identifier (lower case Arabic letter). When working with Rule 4, you will also be asked for the relationship between m2 and m3; this further defines market action into **Categories** (Roman numerals). Finally, these **Rules, Conditions** and **Categories** will, in the "Pre-Constructive Rules of Logic" section, be *transformed* into Structure labels which reveal the internal (unseen) make-up of the monowaves under analysis.

When you have decided which <u>specific</u> Rule applies to each situation, place those markings at the end of m1. Write it on your chart in pencil or photocopy the original and write on the copy. An even better approach requires the use of plastic acetate placed over your chart and drawing on the acetate with a china marker. When you place a Rule at the end of m1, abbreviate the Rule and its sub-sections to prevent clutter on your chart (for example, Rule 4, Condition "a," Category "i" should be abbreviated R-4a-i).

RULE IDENTIFICATION:

If m2's relation to m1 is: **less than 38.2%**, refer to Rule 1 (below);

        at least 38.2%, but less than 61.8%, refer to Rule 2 (page 3-24);

        exactly 61.8%, refer to Rule 3 (page 3-25);

        between 61.8% and 100% (exclusive), refer to Rule 4; (page 3-26)

        at least 100%, but less than 161.8%, refer to Rule 5; (page 3-28)

        between 161.8% and 261.8% (inclusive), refer to Rule 6; (page 3-29)

        more than 261.8%, refer to Rule 7; (page 3-30).

Rule 1 (m2 is less than 38.2% of m1)

If m1 is retraced <u>less than 38.2%</u> by m2, Rule 1 applies; note that fact at the end of m1 (see Figure 3-23). Next, measure the relation of m0 to m1 as previously described, then find that ratio in the list below. That will provide you with the "Condition" Identifier (alphabetic companion) which is to be placed to the **right** of the Rule Identifier.

Figure 3-23

Rule 1
(activation requirement)

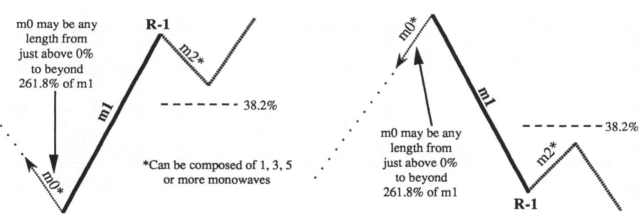

Conditions for Rule 1

Condition "a" -- If **m0 is <u>less than 61.8%</u> of m1**, m1 falls under the jurisdiction of Rule 1a (progress to Rule 1, Condition "a" of the Pre-Constructive Rules of Logic*).

Condition "b" -- If **m0 is <u>at least 61.8%, but less than 100%</u> of m1**, the properties of Rule 1b take effect (move to Rule 1, Condition "b" of the Pre-Constructive Rules of Logic*).

Condition "c" -- If **m0 is <u>between 100% and 161.8%</u> of m1** (inclusive), Rule 1c must be applied (advance to Rule 1, Condition "c" of the Pre-Constructive Rules of Logic*).

Condition "d" -- If **m0 is <u>more than 161.8%</u> of m1** (inclusive), Rule 1d should be considered (jump to Rule 1, Condition "d" of the Pre-Constructive Rules of Logic*).

The Pre-Constructive Rules of Logic begin on page 3-32

- - - - -

Rule 2 (m2 is at least 38.2%, but less than 61.8% of m1)

If m2 retraces <u>at least 38.2%, but less than 61.8%</u> of m1, Rule 2 takes effect (see Figure 3-24). Next, observe the relation of m0 to m1 to decide which alphabetic Condition Identifier applies. Place that letter to the right of Rule 2.

Figure 3-24

Rule 2
(activation requirement)

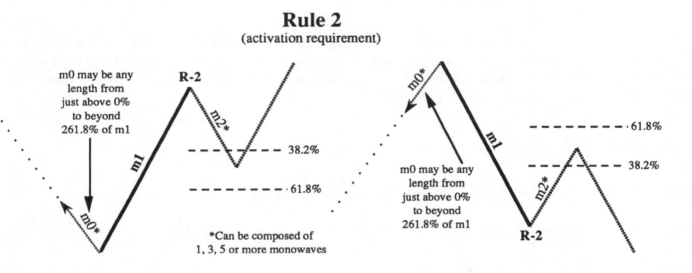

*Can be composed of
1, 3, 5 or more monowaves

Conditions for Rule 2

Condition "a" -- If **m0 is <u>less than 38.2%</u> of m1**, m1 falls under the jurisdiction of Rule 2a (progress to Rule 2, Condition "a" of the Pre-Constructive Rules of Logic*).

Condition "b" -- If **m0 is <u>at least 38.2%, but less than 61.8%</u> of m1**, the properties of Rule 2b take effect (move to Rule 2, Condition "b" of the Pre-Constructive Rules of Logic*).

Condition "c" -- If **m0 is <u>at least 61.8%, but less than 100%</u> of m1**, Rule 2c must be applied (advance to Rule 2, Condition "c" of the Pre-Constructive Rules of Logic*).

Condition "d" -- If **m0 is <u>between 100% and 161.8%</u> of m1** (inclusive), Rule 2d should be considered (jump to Rule 2, Condition "d" of the Pre-Constructive Rules of Logic*).

Condition "e" -- If **m0 is <u>more than 161.8%</u> of m1**, then Rule 2e is the appropriate (proceed to Rule 2, Condition "e" of the Pre-Constructive Rules of Logic*).

*The Pre-Constructive Rules of Logic begin on page 3-32

- - - - -

Rule 3 (m2 is exactly 61.8% of m1)

An <u>exact 61.8%</u> retracement of m1 by m2 (see Figure 3-25) establishes the dominance of Rule 3. When market action activates Rule 3, the *structure* of m1 is the most difficult to decipher since the retracement of m1 is right on the border line of indicating m1 is more likely corrective or more likely Impulsive. To further clarify the market environment surrounding m1, calculate the relationship of m0 to m1 and find that relationship in the listing below to obtain a Condition Identifier for Rule 3.

Figure 3-25

Rule 3
(activation requirement)

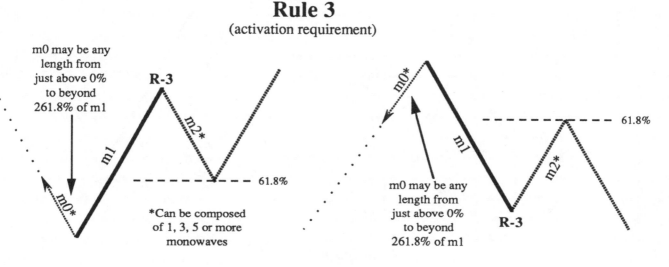

Conditions for Rule 3

Condition "a" -- If **m0 is <u>less than 38.2%</u> of m1**, m1 falls under the jurisdiction of Rule 3a (progress to Rule 3, Condition "a" of the Pre-Constructive Rules of Logic*).

Condition "b" -- If **m0 is <u>at least 38.2%, but less than 61.8%</u> of m1**, the properties of Rule 3b take effect (move to Rule 3, Condition "b" of the Pre-Constructive Rules of Logic*).

Condition "c" -- If **m0 is <u>at least 61.8%, but less than 100%</u> of m1**, Rule 3c must be applied (advance to Rule 3, Condition "c" of the Pre-Constructive Rules of Logic*).

Condition "d" -- If **m0 is <u>at least 100%, but less than 161.8%</u> of m1**, then Rule 3d should be considered (jump to Rule 3, Condition "d" of the Pre-Constructive Rules of Logic*).

Condition "e" -- If **m0 <u>is between 161.8% and 261.8%</u> of m1** (inclusive), Rule 3e is appropriate (proceed to Rule 3, Condition "e" of the Pre-Constructive Rules of Logic*).

Condition "f" -- If **m0 is <u>more than 261.8%</u> of m1**, Rule 3f exerts its influence (proceed to Rule 3, Condition "f" of the Pre-Constructive Rules of Logic*).

The Pre-Constructive Rules of Logic begin on page 3-32

- - - - -

Rule 4 (m2 is between 61.8% and 100% [exclusive])

Rule 4 applies when m2 retraces <u>more than 61.8%, but less than 100%</u> of m1. Figure 3-26 gives an example of this event. Calculate the relation of m0 to m1; then, using the list below, locate a Condition Identifier. Next, measure the percentage retracement of m2 by m3 to obtain a Category Identifier.

Figure 3-26

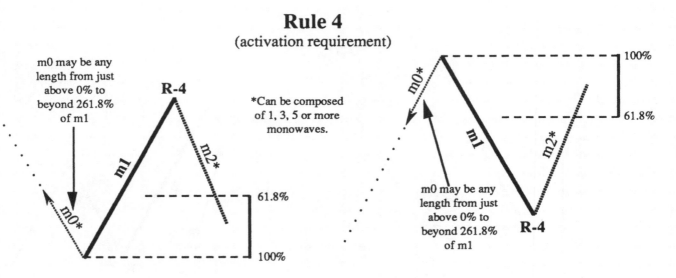

Rule 4
(activation requirement)

Conditions for Rule 4
Condition **"a"** -- If **m0 is <u>less than 38.2%</u> of m1,** m1 falls under the unique jurisdiction of Rule 4a (see Rule 4, Condition "a" of Figure 3-27).

Condition **"b"** -- If **m0 is <u>at least 38.2%, but less than 100%</u> of m1**, then the unusual properties of Rule 4b take affect (see Rule 4, Condition "b" of Figure 3-27).

Condition **"c"** -- If **m0 is <u>at least 100%, but less than 161.8%</u> of m1**, Rule 4c must be applied (see Rule 4, Condition "c" of Figure 3-27).

Condition **"d"** -- If **m0 is <u>between 161.8% and 261.8%</u> of m1** (inclusive), then Rule 4d should be considered (see Rule 4, Condition "d" of Figure 3-27).

Condition **"e"** -- If **m0 is <u>more than 261.8%</u> of m1**, then Rule 4e is the most appropriate (see Rule 4, Condition "e" in Figure 3-27).

Categories for Conditions "a" through "e" of Rule 4
Category i -- If **m3 is <u>at least 100%, but less than 161.8%</u> of m2**, then Rule 4?-i is appropriate (proceed to Rule 4, Condition "?" [the one currently in effect] , Category "i" of the Pre-Constructive Rules of Logic*).

Category ii -- If **m3 is between <u>161.8% and 261.8%</u> of m2** (inclusive), then Rule 4?-ii takes effect (proceed to Rule 4, Condition "?" [the one currently in effect] , Category "ii" of the Pre-Constructive Rules of Logic*).

Category iii -- If **m3 is <u>more than 261.8%</u> of m2**, Rule 4?-iii applies (move to Rule 4, Condition "?" [the one currently in effect], Category "iii" of the Pre-Constructive Rules of Logic*).

The Pre-Constructive Rules of Logic begin on page 3-32

- - - - -

Figure 3-27

Rule 4a

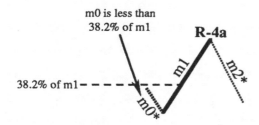

m0 is less than
38.2% of m1

R-4a

38.2% of m1 — — — —

m1

m2*

m0*

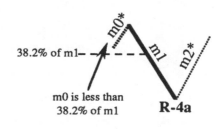

m0*

38.2% of m1—

m1

m2*

m0 is less than
38.2% of m1

R-4a

Rule 4b

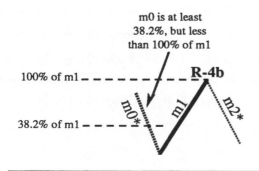

m0 is at least
38.2%, but less
than 100% of m1

100% of m1 — — — — —

R-4b

38.2% of m1 — — — —

m0*

m1

m2*

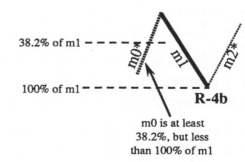

38.2% of m1 — — — —

m0*

m1

m2*

100% of m1 — — —

R-4b

m0 is at least
38.2%, but less
than 100% of m1

Rule 4c

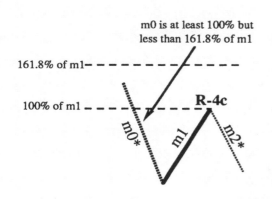

m0 is at least 100% but
less than 161.8% of m1

161.8% of m1 — — — — — — — —

100% of m1 — — — — —

R-4c

m0*

m1

m2*

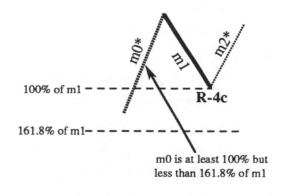

m0*

m1

m2*

100% of m1 — — — —

R-4c

161.8% of m1 — — — — — — —

m0 is at least 100% but
less than 161.8% of m1

Rule 4d

Figure 3-27 continued on next page

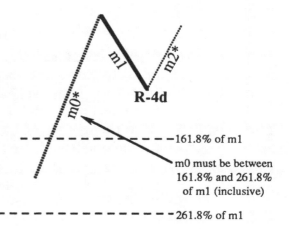

— — — — — — 261.8% of m1

m0 must be between
161.8% and 261.8%
of m1 (inclusive)

— — — — — 161.8% of m1

m0*

R-4d

m1

m2*

> *In all diagrams
> on this page, m0
> and m2 can be
> composed of 1,
> 3, 5, or more
> monowaves

m1

m2*

m0*

R-4d

— — — — — 161.8% of m1

m0 must be between
161.8% and 261.8%
of m1 (inclusive)

— — — — — — — 261.8% of m1

Figure 3-27, continued from page 3-27

Rule 4e

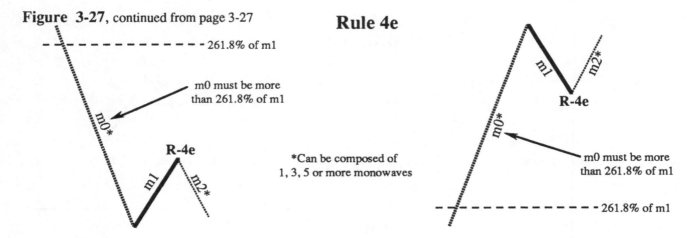

*Can be composed of
1, 3, 5 or more monowaves

Rule 5 (m2 is at least 100%, but less than 161.8% of m1)

For Rule 5 to be activated, m1 must be retraced <u>at least 100%, but less than 161.8%</u> by m2 (see Figure 3-28). Once that requirement has been met, you must observe the relationship of m0 to m1. Find that relationship in the list below to determine the *Condition* Identifier which should be attached to Rule 5.

Figure 3-28

Rule 5
(activation requirement)

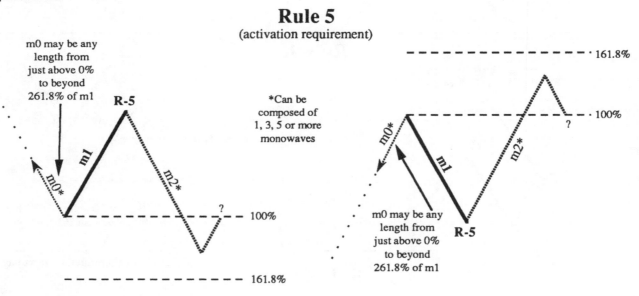

*Can be composed of 1, 3, 5 or more monowaves

Conditions for Rule 5

Condition **"a"** -- If **m0 is <u>less than 100%</u> of m1**, then Rule 5a is activated; make note of that at the end of m1 (progress to Rule 5, Condition "a" of the Pre-Constructive Rules of Logic*).

Condition **"b"** -- If **m0 is <u>at least 100%, but less than 161.8%</u> of m1**; then the properties of Rule 5b take effect (move to Rule 5, Condition "b" of the Pre-Constructive Rules of Logic*).

Condition **"c"** -- If **m0 is <u>between 161.8% and 261.8%</u> of m1** (inclusive), Rule 5c must be applied (advance to Rule 5, Condition "c" of the Pre-Constructive Rules of Logic*).

Condition **"d"** -- If **m0 is <u>more than 261.8%</u> of m1**, the unique circumstances (explained later in this chapter) implied by Rule 5d should be upheld (jump to Rule 5, Condition "d" of the Pre-Constructive Rules of Logic*).

The Pre-Constructive Rules of Logic begin on page 3-32

- - - - -

Rule 6 (m2 is between 161.8% and 261.8% of m1 [inclusive])

For Rule 6 to be activated, m1 must be retraced <u>between 161.8% and 261.8%</u> (inclusively) by m2 (Figure 3-29). Next, measure the relationship of m0 to m1; making use of the table below, this will allow you to decide the "Condition" of current market action.

Figure 3-29

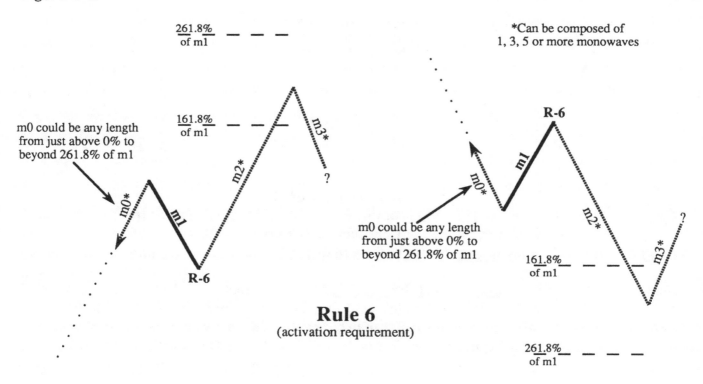

Rule 6
(activation requirement)

Conditions for Rule 6

Condition "a" -- If **m0 is <u>less than 100%</u> of m1**, Rule 6a should be implemented (progress to Rule 6, Condition "a" of the Pre-Constructive Rules of Logic*).

Condition "b" -- If **m0 is <u>at least 100%, but less than 161.8%</u> of m1,** then the properties of Rule 6b take effect (move to Rule 6, Condition "b" of the Pre-Constructive Rules of Logic*).

Condition "c" -- If **m0 is <u>between 161.8% and 261.8%</u> of m1** (inclusive), Rule 6c must be applied (advance to Rule 6, Condition "c" of the Pre-Constructive Rules of Logic*).

Condition "d" -- If **m0 is <u>more than 261.8%</u> of m1**, then Rule 6d is activated; make note of that at the end of m1 (jump to Rule 6, Condition "d" of the Pre-Constructive Rules of Logic*).

    The Pre-Constructive Rules of Logic begin on page 3-32

- - - - -

Rule 7 (m2 is more than 261.8% of m1)

For Rule 7 to be activated, m1 must be retraced <u>more than 261.8%</u> by m2 (Figure 3-30). Next, measure the relationship of m0 to m1, studying the table on the next page, you can decide the "Condition" of current market action (also, see Figure 3-30 on the next page).

Figure 3-30

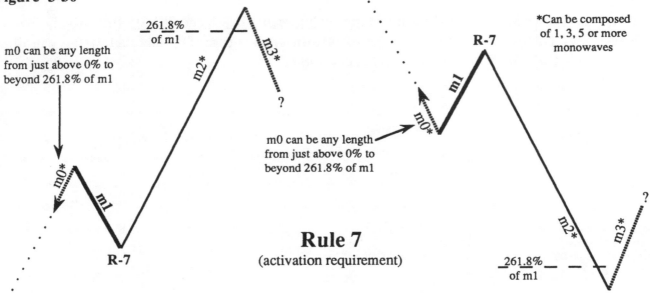

Rule 7
(activation requirement)

Conditions for Rule 7

Condition "a" -- If **m0 is <u>less than 100%</u> of m1**, Rule 7a should be implemented (progress to Rule 7, Condition "a" of the Pre-Constructive Rules of Logic*).

Condition "b" -- If **m0 is <u>at least 100%, but less than 161.8%</u> of m1**, then the properties of Rule 7b take effect (move to Rule 7, Condition "b" of the Pre-Constructive Rules of Logic*).

Condition "c" -- If **m0 is <u>between 161.8% and 261.8%</u> of m1** (inclusive), Rule 7c must be applied (advance to Rule 7, Condition "c" of the Pre-Constructive Rules of Logic*).

Condition "d" -- If **m0 is <u>more than 261.8%</u> of m1**, then Rule 7d is activated; make note of that at the end of m1 (jump to Rule 7, Condition "d" of the Pre-Constructive Rules of Logic*).

*The Pre-Constructive Rules of Logic begin on page 3-32

When you have completed the **Rule** identification process for the <u>current</u> m1, move to the next monowave in the sequence (which previously was m2), make that monowave the new m1 and restart the Rule identification process from that new point. If you have identified all of the monowaves on your chart which you feel are important at this time, proceed to the "Pre-Constructive Rules of Logic" heading and begin *transforming* the Rules at the end of each monowave into usable Structure labels.

Graphical Summary of Retracement Rules

After becoming familiar with the ideas presented in this section, the diagrams on the next two pages should simplify the **Rule** and **Condition** identification process for you. To implement, first measure m2's retracement of m1 in percentage terms, then find that retracement value within one of the relational ranges along m2's ray under the **Rule** Identification diagram (since m2 occurs after m1 in time, m2's ray extends to the right). Jot down the abbreviated Rule which is activated by that retracement range. Once a Rule has been identified, the conditions which surround m1 should be clarified. Move to the **Condition** Identification diagram specifically designed for the **Rule** currently in effect. This time the ray will represent m0 (the wave before m1), therefore it extends to the left. Locate the alphabetic **Condition** associated with the relational value you measured between m0 and m1. Once this information has been obtained, move to the section possessing the same name under the Pre-Constructive Rules of Logic.

Rule Identification

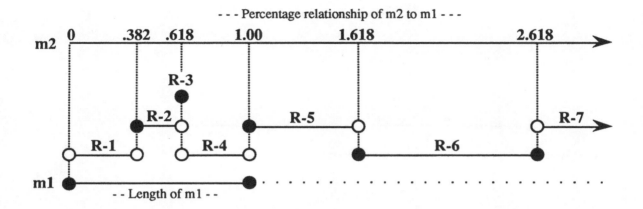

- - - Percentage relationship of m2 to m1 - - -

Condition Identification for Rule 1

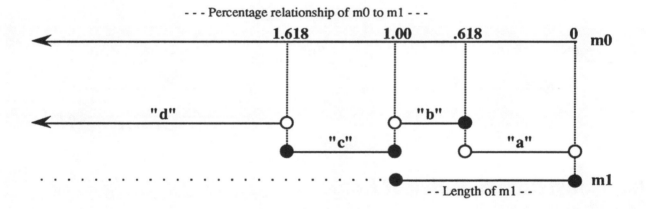

- - - Percentage relationship of m0 to m1 - - -

Condition Identification for Rules 2 and 3

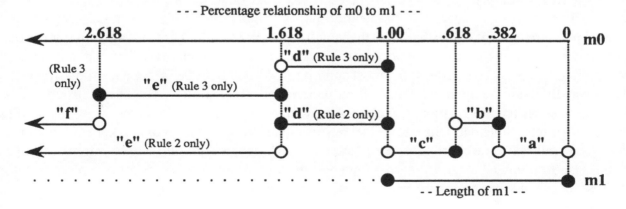

- - - Percentage relationship of m0 to m1 - - -

Condition Identification for Rule 4

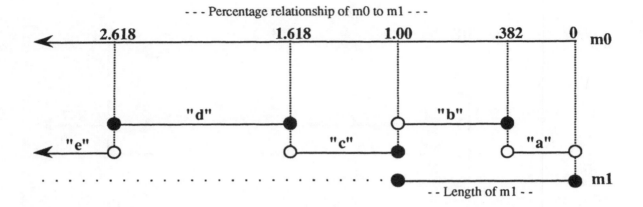

- - - Percentage relationship of m0 to m1 - - -

Condition Identification for Rules 5, 6 and 7

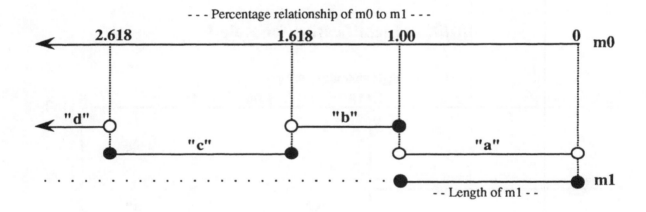

- - - Percentage relationship of m0 to m1 - - -

Pre-Constructive Rules of Logic

The purpose of the "Pre-Constructive Rules of Logic" section is to provide **Structure Labels** to replace the **Rules** you established earlier for each analyzed monowave. All Rules (including their Condition Identifiers) are listed over the next 26 pages. Under each Rule are numerous paragraphs separated into retracement values based on the length of m0, m2 or m3. These paragraphs query you on circumstances unfolding in real-time on your chart. Based on your input regarding market conditions, specific suggestions will be made on the most appropriate Structure Label(s) for each monowave and a guess is usually made on the most likely Elliott pattern unfolding around m1. These Structure labels indicate all possible locations a monowave can occur within an Elliott Wave pattern (simple or complex). Sometimes it is recommended a Structure label be surrounded by parentheses or brackets. If this situation develops on your real-time chart, those Structure labels in <u>brackets</u> represent rare or extremely unlikely possibilities. Structure labels surrounded by <u>parentheses</u> simply have a lower probability of being correct than those which possess no encasement (i.e., no parentheses or no brackets).

To understand and make complete use of Chapter 3, at least a cursory knowledge of all chapters from 1 through 8 (inclusive) is required. If you are still in the early stages of your Elliott Wave learning curve, many concepts and techniques (not as yet explained) will be suggested for use in this section which you do not understand completely or have no working knowledge of. For this reason, become familiar with these other chapters before attempting any analysis which is backed by capital investment. The Elliott Wave Theory is such a maze of interlocking, interdependent concepts that an understanding of only portions of the Theory is insufficient for accurate forecasting. The inherently complex nature of the Wave Theory makes it impossible to confidently apply the prognosticative techniques discussed in this section without some awareness of the general principles which guide wave pattern development.

After reading the book through the end of Chapter 8, return to Chapter 3 and begin the analysis process; you will then be better equipped to understand and apply the ideas presented. For those still becoming familiar with general Elliott Wave concepts, this section is designed to speed your learning curve by providing you with expert counseling on Wave analysis until you become proficient at the process. Furthermore, applying the **Pre-Constructive Rules of Logic** is a great way to learn of the numerous observations and measurements required to properly dissect market action.

It is not recommended you begin your analytical "baptism of fire" by constructing a highly detailed chart and attempting to analyze every wave. Construct a short-term chart and begin with the current day, applying the rules one wave at a time and updating your chart and analysis regularly with daily data. This approach will allow careful consideration of each Rule as it applies to specific situations. Applying the **Pre-Constructive Rules of Logic** in this fashion should minimize the tedium of repetitious application and the chances of analytical error possible when dealing with such extensive relational calculations.

WARNING: This section should NOT be read straight through. The instructions which follow will only make sense, and be useful, when applied in conjunction with real-time market action. If you currently do not have a chart to analyze or this is your first time reading this section, scan through it then proceed to - and scan - the "Implementation of Position Indicators" (page 3-60).

Transformation of Rule Identifiers

The reliable application of Structure labels to replace Rule Identifiers usually requires the study of price action before and after m1. If m1 is the first monowave on your chart, there is no previous market action to help you indirectly decide on the internal Structure of m1. Therefore, simply place the entire Structure list, which appears immediately to the right of each new Rule section, at the end of this first monowave. For example, to the right of Rule 1 (turn page to see listing) appears the Structure list {:5/ (:c3)/(x:c3)/[:sL3]/[:s5]}. If m1 is the first wave on your chart, place that entire list at the end of m1.

Warning: Do not apply these Rules to *Compacted* patterns which exceed their own beginning; retain only the pattern's base Structure. Refer to the top of page 3-68 for a full discussion of this concept. Figure 3-36 on the same page illustrates this behavior. If conditions on your chart are not anticipated under the appropriate Rule section, please verify you are working with the correct Rule and then make sure you reread the conditions necessary to activate that Rule. If m1 is a monowave and you still find there are no stipulations for the conditions on your chart, use the entire Structure list next to the Rule heading of each section. If m1 is a compacted polywave (or higher) pattern **and** you, again, find no stipulations for the conditions on your chart, move on to the section entitled "Implementation of Position Indicators" (page 3-60) and use surrounding Structure labels to decide which Position Indicator - if any - belongs in front of m1's compacted Structure label (i.e., in front of ":3" or ":5").

Remember, all time and price calculations should be made from termination points (the **dots** used earlier to identify the end of each wave); this applies whether working with monowaves or compacted wave groups. If measurement is required to test the validity of a statement and the statement does not specifically mention whether to measure price or time, price is always implied. Furthermore, **do not** take these relationships so seriously that you eliminate a possibility just because a particular wave was 60% of m1 when it was supposed to be" at least 61.8% of m1." All specific ratios mentioned in this text (such as 61.8%, 161.8%, etc.) must be given a slight range of error. When two patterns relate by 61.8%, that means the smaller of the two may be anywhere from approximately 58% to 66% of the larger wave. This provides about four (4) percent leeway either side of an ideal Fibonacci relationship. Finally, the phrase "almost" and "close to" used throughout this section translate to "within 10% of the ratio mentioned."

For the same reasons that relationships between waves cannot always be expected to behave perfectly, when following the guidelines listed in this section, if all conditions except one are scrupulously adhered to (and the one which is not followed is off by just a little), you should probably assume the rules were correctly followed and place the suggested Structure label or list at the end of m1. This is all part of the **Exception Rule** discussed in Chapter 9, page 7.

On the other hand, time calculations which involve retracement of previous patterns must be taken precisely as written. For example, many times throughout this section you will find something like this; "if m1 (plus one time unit) is retraced completely in the same amount of time that it took to form (or less)," that statement must be taken literally. In this case, you would add one standard time unit (the time consumed from one data point to the next) to the number of time units consumed by m1, then compare that larger amount to the number of time units taken by m2 to completely retrace the length of m1. If the two numbers are equal or if m1 is a larger number, the statement above is true. If m2, after such proceedings, is a larger number than m1, the statement above would be false.

The greatest technical and analytical value is derived from this section when the Rules are applied to monowaves. When returning to this section for **reassessment** of a confirmed, Compacted Elliott pattern, the primary purpose in applying the Rules is to detect the presence of "missing" waves. During this reassessment, if none of the conditions listed (under the appropriate section) agree with the base Structure of the compacted pattern, the compacted pattern *might* be missing a wave (page 12-34 discusses this phenomenon). Encase those choices which do not agree with the base Structure of the compacted pattern within squares. This provides you with a **warning** which should only be taken seriously if it is later found that the compacted pattern cannot be adjoined acceptably to any surrounding monowaves or wave groups. During reassessment, if you find that the Structure labels suggested by this section do not allow you to connect the compacted pattern with surrounding price action, assume the pattern is part of a Complex formation. If trying to reassess a compacted pattern and you find none of the conditions listed under the appropriate section apply to the pattern, place all Structure labels listed next to the Rule heading and encase those labels which disagree with the base Structure of the pattern within squares.

At the beginning of each Rule section, the chart used during the Rule Identification process earlier has been duplicated for your convenience. It will probably help to glance at these diagrams as you read the directions, comparing your real-time chart with the prescribed action. The first sentence of each of the paragraphs to follow is the most important. If the conditions described in the first sentence are not reflected in the actual market action, move on to the next paragraph (unless otherwise directed). If the first sentence of a paragraph accurately describes the market action you are analyzing, always read the rest of the entire paragraph. Some of the sentences which follow are very lengthy and contain many conditions. Usually, all parameters enumerated in one sentence must be adhered to for the Structure list to apply. Carefully read each sentence while comparing the statements to your chart and apply those Structure labels to the end of those monowaves which it directs you to. Sometimes you will be trying to analyze m1, but the section will direct you to place Structure or Progress labels or both at the end of m0, m2 or other waves other than m1; those directions are not misprints, place those labels as directed.

Rule 1
(activation requirement)

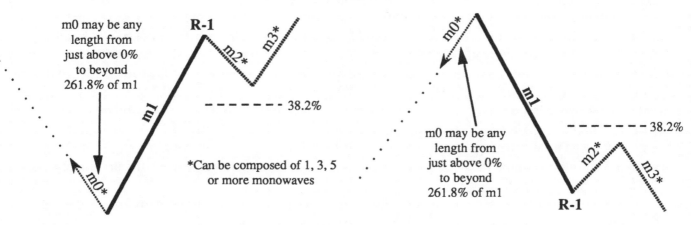

R-1

m0 may be any length from just above 0% to beyond 261.8% of m1

m1

m2*

m3*

38.2%

m0*

*Can be composed of 1, 3, 5 or more monowaves

m0*

m1

m2*

m3*

38.2%

m0 may be any length from just above 0% to beyond 261.8% of m1

R-1

__Rule 1__ {:5/(:c3)/(x:c3)/[:sL3]/[:s5]}
Condition "a" -- if m0 is less than 61.8% of m1

If m2 takes the same amount of time (or more) as m1 OR m2 takes the same amount of time (or more) as m3, place a ":5" at the end of m1. If the length of m(-1) is between 100-161.8% (inclusive) of m0 **and** m0 is very close to 61.8% of m1 **and** m4 does not exceed the end of m0, m1 may complete a Flat pattern within a Complex formation where m2 is an x-wave (x:c3); place ":s5" at the end of m1. Read through the rest of this section to see what other possibilities could be unfolding.

If m0 is composed of more than three monowaves **and** m1 retraces all of m0 in the same amount of time (or less) that m0 took to form, m0 is probably the end of an important Elliott pattern; note on chart.

If m0 and m2 are approximately equal in price and time (or related by 61.8% in either case) **and** m(-1) is 161.8% (or more) of m1 **and** m3 (or m3 through m5) achieves a price length equal to or greater than m(-1) within a time frame equal to or less than that of m(-1), a Running Correction (any variation) is probably unfolding; take note of that fact and add "[:c3]" after the ":5" already at the end of m1. If the Running correction is a simple variation, it most likely started at the beginning of m0 and concluded at the end of m2 with m1 the "b-wave" of the correction. For the Running Correction to be of the complex Double Three variety, m(-2) must be shorter than m(-1); in that case, the formation probably started with m(-2) and concluded with m4 making m1 the "x-wave" of the formation (x:c3). Read the next paragraph to be aware of any additional or unusual circumstances which could evolve under these conditions.

If m0 and m2 are approximately equal in price and time (or related by 61.8% in either case) **and** m(-1) is less than 161.8% of m1 **and** m(-1) is larger than m0 **and** m3 <u>or</u> m5 is 161.8% of m1 (or more), a Running Correction (any variation), which concludes more than one pattern, might be under formation; note that fact and add ":c3" after the ":5" already at the end of m1. If m(-2) is longer than m(-1), go back to what is currently m(-1) and add ":sL3" to its Structure list. If the Running correction is a simple variation, it most likely started at the beginning of m0 and concluded at the end of m2 with m1 the "b-wave" of the correction. For the Running Correction to be of the complex Double Three variety, m(-2) must be shorter than m(-1) **and** m3 must not be more than 161.8% of m1; under those specific conditions, the formation probably started with m(-2) and concluded with m4 making m1 the "x-wave" of the formation (add an "x" in front of the ":c3").

If m0 and m2 are approximately equal in price and time (or related by 61.8% in either case) **and** m3 is less than 161.8% of m1 **and** m3 (plus one time unit) is completely retraced in the same amount of time (or less) that it took to form, m1 may be part of a Complex Correction which will necessitate the use of an "x-wave" Progress label. The x-wave would be in one of two places; the end of m0 **or** hidden from sight (i.e., invisible or "missing") in the center of m1. This concept of "missing waves" is discussed in Chapter 12, page 12-34. To warn of these two possibilities, take a pencil and place an "x:c3?" at the end of m0. Also, circle the center of m1 and, to the right of the circle, place "x:c3?," to the left place ":s5." If m(-2) is longer than m(-1), the x-wave is not at the end of m0, drop that as a possibility. If the length of m3 is less than 61.8% of m1, the probabilities increase dramatically that the x-wave is hidden in the center of m1. These warnings will come in helpful as you group monowaves in Chapter 4 and as you finalize your interpretation throughout the analysis process. If the x-wave is used, the previously placed ":5" Structure label applies.

If m0 and m2 are obviously different from each other in price or time or both **and** m0 and m2 do not share any similar price range **and** m1, when compared to m(-1) and m3, is not the shortest of the three, m1 may be part of a larger Trending Impulse pattern; if so, the previously placed ":5" is used.

Condition "b" -- if m0 is at least 61.8%, but less than 100% of m1

Place a ":5" at the end of m1. If the length of m(-1) is between 100-161.8% (inclusive) of m0 **and** m4 does not exceed the end of m0, m1 may complete a Flat pattern within a Complex formation where m2 is an x-wave; place ":s5" at the end of m1 and "x:c3?" at the end of m2. Additional Structure labels may be needed if certain behavior is exhibited by the price action. Read the below descriptions to decide if any additional Structure labels should be added.

If m0 is composed of more than three monowaves **and** m1 (minus one time unit) retraces all of m0 in the same amount of time (or less) m0 took to form, m0 might conclude an important Elliott pattern.

If part of m2's price range is shared by m0 **and** m3 is longer and more vertical than m1 during a time span equal to (or less than) m1 **and** m(-1) is longer than m1, add ":sL3" to m1's Structure list.

If part of m2's price range is shared by m0 **and** m3 is longer and more vertical than m1 during a time span equal to (or less than) m1 **and** m(-1) is shorter than m1 **and** m0 and m2 are obviously different in price or time or both **and** m4 (or m4 through m6) returns to the beginning of m1 in a time period 50% of that consumed by m1 through m3, a 5th Extension Terminal pattern may have completed with m3; add ":c3" to m1's Structure list.

If m3 is shorter than m1 **and** part of m2's price range is shared by m0 **and** m(-1) is longer than m0 **and** m1, when compared to m(-1) and m3, is not the shortest of the three **and** from the end of m3, the market returns to the beginning of m1 (or beyond) in a span of time which is 50% or less of that consumed by m(-1) through m3; place ":c3" at the end of m1.

Condition "c" -- if m0 is between 100% and 161.8% of m1 *(inclusive)*

Place a ":5" at the end of m1. Read the below descriptions to decide if any additional Structure labels should be added.

If m0 and m1 are approximately equal (allow for 10% leeway) in price **and** m0 and m1 are equal or related by 61.8% in time **and** m3 is longer and more vertical than m1 **and** m2 takes at least as much time as m0 or m1 **and** m2 is very close to 38.2% of m1 **and** one of m0's Structure label choices is ":F3," then add "[:c3]" to the Structure list for m1. For ":c3" to be a good choice, it is preferable m2 terminate at an important Fibonacci price level to m0, m1 or the previous Impulsive wave (monowave or higher). Keep in mind that a ":c3," even under these conditions, is still a risky choice (that is the reason for the brackets which surround it).

If m3 is longer and more vertical than m1 **and** either m3 is completely retraced OR m3 is retraced no more than 61.8% **and** m2 is very close to 38.2% of m1 **and** one of m0's Structure label choices is ":c3" **and** m(-3) is longer than m(-2) **and** m(-2) or m(-1) is longer than m0, m1 may be the second-to-last leg of a Contracting Triangle; add "(:sL3)" to the Structure list. Keep in mind that an ":sL3," even under these ideal conditions, is still a less likely Structure label than the ":5" placed earlier (that is the reason for the brackets which surround it).

Condition "d" -- if m0 is more than 161.8% of m1

There is only one possibility under these circumstances, place a ":5" at the end of m1.

Rule 2
(activation requirement)

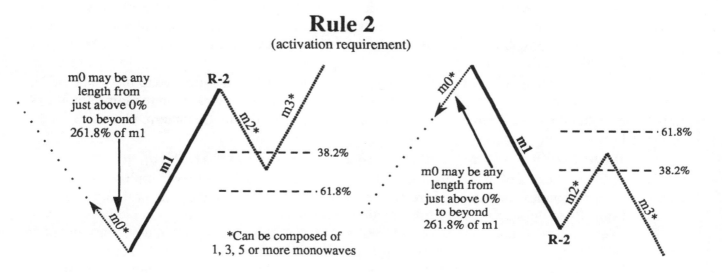

*Can be composed of
1, 3, 5 or more monowaves

Rule 2 {:5/(:sL3)/[:c3]/[:s5]}

Condition "a" -- if m0 is less than 38.2% of m1

Place a ":5" at the end of m1. If m4 does not exceed the end of m0, m1 may be completing a Corrective pattern within a Complex formation where m2 is an x-wave; place ":s5" at the end of m1 and "x:c3?" at the end of m2. When comparing m(-1), m1 and m3, if m1 is not the shortest of the three **and** the longest of the three is close to (or greater than) 161.8% of the next longest **and** m3 is retraced at least 61.8%, the market may be forming an Impulse pattern with m1 the center phase (wave-3). For additional Structure label possibilities which may pertain to m1, read the rest of this section.

If m0 is composed of more than three monowaves **and** m1 retraces all of m0 in the same amount of time (or less) that m0 took to form, m0 is probably the end of an important Elliott pattern.

If m0 and m2 are related by 61.8% in price **and** equal or related (by 61.8%) in time **and** m(-1) is 161.8% or more of m1 **and** m3 (and any additional monowaves required) achieves a price length longer than m(-1) within a time frame equal to or less than m(-1), a Running Correction (any variation) is probably unfolding; take note of that fact and add "[:c3]" after the ":5" already at the end of m1. The Running correction most likely started at the beginning of m0 and concluded at the end of m2. When you begin grouping your Structure labels in Chapter 4, work on the Running Correction as the first possibility with the ":c3" either wave-b of a Running Correction or wave-x of a Double Three Running Correction. Read on to be aware of additional circumstances possible at this juncture.

If m0 and m2 are approximately equal in time **and** m3 is less than 161.8% of m1 **and** m(-1) is longer than m0, m1 might be part of a Complex Correction which will necessitate the use of an "x-wave" Progress label. The x-wave will be in one of three places; if m(-2) is shorter than m(-1), the x-wave might be at the end of m0; if m4 is no more than 161.8% of m3, the x-wave may be at the end of m2; **or**, if m0 is no more than 50% of m1 **and** m1 is the longest when compared to m(-1) and m3, the x-wave may be hidden from sight (i.e., invisible or missing) in the center of m1. To warn of the three possibilities, take a pencil and place an "x:c3?" at the end of m0, the end of m2 and, if appropriate, the center of m1 (using a circle to mark the center of m1). The "missing" x-wave in the center of m1 is the least likely of the three choices under these circumstances. [The concept of "missing waves" is discussed in Chapter 12, page 12-34.] **NOTE**: The x-wave can occur in only one of the three places mentioned. If the x-wave concept is used at one point, erase the other two possibilities. These warnings will come in helpful as you group monowaves together in Chapter 4 and as you finalize your interpretations throughout the analysis process. The x-wave possibility makes use of the ":5" Structure label which should already be in place at the end of m1.

If m(-1) is longer than m0 **and** m0 is shorter than m1 **and** m1, when compared to m(-1) and m3, is not the shortest of the three **and** m3 (plus one time unit) is completely retraced in the same amount of time (or less) that it took to form, m1 may be wave-3 of a Terminal Impulse pattern; place ":c3" at the end of m1.

Condition "b" -- if m0 is at least 38.2%, but less than 61.8% of m1

Place a ":5" at the end of m1, if m4 does not exceed the end of m0, m1 may complete a Corrective pattern within a Complex formation where m2 is an x-wave; place ":s5" at the end of m1 and "x:c3?" at the end of m2. Read through the rest of this section to see what other possibilities could be unfolding.

If m0 is composed of more than three monowaves **and** m1 retraces all of m0 in the same amount of time (or less) that m0 took to form, m0 is probably the end of an important Elliott pattern.

If m0 and m2 are approximately equal in price and time (or related by 61.8% in either case) **and** the monowave before m0 is 161.8% or longer than m1 **and** if m3 (and any additional monowaves that are required) achieves a price length longer than m(-1) within a time frame equal to or less than m(-1), **then** a Running Correction (any variation) is probably unfolding; take note of that fact and add "[:c3]" after the ":5" already at the end of m1. The Running correction would most likely start at the beginning of m0 and conclude at the end of m2. When you begin grouping your Structure labels in Chapter 4, work on the Running Correction as the first possibility with the ":c3" either wave-b of a Running Correction or wave-x of a Double Three Running Correction. Read the next paragraph to be aware of any additional or unusual circumstances which could evolve under these conditions.

If m0 and m2 are approximately equal in price and time **and** m3 is not 161.8% (or more) of m1 **and** m3 (plus one time unit) is retraced completely by m4 faster than m3 took to form, m1 may be part of a complex correction which will necessitate the use of an "x-wave" Progress label. The x-wave will be in one of three places; the end of m0, the end of m2 **or** hidden from sight (i.e., invisible or missing) in the center of m1. [The concept of "missing waves" is discussed in Chapter 12, page 12-34.] To warn of these three possibilities, take a pencil and place an "x:c3?" at the end of m0, the end of m2 and the center of m1 (using a circle to mark the center of m1). As the length of m3 drops below 61.8% of m1, the probabilities increase dramatically that the x-wave (if it exists) is hidden in the center of m1. These warnings will come in helpful as you group monowaves together in Chapter 4 and as you finalize your interpretation throughout the analysis process.

If part of the price range of m2 is shared by m0 **and** m0 and m2 differ by at least 61.8% in time **and** if m1 is not the shortest wave when compared to m3 and m(-1) **and** after m3, the market quickly returns to the beginning of m1, there is a chance m1 is an ":sL3" which is part of a Terminal pattern; jot down that fact and add ":sL3" to the Structure list.

Condition "c" -- if m0 is at least 61.8%, but less than 100% of m1

Under all circumstances, place ":5" at the end of m1. If m4 <u>does not</u> exceed the end of m0, m1 may complete a Flat pattern within a Complex formation where m2 is an x-wave; place ":s5" at the end of m1 and "x:c3?" at the end of m2. Other possibilities could be unfolding, read the rest of this section to make sure no additional possibilities are missed. If the ":5" is used as the preferred Structure label, it may be the end of a Zigzag within a Running or Irregular Failure Flat correction **OR** part of a Complex correction where wave-x is at the end of m0 or m2 (add "x:c3?" to the end of m0). If considering the Complex scenario, for the x-wave to work in the m0 position, m(-2) must be shorter than m(-1) **and** it is extremely likely m(-4) would be larger than m(-3). For the x-wave to work in the m2 position, it is extremely likely m(-2) would be longer than m(-1). Furthermore, for the x-wave to work in the m2 position, m1 must be at least 38.2% of m(-1) with it preferable m1 be 61.8% or more of m(-1). Record these facts and classify them based on their proper Chronological label.

If m(-1) is larger than m0, but less than 261.8% of m1 **and** m3 is shorter than m1 **and** after m3, the market quickly returns to the beginning of m1 (or beyond), a Terminal may have completed with m3; place ":c3" at the end of m1.

If m0 is composed of more than three monowaves **and** m1 retraces all of m0 in the same amount of time (or less) that m0 took to form, m0 probably ends an important Elliott pattern; note that on chart.

If m2 (plus one time unit) is completely retraced in the same amount of time (or less) that it took to form **and** m3 is longer and more vertical than m1 **and** m(-1) is not more than 161.8% of m1, a Running Triangle may have terminated with m2; place ":sL3" at the end of m1. If the thrust out of the Triangle (m3) is completely retraced by m4 faster than it took to form, the Triangle is of the Limiting variety. If m3 <u>is not</u> completely retraced by m4 **OR** m4's slope is much less than m3's **and** m4 is completely retraced, the Triangle is probably of the Non-Limiting variety **or** m3 will become part of a five segment Terminal formation.

If m3 and m(-1) are both at least 161.8% of m1, an Irregular Failure may have completed with m2; place ":c3" at the end of m1.

Condition "d" -- if m0 is between 100% and 161.8% of m1 (inclusive)

If m2 takes the same amount of time (or more) as m1 **OR** m2 takes the same amount of time (or more) as m3, place a ":5" at the end of m1; other possibilities could be unfolding, read the rest of this section to make sure no possibilities are overlooked.

If m2 (plus one time unit) is completely retraced in the same amount of time (or less) that it took to form **and** m3 is longer and more vertical than m1 **and** m0 and m1 consume a similar amount of <u>time</u> (within a 61.8% tolerance) **and** m2 is <u>at least</u> 61.8% of the time of m0 or m1 **and** m0 is no more than 138.2% of m1, then a severe "C-Failure" Flat may have concluded with m2; place ":c3" at the end of m1.

If m3 is longer and more vertical than m1 **and** either m3 is completely retraced OR m3 is retraced no more than 61.8% **and** one of m0's Structure label choices is ":c3" **and** m(-3) is longer than m(-2) **and** m(-2) or m(-1) is longer than m0, m1 may be the second-to-last leg of a Contracting Triangle; add "(:sL3)" to the Structure list.

If m3 is shorter than m1 **and** m3 is retraced at least 61.8% **and** m1 takes less time than m0 **and** m2 takes just as much time (or more) as m1, then m1 is probably part of a Zigzag which concludes with m3; put a ":5" at the end of m1.

Condition "e" -- if m0 is more than 161.8% of m1

No matter what the circumstances, a ":5" is a very likely Structure label for m1; place ":5" at the end of m1. If m3 is shorter and less vertical than m1, the ":5" is the **only** plausible choice.

If m2 (plus one time unit) is completely retraced in the same amount of time (or less) that it took to form **and** m3 is longer and more vertical than m1 **and** m(-1) does not share any similar price territory with m1, the market may have completed a Complex Correction at m2 with a "missing" x-wave in the middle of m0; add ":c3" to the current Structure list of m1, place a dot in the center of m0 (around the same price level m1 completed) and place "x:c3?" to the right of the dot and ":5" to the left of the dot.

Rule 3
(activation requirement)

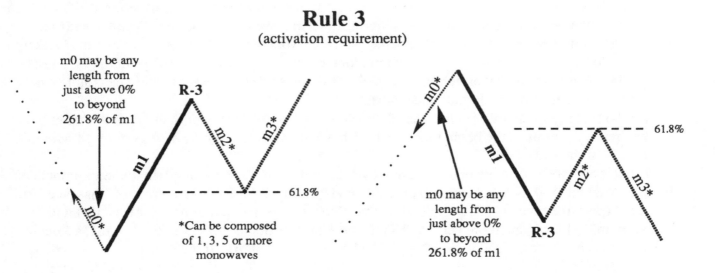

m0 may be any length from just above 0% to beyond 261.8% of m1

R-3

61.8%

*Can be composed of 1, 3, 5 or more monowaves

m0 may be any length from just above 0% to beyond 261.8% of m1

R-3

61.8%

Rule 3 {:F3/:c3/:s5/:5/(:sL3)/[:L5]}

Condition "a" -- if m0 is less than 38.2% of m1

If m3 is <u>more than 261.8%</u> of m1, m1 most likely is the center portion of a Running Correction, but it could also be the end of a Zigzag within a Complex Correction; place ":c3/(:s5)" at the end of m1 to show these two possibilities respectively. If m1 is the longest when compared to m(-1) and m(-3) **and** m2 breaks a trendline drawn across the low of <u>m(-2) and m0</u> in a period of time equal to or less than that taken by m1, m1 may be the 5th wave of a 5th Extension pattern; add [:L5] to the end of m1. If m(-1) is more than 161.8% of m1, drop ":s5" from the list. If m(-1) is less than 61.8% of m3, more than one Elliott pattern (each of a larger magnitude) may have completed at m2.

If m3 is <u>between 161.8 and 261.8%</u> (inclusive) of m1, m1 may be the center portion of an Impulse pattern with a 5th wave Extension, the center portion of a Running Correction, or the first leg of an Elliott pattern within a Complex Correction; place ":s5/:c3/:F3" at the end of m1 to list these three possibilities in their respective order. If m1 is the longest when compared to m(-1) and m(-3) **and** m2 breaks a trendline drawn across the low of <u>m(-2) and m0</u> in a period of time equal to or less than that taken by m1, m1 may be the 5th wave of a 5th Extension pattern; add [:L5] to the end of m1. If m(-1) is longer than m3, drop ":c3" from the Structure list. If m(-1) is longer than m1, the ":s5" (if used for m1) could only be the c-wave of a Zigzag within a Complex Correction; m2 would be an x-wave which is then most likely followed by wave-a of a Contracting Triangle.

If m3 is <u>at least 100, but less than 161.8%</u> of m1, m1 may be the first leg of a Standard Elliott pattern within a Complex Correction, wave-3 of an Impulse pattern with a 5th-wave Extension **or** the c-wave of a Zigzag within an ongoing Complex Correction; place ":F3/:5/:s5" at the end of m1 to show these three possibilities respectively. If m1 is the longest when compared to m(-1) and m(-3) **and** m2 breaks a trendline drawn across the low of <u>m(-2) and m0</u> in a period of time equal to or less than that taken by m1, m1 may be the 5th wave of a 5th Extension pattern; add [:L5] to the end of m1. If m4 is smaller than m3, drop ":F3" from the Structure list. If m0 simultaneously takes less time than m(-1) and m1, drop ":s5" from the list. If m(-1) is longer than m1 and the ":s5" is used, m1 could **only** be the c-wave of a Zigzag within a Complex Correction; m2 would be an x-wave.

If m3 is <u>shorter than m1</u> **and** m3 is completely retraced <u>faster</u> than it took to form, an Impulsive or Complex Corrective pattern may have concluded with m3; place ":5/:F3" at the end of m1. If m1 is the longest when compared to m(-1) and m(-3) **and** m2 breaks a trendline drawn across the low of <u>m(-2) and m0</u> in a period of time equal to or less than that taken by m1, m1 may be the 5th wave of a 5th Extension pattern; add [:L5] to the end of m1.

If m3 is <u>shorter than m1</u> **and** m3 is completely retraced <u>slower</u> than it took to form, m1 concludes a Zigzag which is part of a Complex Corrective pattern; place ":s5" at the end of m1. If m1 is the longest when compared to m(-1) and m(-3) **and** m2 breaks a trendline drawn across the low of <u>m(-2) and m0</u> in a period of time equal to or less than that taken by m1, m1 may be the 5th wave of a 5th Extension pattern; add [:L5] to the end of m1.

If m3 is <u>shorter than m1</u> **and** m4 is shorter than m3, m1 may have completed a Zigzag which is part of a Complex Correction **or** is part of a Terminal Impulse pattern; place ":s5/:F3" at the end of m1 to show these two possibilities, respectively. If m1 is the longest when compared to m(-1) and m(-3) **and** m2 breaks a trendline drawn across the low of <u>m(-2) and m0</u> in a period of time equal to or less than that taken by m1, m1 may be the 5th wave of a 5th Extension pattern; add [:L5] to the end of m1. If m5 is longer than m3, drop ":F3" as a possible Structure label.

Condition "b" -- if m0 is at least 38.2%, but less than 61.8% of m1

If m3 is <u>more than 261.8%</u> of m1, m1 most likely is the center portion of an Irregular Failure, but it could also be the end of a Zigzag within a Complex Correction; place ":c3/(:s5)" at the end of m1 to show these two possibilities respectively. If m(-1) is more than 161.8% of m1, drop ":s5" from the list. If m(-1) is less than 61.8% of m3, more than one Elliott pattern may have completed at m2 (each of a larger magnitude).

If m3 is <u>between 161.8 and 261.8%</u> (inclusive) of m1, m1 may be the center portion of an Irregular Failure, the c-wave of a Zigzag within a Complex Correction **or** the center portion of a 5th Extension Terminal Impulse pattern; place ":c3/:s5" at the end of m1. The Running Correction and Terminal Impulse both make use of the ":c3" while the ":s5" would be used for the Zigzag scenario. If m(-1) is longer than m1, forget the Terminal scenario.

If m3 is <u>at least 100, but less than 161.8%</u> of m1, m1 may be the <u>first</u> or <u>last leg</u> of a Zigzag within a Complex Correction **or** the center portion of a 5th Extension Terminal Impulse pattern; place ":5/:s5/:c3" at the end of m1. If m(-1) is longer than m1, drop ":c3" from the list of possibilities. If m(-1) is longer than m1 **and** the ":s5" is used for m1, m1 would be wave-c of a Zigzag within a Complex Correction; place "x:c3?" at the end of m2. If m4 is smaller than m3, drop ":5" from the Structure list. If m3 (plus one time unit) is completely retraced faster than it took to form, drop ":s5" from the Structure list.

If m3 is <u>shorter than m1</u> **and** m3 (plus one time unit) is completely retraced <u>faster</u> than it took to form, a Complex Correction may have concluded with m3; place ":5" at the end of m1. If m4 returns to the beginning of m(-1) within a period of time equal to 50% (or less) of that consumed by m(-1) through m3 **and** m(-1) is not more than 261.8% of m1, m1 may be part of a Terminal Impulse pattern; add ":c3" to m1.

If m3 is <u>shorter than m1</u> **and** m3 is completely retraced <u>slower</u> than it took to form, m1 concludes a Zigzag which is part of a Complex Corrective pattern; place ":s5" at the end of m1.

If m3 is <u>shorter than m1</u> **and** m4 is shorter than m3, m1 may have completed a Zigzag which is part of a Complex Correction **or** is part of a Terminal Impulse pattern; place ":s5/:F3" at the end of m1 to show these two possibilities, respectively. If m5 is longer than m3, drop ":F3" as a possible Structure label.

Condition "c" -- if m0 is at least 61.8%, but less than 100% of m1

If m3 is <u>more than 261.8%</u> of m1, m2 probably completed an Irregular Failure Flat **or** a Non-Limiting Triangle; place ":c3/:sL3" at the end of m1. If m(-1) is more than 161.8% of m1, drop ":sL3" from the list of Structure possibilities. If m(-1) is not more than 161.8% of m1 **and** m(-2) is at least 61.8% of m(-1), drop ":c3" from the Structure list.

If m3 is <u>between 161.8% and 261.8%</u> (inclusive) of m1, m1 may be the center portion of an Irregular Failure Flat, the second-to-last leg of a Contracting Triangle **or** part of a Complex Correction; place ":F3/:c3/:sL3/:s5" at the end of m1. If m3 (plus one time unit) is completely retraced faster than it took to form, drop ":s5" from the above list. If m(-1) is more than 161.8% of m1, drop ":sL3" from the above list. If m(-1) is not more than 161.8% of m1 **and** m(-1) is retraced at least 61.8%, drop ":c3" from the above list. If m4 is shorter than m3, drop ":F3" from the list.

If m3 is <u>at least 100%, but less than 161.8%</u> of m1, m1 may be the center portion of an Irregular Failure Flat, the second-to-last leg of a Contracting Triangle, the center leg of a 5th Extension Terminal pattern **or** one of the legs of a Complex Correction; place ":F3/:c3/:sL3/:s5" at the end of m1. If m4 is shorter than m3, drop ":F3" from the list and forget about the Terminal scenario. If m3 (plus one time unit) is completely retraced faster than it took to form, drop ":s5" from the above list. If m(-1) is more than 161.8% of m1, drop ":sL3" from the above list. If m(-1) is not more than 161.8% of m1 **and** m(-1) is retraced at least 61.8%, drop ":c3" from the above list.

If m3 is <u>shorter than m1</u> **and** m3 (plus one time unit) is completely retraced <u>faster</u> than it took to form, a Terminal Impulse or Complex Correction may have concluded with m3; place ":c3/:F3" at the end of m1. If m(-1) is less than 138.2% **or** more than 261.8% of m1, the ":c3" becomes very improbable; place brackets around it - "[:c3]."

If m3 is <u>shorter than m1</u> **and** m3 (plus one time unit) is completely retraced <u>slower</u> than it took to form, m1 may be wave-a of a Zigzag or the c-wave of a Zigzag within a Complex Correction; place ":F3/(:s5)" at the end of m1. If m5 (plus one time unit) is completely retraced by m4 faster than m4 took to form, drop "(:s5)" from the list.

If m3 is <u>shorter than m1</u> **and** m4 is shorter than m3, m1 may be the <u>last leg</u> of a Zigzag or Flat within a Complex Correction, one of the center legs of a Running Contracting Triangle **OR** the first leg of a Terminal Impulse pattern; place ":s5/:c3/(:F3)" at the end of m1. If m5 is longer than m3, drop "(:F3)" from the Structure list. If m(-1) is longer than 261.8% of m1, drop ":s5" from the list.

Condition "d" -- if m0 is at least 100%, but less than 161.8% of m1

If m3 is <u>more than 261.8%</u> of m2, m1 may be the first leg of a Zigzag, the center section of a C-Failure Flat **or** the second-to-last leg of a Triangle; place ":5/:c3/(:sL3)" at the end of m1. If m(-1) is less than 61.8% or more than 161.8% of m0, drop "(:sL3)" from the Structure list. If m2 is retraced slower than it took to form, drop "(:sL3)" and ":c3" from the list. If m3 is more than 161.8% of m1, drop ":5" from the list.

If m3 is <u>between 161.8 and 261.8%</u> (inclusive) of m2, m1 may be the center section of a C-Failure Flat, the second-to last leg of a Contracting Triangle **or** the first leg of a Zigzag; place ":c3/:sL3/:5" at the end of m1 to show these three possibilities respectively. If m(-1) is less than 61.8% of m0 <u>OR</u> more than 161.8% of m0, then check to see if m1 is less than 38.2% of m(-3) through m0; if so, drop ":sL3" from the list; if m1 is more than 38.2%, but less than 61.8% of m(-3) through m(-1), place parentheses around ":sL3" to show that, though it is possible, it is not the preferred choice. If m(-1) falls between 61.8-161.8% of m0, drop ":c3" from the list. If m4 is less than 61.8% of m0, place parentheses around ":5" to indicate its lower probability.

If m3 is <u>at least 100, but less than 161.8%</u> of m2, m1 is probably the first leg of a Zigzag, but it may be in a Triangle; place ":5/(:c3)/[:F3]" at the end of m1. If m4 is larger than m3, drop "(:c3)" and "[:F3]" from the list of possibilities. If m4 is shorter than m3 **and** m5 retraces m4 faster than m4 took to form **and** m5 is equal to (or longer) and more vertical than m1, drop ":5" from the list.

Condition "e" -- if m0 is between 161.8% and 261.8% of m1 *(inclusive)*

If m3 is <u>more than 261.8%</u> of m2, m1 may be the first leg of a Zigzag, the center section of a C-Failure Flat which concludes a Complex Correction (with a "missing" x-wave in the middle of m0) **or** the second-to-last leg of a Triangle; place ":5/:c3/(:sL3)" at the end of m1. If m(-1) is less than 61.8% or more than 161.8% of m0, drop "(:sL3)" from the Structure list. If m2 is retraced slower than it took to form, drop "(:sL3)" and ":c3" from the list. If m3 is more than 161.8% of m1, drop ":5" from the list.

If m3 is <u>between 161.8 and 261.8%</u> (inclusive) of m2, m1 may be the first leg of a Zigzag **or** the center section of a C-Failure Flat which concludes a Complex Correction (with a "missing" x-wave in the middle of m0); place ":5/:c3" at the end of m1 and put a dot in the middle of m0 with "x:c3?" to the right and ":s5?" to the left of the dot. If m2 is retraced slower than it took to form, drop ":c3" from the list. If m3 is more than 161.8% of m1, drop ":5" from the list.

If m3 is <u>at least 100, but less than 161.8%</u> of m2, m1 may be the first leg of a Zigzag or the first leg of a Triangle; place ":5/(:F3)" at the end of m1. If m4 is a monowave **and** m4 is longer than m3, drop "(:F3)" from the Structure list.

Condition "f" -- if m0 is more than 261.8% of m1

If m3 is <u>more than 261.8%</u> of m2, m1 may be the first leg of a Zigzag **or** the center section of a C-Failure Flat which concludes a Complex Correction (with a "missing" x-wave in the middle of m0); place ":5/(:c3)" at the end of m1. If m2 is retraced slower than it took to form, drop ":c3" from the list. If m3 is more than 161.8% of m1, drop ":5" from the list. If the "(:c3)" Structure label is used for m1 and m(-1) shares no similar price territory with m1, mark the middle of m0 with a dot and place "x:c3?" to the right of it and ":s5" to the left of the dot to represent m0's "missing" x-wave possibility.

If m3 is <u>between 161.8 and 261.8%</u> (inclusive) of m2, m1 may be the first leg of a Zigzag **or** the center section of a C-Failure Flat which concludes a Complex Correction (with a "missing" x-wave in the middle of m0); place ":5/(:c3)" at the end of m1. If m3 is longer than m2, drop "(:c3)" from the list. If m2 is retraced slower than it took to form, drop "(:c3)" from the list. If m3 is more than 161.8% of m1, drop ":5" from the list. If the "(:c3)" Structure label is used for m1 and m(-1) shares no similar price territory with m1, mark the middle of m0 with a dot and place "x:c3?" to the right of it and ":s5" to the left of the dot to represent m0's "missing" x-wave possibility.

If m3 is <u>at least 100, but less than 161.8%</u> of m2, m1 may be the first leg of a Zigzag or the first leg of a Triangle; place ":5/(:F3)" at the end of m1. If m4 is a monowave **and** m4 is longer than m3, drop "(:F3)" from the Structure list

Rule 4a

m3 may be any length from
100% to beyond 261.8% of m2

R-4a

m1

*m0**

*m2**

*m3**

38.2% of m1 - - - - - -

*Can be composed of 1,
3, 5 or more monowaves

*m0**

38.2% of m1 - - - - - *m1*

*m2**

*m3**

R-4a

m3 may be any length from
100% to beyond 261.8% of m2

Rule 4

Condition "a" {:F3/:c3/:s5/[:sL3]}

Category "i" -- if m3 is at least 100%, but less than 161.8% of m2

If m3 (plus one time unit) is completely retraced **slower** than it took to form, m1 should be the first leg of a correction which follows an x-wave (m0) **or** the end of a corrective phase which is part of a larger Standard or Non-Standard formation; place ":F3/:s5" at the end of m1. If ":F3" is chosen, m1 is wave-a of a Flat correction. If ":s5" is appropriate, m1 is the end of a Zigzag pattern. If m1 is less than 61.8% of m(-1), drop ":s5" from the Structure list. If m0 simultaneously takes less time than m(-1) and m1, drop ":s5" from m1's list.

If m3 is completely retraced in the same amount of time (or less) that it took to form, there is virtually no chance m1 completed an Elliott pattern; place only ":F3/:c3" at the end of m1. If m1 is retraced <u>no more</u> than 70% **and** none of m2's price range is shared by m0 **and** m3 is almost 161.8% of m1 **and** m0 takes more time than m(-1) <u>or</u> m0 takes more time than m1, add ":s5" to the Structure list. If none of m2's price range is shared by m0, drop ":c3" as a choice. If ":F3" is used, m1 would be wave-a of a correction within a larger complex formation; m0 would be an x-wave. If ":c3" is still a possibility, m1 may be part of an Expanding Triangle or a Terminal Impulse pattern. If ":s5" is a possibility, it would be wave-3 of a 5th Extension Impulse pattern.

If m3 is retraced less than 100%, place ":F3/:s5" at the end of m1. If m2 is composed of more than three monowaves **and** m2 (plus one time unit) is completely retraced faster than it took to form **and** m2 takes more time than m1 **and** m2 breaks a line draw across the end of <u>m(-2) and m0</u> faster than m1 took to form, <u>then</u> m1 may be the end of a Zigzag contained within an Irregular or Running Correction; add ":L5" to the Structure list at the end of m1. The ":L5" in this case was justified based on the two stages of confirmation possible when dealing with polywave patterns (see Chapter 6 for further details). If m0 simultaneously takes less time than m(-1) and m1, drop ":s5" from list. If ":F3" is chosen, m1 is wave-a of a Flat or Triangle correction. If ":s5" is appropriate, m1 is the end of a Zigzag pattern.

Category "ii" -- if m3 is between 161.8% and 261.8% of m2 (inclusive)

If m(-1) is more than 261.8% of m1, there is virtually no chance m1 is the end of any Elliott formation; place <u>only</u> ":F3" at the end of m1.

If m4 is longer than m3, it is extremely unlikely m1 is the end of any Elliott pattern, place only ":F3" at the end of m1.

If m3 is retraced less than 100%, place ":s5" at the end of m1 then follow the below directions to decide what type of Elliott pattern is forming.

1. *If m1 is retraced no more than 70% and m1 is from 101-161.8% of m(-1) and none of the price range covered by m2 is shared by m0 and m(-2) is longer than m(-1), m1 may conclude the 3rd wave of* a Trending 5th Extension Impulse pattern. If m1 is between 161.8-261.8% of m(-1), it becomes more likely that m1 is the end of a Zigzag inside of a Complex Correction with m2 the end of an "x-wave." But, the 3rd wave idea is still possible as long as you identify the 3rd wave as part of a Double Extension Impulse pattern (see diagram; Chapter 12, page 12) with the 5th wave the longest. If m1 is more than 261.8% of m(-1), the Complex corrective scenario is the only proper choice.

   2. If m1 is retraced no more than 70% **and** m1 is at least 100, but less than 161.8% of m(-1) **and** some of the price range covered by m2 is shared by m0 **and** m(-2) is longer than m(-1), then m1 may conclude the 3rd wave of a Terminal 5th Extension Impulse pattern. If m1 is between 161.8-261.8% (inclusive) of m(-1), it becomes more likely m1 is the end of a Zigzag inside of a Complex Correction with m2 the end of an "x-wave." But, the 3rd wave idea is still possible as long as you identify the 3rd wave as part of a Double Extension Terminal Impulse pattern with the 5th wave the longest. If m1 is more than 261.8% of m(-1), the Complex Corrective scenario is the only proper choice.

   3. If m1 is retraced no more than 70% **and** m1 is smaller than m(-1), m1 can only be part of a Zigzag pattern.

   4. If m1 is retraced more than 70%, it is most likely m1 terminated a Zigzag. But, if the price ranges of m0 and m2 share some common territory **and** m3 is retraced faster than it took to form, m1 may be the 3rd wave of a 5th Extension Terminal Impulse pattern.

   Make notes of these facts on your chart.

Category "iii" -- if m3 is more than 261.8% of m2

   If m(-1) is more than 261.8% of m1, there is virtually no chance m1 is the end of any Elliott formation; place <u>only</u> ":F3" at the end of m1.

   If m3 is retraced <u>completely</u> by m4, it is extremely unlikely m1 is the end of any Elliott pattern, place only ":F3" at the end of m1.

   If m3 is retraced <u>less</u> than 100%, it is extremely unlikely m1 is the beginning of any Elliott pattern, place only ":s5" at the end of m1.

Rule 4b

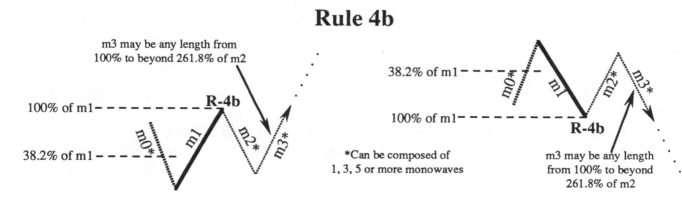

Condition "b" {:F3/:c3/:s5/(:sL3)/(x:c3)/[:L5]}
Category "i" -- if m3 is at least 100%, but less than 161.8% of m2

   If m3 (plus one time unit) is <u>completely retraced</u> in the same amount of time (or less) that it took to form, there is virtually no chance m1 completed an Elliott pattern; place only ":F3/:c3" at the end of m1. If later it is found that ":c3" is the preferred choice, m1 may be part of a Terminal Impulse pattern. If the end of m3 is exceeded before the end of m0 **and** m1 is the longest when it is compared to m(-1) and m(-3) **and** m2 breaks a trendline drawn across the low of <u>m(-2) and m0</u>, in a period of time equal to or less of that taken by m1, m1 may be the 5th wave of a 5th Extension pattern; add [:L5] to the end of m1.

If m3 is <u>completely retraced slower</u> than it took to form; place ":F3/:c3/:s5" at the end of m1. If the end of m1 is exceeded during the formation of m2, add an "x" in front of ":c3." If the end of m3 is exceeded before the end of m0 **and** m1 is the longest when compared to m(-1) and m(-3) **and** m2 breaks a trendline drawn across the low of <u>m(-2) and m0</u> in a period of time equal to or less than that taken by m1, m1 may be the 5th wave of a 5th Extension pattern; add [:L5] to m1's Structure list. If m1 is less than 61.8% of m(-1), drop ":s5" from the list. If m(-1) is 161.8% (or more) of m1 **and** m3 is retraced less than 61.8%, drop ":F3" as a possibility. If m0 (plus one time unit) simultaneously takes less time than m(-1) and m1, drop ":s5" from list.

If m3 is <u>retraced less than 100%</u>, it is extremely unlikely m1 is the beginning of any Elliott pattern; place ":c3/:s5" at the end of m1. If the end of m1 is exceeded during the formation of m2, add an "x" in front of ":c3." Even if this next condition is not met, read this entire paragraph. If m2 is composed of more than three monowaves **and** m2 is completely retraced faster than it took to form **and** m2 takes more time than m1 **and** m(-1) is at least 161.8% of m0 **and** m2 breaks a line which is draw across the end of <u>m(-2) and m0</u> faster than m1 took to form, <u>then</u> m1 may be the end of a Zigzag contained within an Irregular or Running Correction; add ":L5" to the Structure list at the end of m1. The ":L5" in this case was justified based on the two stages of confirmation possible when dealing with polywave patterns (see Chapter 6 for further details). If m0 (plus one time unit) takes less time than m(-1) and less time than m1, drop ":s5" from list. If m(-2) is longer than m(-1) **and** ":c3" in the list is not preceded by an "x," drop ":c3" from the list. If m5 is not completely retraced as fast as it took to form, drop ":c3" from the list. If ":c3" is <u>still</u> a possibility, m1 may be the x-wave of a Complex Correction; add "x:c3?" as a additional Structure label to m1's list.

If m3 is <u>retraced less than 61.8%</u>, it is extremely unlikely m1 is the beginning of any Elliott pattern; place ":c3/:sL3/:s5" at the end of m1. If the end of m1 is exceeded during the formation of m2, add an "x" in front of ":c3." If m1 is the longest when compared to m(-1) and m(-3) **and** m2 breaks a trendline drawn across the low of <u>m(-2) and m0</u> in a period of time equal to or less than that taken by m1, m1 may be the 5th wave of a 5th Extension pattern; add [:L5] to the end of m1. If m3 through m5 do not achieve a price distance of 161.8% of m1 (or more), drop ":sL3" from the list. If m2 (plus one time unit) is not completely retraced in the same amount of time that it took to form (or less), drop ":sL3" from the list. If m0 (plus one time unit) simultaneously takes less time than m(-1) and m1, drop ":s5" from list. If m(-2) is longer (in price) than m(-1) **and** ":c3" is not preceded by an "x," drop ":c3" from the list. If ":c3" is still a possibility, m1 may be the x-wave of a Complex Correction; add "x:c3?" as an additional Structure label to m1's list.

Category "ii" -- if m3 is between 161.8% and 261.8% of m2 (inclusive)

If m(-1) is more than 261.8% of m1, there is virtually no chance m1 is the end of any Elliott formation; place <u>only</u> ":F3/:c3" at the end of m1. If the end of m1 is exceeded during the formation of m2, add an "x" in front of ":c3."

If m1 is the longest wave when compared to m(-1) and m(-3) **and** m2 (plus one time unit) breaks a trendline drawn across the low of <u>m(-2) and m0</u> in a period of time equal to or less than that taken by m1, m1 may be the 5th wave of a 5th Extension pattern; add [:L5] to the end of m1.

If m3 is retraced less than 61.8%, it is extremely unlikely m1 is the beginning of any Elliott pattern, place only ":c3/(:sL3)/(:s5)" at the end of m1. If the end of m1 is exceeded during the formation of m2, add an "x" in front of ":c3." If m3 through m5 do not achieve a price distance of 161.8% (or more) of m1, drop ":sL3" from the list. If m0 (plus one time unit) simultaneously takes less time than m(-1) and m1, drop ":s5" from list. If m2 is completely retraced slower than it took to form, drop ":sL3" from the list. NOTE: if ":sL3 is used, the Triangle (which concludes with m2) is Non-Limiting.

If neither of the conditions above applies, place ":F3/:c3/:sL3/:s5" at the end of m1. If the end of m1 is exceeded during the formation of m2, add "x:c3" to the Structure list. If m1 is the longest when compared to m(-1) and m(-3) **and** m2 breaks a trendline drawn across the low of m(-2) and m0 in a period of time equal to or less than that taken by m1, m1 may be the 5th wave of a 5th Extension pattern; add [:L5] to the end of m1. When comparing m(-1), m1 an m3, if m1 is the shortest of the three **and** m3 (plus one time unit) is completely retraced faster than it took to form, drop ":c3" as a possibility. If m1 is less than 61.8% of m(-1), drop ":s5" from the list. If m3 is retraced less than 61.8%, drop ":F3" as a possibility. If m0 simultaneously takes less time than m(-1) and m1, drop ":s5" from the list. If m2 (plus one time unit) **is** retraced slower than it took to form, drop ":sL3" from the list.

Category "iii" -- if m3 is more than 261.8% of m2

If m(-1) is more than 261.8% of m1, there is virtually no chance m1 is the end of any Elliott formation; place <u>only</u> ":c3/(:F3)" at the end of m1. If the end of m1 is exceeded during the formation of m2, add an "x" in front of ":c3."

If m(-1) is at least 161.8% of m1 **and** m0 is retraced slower than it took to form **and** m1 takes 161.8% of the time of m0 (or more), it is almost certain m0-m2 form an Irregular Failure Flat, place ":c3/(:F3)" at the end of m1. If the end of m1 is exceeded during the formation of m2, add an "x" in front of ":c3."

If m1 is the longest when compared to m(-1) and m(-3) **and** m2 breaks a trendline drawn across the low of m(-2) and m0 in a period of time equal to or less than that taken by m1, m1 may be the 5th wave of a 5th Extension pattern; add [:L5] to the end of m1.

If m3 is retraced less than 61.8%, it is extremely unlikely m1 is the beginning of any Elliott pattern, place only ":F3/:c3/(:s5)" at the end of m1. If ":F3" is used, an Elongated Flat begins with the start of m1. If m1 is the longest when it is compared to m(-1) and m(-3) **and** m2 breaks a trendline drawn across the low of m(-2) and m0 in a period of time equal to or less than that taken by m1, m1 may be the 5th wave of a 5th Extension pattern; add [:L5] to the end of m1. If m0 simultaneously takes less time than m(-1) and m1, drop ":s5" from list. If the end of m1 is exceeded during the formation of m2, add an "x" in front of ":c3."

If none of the conditions above apply, place ":F3/:c3/:sL3/:s5" at the end of m1. If m1 is the longest when compared to m(-1) and m(-3) **and** m2 breaks a trendline drawn across the low of m(-2) and m0 in a period of time equal to or less than that taken by m1, m1 may be the 5th wave of a 5th Extension pattern; add [:L5] to the end of m1. If m1 is less than 61.8% of m(-1), drop ":s5" from the list. If m3 is retraced less than 61.8%, drop ":F3" as a possibility. If m0 simultaneously takes less time than m(-1) and m1, drop ":s5" from list. If the end of m1 is exceeded during the formation of m2, add an "x" in front of ":c3."

Rule 4c

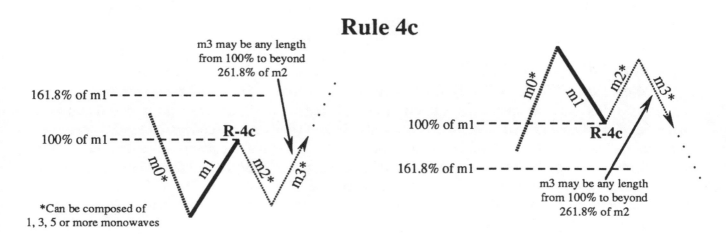

m3 may be any length from 100% to beyond 261.8% of m2

161.8% of m1

100% of m1

R-4c

*Can be composed of 1, 3, 5 or more monowaves

100% of m1

R-4c

161.8% of m1

m3 may be any length from 100% to beyond 261.8% of m2

Condition "c" {:c3/(:F3)/(x:c3)}

Category "i" -- if m3 is at least 100%, but less than 161.8% of m2

The subtleties of this situation do not allow for specific quantifications, just place an ":F3/:c3" at the end of m1. If the end of m1 is exceeded during the formation of m2, add an "x" in front of ":c3." The interaction of m1 with surrounding Structure will help to decide which of the two Structure labels best fits current conditions.

Category "ii" -- if m3 is between 161.8% and 261.8% of m2 (inclusive)

If m2 (plus one time unit) is completely retraced in the same amount of time (or less) that it took to form **and** m3 is more than 161.8% of m1, the chances are better that m1 was the center section of a C-Failure Flat or of a Contracting Triangle; place ":c3/(:F3)" at the end of m1. If the end of m1 is exceeded during the formation of m2, add an "x" in front of ":c3." If the lower probability "(:F3)" is correct, m1 is part of an Elongated Flat.

Under all other situations, place ":F3/:c3/x:c3" at the end of m1.

Category "iii" -- if m3 is more than 261.8% of m2

If m2 (plus one time unit) is completely retraced in the same amount of time (or less) that it took to form, it is almost certain m1 is the center section of a C-Failure Flat or a Non-Limiting, Contracting Triangle; place ":c3/[:F3]" at the end of m1. If the end of m1 is exceeded during the formation of m2, add an "x" in front of ":c3." The only way "[:F3]" should be considered is if m3 is retraced more than 61.8% within a period of time equal to or less than m3.

Rule 4d

m3 may be any length from 100% to beyond 261.8% of m2

261.8% of m1

161.8% of m1

R-4d

*Can be composed of 1, 3, 5 or more monowaves

261.8% of m1

Condition "d" {:F3/(:c3)/(x:c3)}

Categories "i" & "ii" -- if m3 is between 100% and 261.8% of m2 (inclusive)

If m2 (plus one time unit) is completely retraced in the same amount of time (or less) that it took to form **and** m3 is not retraced more than 61.8% **and** m3 (or m3 through m5) consumed 161.8% (or more) of m1 in price within the same amount of time (or less) as m1, m1 may be part of a complex correction in which m0 contains a "missing" x-wave toward its center; place ":F3/[:c3]" at the end of m1 and mark (with a pencil) the center of m0 with a circle and place "x:c3?" to the right of it, "c:5?" to the left of it and ":F3?" at the end of m0. All of the labels with question marks are interdependent; if one is used all must be used, if one is not used, none of them can be used.

If m2 is completely retraced **slower** than it took to form, a Flat or Triangle is probably involved; place ":F3/:c3" at the end of m1. If the end of m1 is exceeded during the formation of m2, add an "x" in front of ":c3."

If m3 (plus one time unit) is completely retraced in the same amount of time (or less) that it took to form, ":F3" is the only reasonable choice for m1.

If m3 is retraced at least 61.8%, but less than 100%, ":F3" is the only reasonable choice for m1.

If m3 is retraced less than 61.8%, place ":F3" at the end of m1. If m5 is not the longest when compared with m1 and m3 **and** m5 (plus one time unit) is completely retraced in the same amount of time (or less) that it took to form, m1 may be part of a Terminal pattern. If m5 is longer than m1 and m3, m1 may be part of a Complex Double Flat pattern in which m4 is an x-wave. Make note of these possibilities. Under both conditions listed in this paragraph, m1 would be marked with an ":F3."

Category "iii" -- if m3 is more than 261.8% of m2

If m3 takes the same amount of time (or less) as m1 **and** m2 (plus one time unit) is completely retraced in the same amount of time (or less) that it took to form, there is a strong possibility of a "missing" x-wave hidden toward the middle of m0; place ":c3" at the end of m1. If the end of m1 is exceeded during the formation of m2, add an "x" in front of the ":c3" at the end of m1. If m3 is retraced at least 61.8%, it is possible m1 is the first leg of a Flat; add ":F3" to the end of m1.

If m3 takes more time than m1, place ":F3/:c3" at the end of m1. If the end of m1 is exceeded during the formation of m2, add an "x" in front of ":c3."

Rule 4e

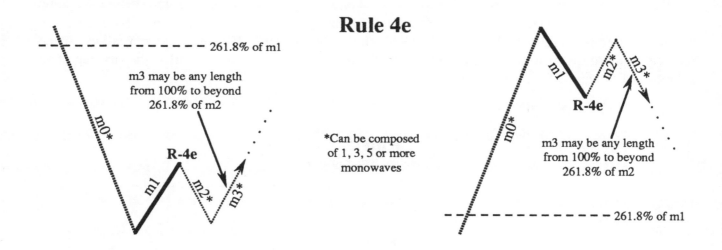

m3 may be any length from 100% to beyond 261.8% of m2

261.8% of m1

R-4e

*Can be composed of 1, 3, 5 or more monowaves

Condition "e" {:F3/(x:c3)/[:c3]}

Categories "i" & "ii" -- if m3 is between 100% and 261.8% of m2 (inclusive)

If m3 (plus one time unit) is completely retraced in the same amount of time (or less) that it took to form, ":F3" is the only reasonable choice, place it at the end of m1.

If m3 is not more than 161.8% of m2 **and** m3 is not completely retraced **and** m4 is retraced faster than it took to form, m1 could be the x-wave of a complex correction; add "x:c3" to the end of m1. In this scenario, if m(-1) is more than 61.8% of m0, m0 would be missing an x-wave toward its center.

If m2 (plus one time unit) is completely retraced in the same amount of time (or less) that it took to form **and** m(-1) is not more than 61.8% of m0 **and** m3 is retraced no more than 61.8% **and** m3 (or m3 through m5) achieved a price distance equal to m1 (or more) within the same amount of time (or less) as m1, m1 may be part of a Complex Correction in which m0 contains a "missing" x-wave toward its center or m1 is the x-wave after a Zigzag; place ":F3/[:c3]" at the end of m1 and mark (with a pencil) the center of m0 with a dot and place "x:c3?" to the right and ":s5" to the left of it. If the end of m1 is exceeded during the formation of m2, m1 may be an x-wave, add "x:c3" to m1's list.

If m2 is retraced **slower** than it took to form **and** m(-1) is not more than 61.8% of m0 **and** m3 is retraced no more than 61.8% **and** m3 through m5 consume 161.8% (or more) of m1 in price within the same amount of time (or less) as m1, m1 may be part of a complex correction in which m0 contains a "missing" x-wave toward its center; place ":F3/[:c3]" at the end of m1 and mark (with a pencil) the center of m0 with a dot and place "x:c3?" to the right and ":s5" to the left of it. If the end of m1 is exceeded during the formation of m2, m1 may be an x-wave, add "x:c3" to m1's list.

If m2 is retraced **slower** than it took to form, a Flat or Triangle is probably involved; place an ":F3" at the end of m1.

If m0 is a polywave (or a monowave with a suspected "missing" wave toward its center), m1 may be an x-wave of a Complex Correction; add "x:c3" to any existing Structure labels at the end of m1.

If m(-1) is no more than 61.8% of m0, m1 may be an x-wave of a Complex Correction; add "x:c3" to any existing Structure labels at the end of m1.

Category "iii" -- if m3 is more than 261.8% of m2

If m3 takes no more time than m1 **and** m2 (plus one time unit) is completely retraced in the same amount of time (or less) that it took to form, there is a strong possibility of a "missing" x-wave hidden toward the middle of m0; put an "x:c3" next to the termination of m1. If m3 through m5 fail to exceed the beginning of m0 **and** m3 is retraced at least 61.8%, it is also very possible m1 is the first leg of an Elongated Flat; add ":F3" to the end of m1.

Rule 5
(activation requirement)

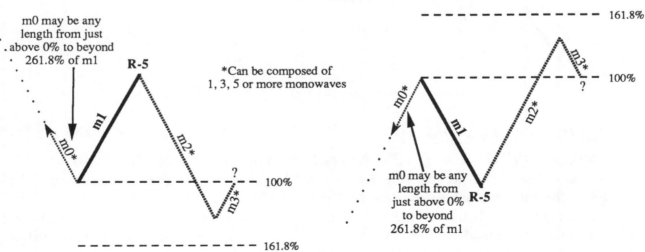

m0 may be any length from just above 0% to beyond 261.8% of m1

R-5

*Can be composed of 1, 3, 5 or more monowaves

161.8%

100%

Rule 5 {:F3/:c3/:5/:L5/(:L3)}

Condition "a" -- m0 is less than 100% of m1

**(if m2 is composed of more than three monowaves [or monowave groups])

If the first three monowaves of m2 **do not** retrace more than 61.8% of m1, a Complex Correction may be unfolding with the first or second monowave (immediately after the end of m1) moving in the opposite direction of m1 representing an x-wave OR m1 contains a "missing" x-wave or "missing" b-wave in its center OR m1 is the 3rd-wave of an 5th-Failure Impulse pattern (Trending or Terminal); place ":5/:s5" at the end of m1. Add ":F3" to the end of m1 **if** m1 is retraced at least 25% by the first 3 monowaves. NOTE: if the "missing" x-wave is used, circle the center of m1 and place ":5 ?" to the left of the circle and "x:c3 ?" to the right of the circle. If the "missing" b-wave is used, circle the center of m1 and place ":5 ?" to the left of the circle and "b:F3 ?" to the right of the circle. If the b-wave is used, the termination of the Complex Correction will be confirmed immediately before the point where the market violently turns (in the opposite direction of m1) and exceeds the 61.8% retracement level of m1.

If the first three monowaves of m2 retrace more than 61.8% of m1, m1 may have completed wave-a of a Flat with a complex b-wave OR m1 may have completed wave-3 of a 5th Failure Impulse pattern; place ":F3/:5" at the end of m1 to show these two possibilities respectively.

**(if m2 is composed of three monowaves [or monowave groups] or less)

If m1 (plus one time unit) is completely retraced in the same amount of time (or less) that it took to form **and** m(-2) and m0 <u>do not</u> share any similar price territory **and** m2 is larger than m(-2) **and** m(-2) and m0 <u>are obviously different</u> in price or time or both **and** when the price lengths of m(-3), m(-1) and m1 are compared, m(-1) is not the shortest, a *Trending* Impulse pattern may have concluded with m1; place ":L5" at the end of m1.

If m1 (plus one time unit) is completely retraced in the same amount of time (or less) that it took to form **and** m2 is longer than m(-2) **and** m(-4) is longer than m(-3), a Zigzag or Flat may have concluded with m1; place ":L5" at the end of m1.

If m1 (plus one time unit) is completely retraced in the same amount of time (or less) that it took to form **and** m2 is longer than m(-2) **and** m(-3) is longer than m(-2) **and** m(-4) is shorter than m(-3), m1 may be the end of a Standard Elliott pattern which is part of a Complex Correction where m(-2) is an x-wave; place ":L5" at the end of m1 and "x:c3?" at the end of m(-2). In the above situation, if m(-1) is at least 161.8% of m0, the Standard Correction unfolding is probably a Zigzag. If m(-1) is at least 100%, but less than 161.8% of m0, the Standard Correction is probably a Flat.

If m1 (plus one time unit) is completely retraced in the same amount of time (or less) that it took to form **and** m2 is smaller than m(-2), a Flat or Zigzag may have concluded with m1; place ":L5" at the end of m1.

If m1 (plus one time unit) is completely retraced in the same amount of time (or less) that it took to form **and** m(-2) is smaller than m(-1) **and** m(-1) is not the shortest when compared to m(-3) and m1 **and** m2 is not retraced more than 61.8% **and** the market approaches (or exceeds) the beginning of m(-3) within a time period which is 50% or less of that consumed by m(-3) through m1 **and** the end of m1 is not exceeded for a period which is four times as long as that consumed by m(-3) through m1 **and** m2-m4's <u>price</u> coverage is at least twice that of m1 **and** one of m(-1)'s possible Structure labels is ":c3," a *Terminal* Impulse may have completed at the end of m1; add ":L3" to the current Structure list for m1.

If m1 (plus one time unit) is completely retraced in the same amount of time (or less) that it took to form **and** m3 is between 61.8-100% of m1, m1 may be part of an Irregular Failure Flat; place ":F3" at the end of m1.

If m1 (plus one time unit) is completely retraced in the same amount of time (or less) that it took to form **and** m3 is longer than m2 **and** m0 is at least 161.8% of m2 **and** m3 is completely retraced in the same amount of time (or less) that it took to form, m1 may be part of an Irregular Flat; place ":F3" at the end of m1.

If m1 is completely retraced **slower** than it took to form **and** m2 did not retrace more than 61.8% of the distance from the beginning of m(-1) to the end of m1 **and** m3 is shorter than m2, the market may be concluding a Complex Correction in which m1 will be the most extreme price obtained by the market for a period at least twice that of the combined times of m0-m2; place an ":F3" at the end of m1.

If m1 is completely retraced **slower** than it took to form **and** m3 is longer than m2 **and** m2 is not more than 61.8% of the price distance from the beginning of m(-1) to the end of m1, <u>then</u> the market could be forming a Complex Correction (where m1 would be the end of one corrective phase of the pattern and m2 would probably be the end of an x-wave) or an Expanding Triangle; place ":F3/:c3/:L5" at the end of m1.

If m1 is completely retraced **slower** than it took to form **and** m2 is smaller than m(-2), a Zigzag (which is part of a Contracting Triangle) may have concluded with m1; place ":L5" at the end of m1.

If m1 is completely retraced **slower** than it took to form **and** m(-1) is at least 61.8% of m1 **and** m3 is smaller than m2 **and** m3 (plus one time unit) is completely retraced in the same amount of time (or less) that it took to form, m1 may be part of a Flat which concludes a larger pattern; place ":F3" at the end of m1.

If m3 is longer than m2 **and** m4 is longer than m3 **and** m0 is less than 61.8% of m1, m1 may have begun an Expanding Triangle; add "(:F3)" to the Structure list at the end of m1.

If m3 is longer than m2 **and** m4 is longer than m3 **and** m0 is between 61.8-100% of m1, the market may be forming an Expanding Triangle; add "(:c3)" to m1.

Condition "b" -- m0 is at least 100%, but less than 161.8% of m1

If m3 is longer than m2 **and** m0 is <u>closer to</u> 100% than to 161.8% of m1, place ":c3" at the end of m1. If m(-1) is longer than m0 **and** the ":c3" is used as the preferred Structure choice, add "b" in front of ":c3" to get "b:c3." This means that m1 is the b-wave of a Flat correction. If m(-1) is less than m0, m1 may be the x-wave of a Complex correction, add "x" in front of the ":c3" instead.

If m3 is longer than m2 **and** m0 is <u>closer to</u> 161.8% of m1 than to 100% of m1, place ":F3" at the end of m1. If m(-1) is longer than m0, m2 is probably the end of a Zigzag. If m(-1) is shorter than m0, m1 may be an x-wave of a Complex Correction which ends with m4. To show these two possibilities respectively, add "b:c3" and "x:c3" to the Structure list at the end of m1.

If m3 is at least 61.8% of m1 **and** m3 completes without any part of it exceeding the end of m2 **and** m2 is <u>close to</u> 61.8% of m0, m1 may be the first leg of an Irregular Failure Flat; add ":F3" to m1's Structure list if it is not already present.

If m2 is retraced less than 100% **and** m3 is at least 61.8% of m1 **and** m3 completes without any part of it exceeding the end of m2, place ":F3" at the end of m1.

If m2 is retraced less than 61.8% **and** m1 and m3 share some of the same price range **and** m4 is no more than 261.8% of m2 **and** m2 is not the shortest wave when compared with m0 and m4 **and** m4 (plus one time unit) is completely retraced faster than it took to form with the market approaching (or exceeding) the beginning of m0 in 50% of the time (or less) that it took to form m0 through m4, <u>then</u> the market possibly completed a Terminal pattern at m4; place ":c3" at the end of m1 (if ":F3" is one of m1's current Structure possibilities, add brackets "[]" around it to indicate ":c3" is a much better choice).

If m1 (plus one time unit) is completely retraced in the same amount of time (or less) that it took to form **and** m2 is almost 161.8% of m1 **and** m2 is retraced less than 61.8% **and** m(-1) is at least 61.8% of m0 **and** m(-2) is between 61.8-161.8% of m(-1) **and** m(-3) is between 61.8-161.8% of m(-2) **and** the combined price coverage of m2 through m4 is longer than m0, m1 may have completed a Contracting Triangle; add ":L3" to any existing Structure list.

If m1 (plus one time unit) is completely retraced in the same amount of time (or less) that it took to form **and** m1 is not more than 161.8% of m(-1) **and** m2 is almost 161.8% of m1 **and** m2 is retraced less than 61.8% **and** the combined price coverage of m2 through m4 is longer than m0, m1 may have completed a Flat pattern; add ":L5" to any existing Structure list.

If m(-1) is shorter than m0 **and** m2 (plus one time unit) is completely retraced in the same amount of time it took to form (or less) **and** m0 is not the shortest when compared to m(-2) and m2 **and** the market approaches (or exceeds) the beginning of m(-2) in a period of time 50% (or less) of that consumed by m(-2) through m2, add ":sL3)" to the Structure list, a Terminal Impulse pattern could have completed at the end of m2.

If none of the above conditions fit your current situation **and** m1 is a monowave, place all Structure Labels at the beginning of this section at the end of m1. If none of the above conditions fit your current situation **and** m1 is a compacted polywave (or higher) pattern, just move on to the section entitled "Implementation of Position Indicators" and use surrounding Structure labels to decide the Position Indicator which belongs in front of m1's compacted Structure label.

Condition "c" -- m0 is between 161.8% and 261.8% of m1 *(inclusive)*

If m3 is between 61.8-161.8% of m1 (inclusive) **and** m2 is less than 61.8% of m0 **and** m4 is at least 100% of m2 **and** m4 (or m4 through m6) is at least 100% of m0, m1 may be the first leg of an Irregular Flat (either variation) or a Running Triangle pattern; place ":F3" at the end of m1.

If m3 is between 101%-161.8% of m2, there is the unlikely possibility an Expanding Triangle is forming; place ":c3" at the end of m1.

If m2 (plus one time unit) is completely retraced in the same amount of time (or less) that it took to form **and** m(-1) is shorter than m0 **and** m0, when compared to m(-2) and m2, is not the shortest of the three **and** m(-1) and m1 share some of the same price range **and** the market approaches (or exceeds) the beginning of m(-2) in a period of time which is half (or less) of that consumed by m(-2) through m2, add ":sL3)" to the Structure list (indicating the possibility of a Terminal pattern completing with m2).

If m2 is retraced less than 61.8% **and** m1 and m3 share some of the same price range **and** m4 is shorter than m2 **and** m4 (plus one time unit) is completely retraced faster than it took to form with the market approaching (or exceeding) the beginning of m0 in a period of time which is 50% (or less) of that taken to form m0 through m4, <u>then</u> the market possibly completed a Terminal pattern at m4; place ":c3" at the end of m1.

If m1 (plus one time unit) is completely retraced in the same amount of time (or less) that it took to form **and** m2 is close to 161.8% of m1 **and** m2 is retraced less than 61.8% **and** m(-1) is at least 61.8% of m0 **and** m(-2) is between 61.8-161.8% of m(-1) **and** m(-3) is between 61.8-161.8% of m(-2) **and** the combined price coverage of m2 through m4 is longer than m0, m1 may have completed a Contracting Triangle; add ":L3" to any existing Structure list.

If m1 (plus one time unit) is completely retraced in the same amount of time (or less) that it took to form **and** m1 is very close to 61.8% of m0, but m1 is no more than 161.8% of m(-1) **and** m2 is very close to 161.8% of m1 **and** m2 is retraced less than 61.8% **and** the combined price coverage of m2 through m4 is longer than m0, m1 may have completed a Flat pattern; add ":L5" to any existing Structure list.

If no Structure label has been placed for m1 during this section, place ":F3" at the end of m1.

Condition "d" -- m0 is more than 261.8% of m1

If m2 is composed of more than three monowaves, place ":F3" at the end of m1.

If m2 (plus one time unit) is completely retraced in the same amount of time (or less) that it took to form **and** m(-2) is shorter than m0 **and** after m2, the market approaches (or exceeds) the beginning of m(-2) within 50% (or less) of the time taken to form m(-2) through m2; add "(:sL3)" to the Structure list (indicating the outside possibility of a 3rd Extension Terminal pattern completing with m2).

If m2 is retraced less than 61.8% **and** m1 and m3 share some of the same price range **and** m4 is shorter than m2 **and** m4 is completely retraced faster than it took to form with the market approaching (or exceeding) the beginning of m0 within 50% (or less) of the time it taken to form m0 through m4, <u>then</u> the market possibly completed a Terminal pattern at m4; place ":c3" at the end of m1.

If m2 is retraced less than 61.8% **and** m2-m4's combined price coverage is larger and their movement more vertical than m0's **and** m(-1) is at least 61.8% of m0, there is a remote chance m1 completed a Contracting Triangle; add "(:L3)" to m1.

If m2 is retraced less than 61.8% **and** m2-m4's combined price coverage is larger and their movement more vertical than m0's **and** m(-1) is approximately equal to m0 in price **and** m1 is equal (or greater) in time to m(-1) **and** m0 takes more time than m(-1) and more time than m1, there is a remote chance m1 completed a "severe" C-Failure Flat; add "[:L5]" to m1.

If m3 is <u>between 61.8-100%</u> of m2 **and** m4 is less than 61.8% of m0 **and** m1 takes less time than m0, m1 may be an "x-wave" within a Complex Correction; jot that down and place "x:c3" at the end of m1.

If m3 is <u>between 61.8-100%</u> of m2 **and** m4 is 61.8% of m0 (or more), put ":F3" at the end of m1.

If m3 is <u>less than 61.8%</u> of m2, place an ":F3/:c3" at the end of m1.

If m3 is at least 61.8% of m1 but less than 100% of m2 **and** m4 is at least as long as m2 **and** m4 (or m4 through m6), without breaking beyond the end of m3, is at least 61.8% of m0, there is a good chance m1 is the first leg of an Irregular Failure Flat; add ":F3" to m1.

If at this time there are still no Structure label(s) present for m1, place ":F3" at the end of m1.

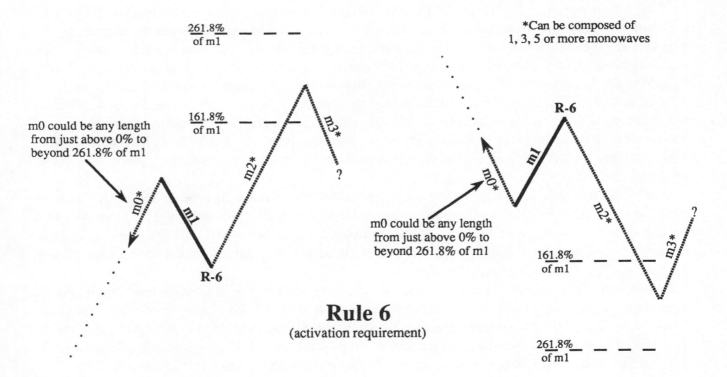

Rule 6
(activation requirement)

Rule 6 { any Structure possible, if no descriptions apply, use Position Indicator Sequences on page 3-61 for guidance }

Condition "a" -- m0 is less than 100% of m1

**(if m2 is composed of more than three monowaves [or monowave groups])

If the first three monowaves of m2 **do not** retrace more than 61.8% of m1, a Complex Correction may be unfolding with the first or second monowave (immediately after the end of m1), moving in the opposite direction of m1, representing the x-wave OR m1 contains a "missing" x-wave in its center OR m1 is the 3rd-wave of a 5th-Failure Impulse pattern (Trending or Terminal); place ":5/:s5" at the end of m1. NOTE: if the "missing" x-wave is used, circle the center of m1 and place ":5 ?" to the left of the circle and ":F3 ?" to the right of the circle, the end of the Complex Correction will be confirmed immediately before the point when the market violently turns (in the opposite direction of m1) and exceeds the 61.8% retracement level of m1.

If the first three monowaves of m2 retrace more than 61.8% of m1, m1 may have completed wave-a of a Flat with a complex b-wave OR m1 may have completed wave-3 of a 5th Failure Impulse pattern; place ":F3/:5" at the end of m1 to show these two possibilities respectively.

**(if m2 is composed of three monowaves [or monowave groups] or less)

If m2 is retraced <u>less than 61.8%</u> by m3, place ":L5" at the end of m1. If m0 and m(-2) share some of the same price territory, add ":L3" to the list.

If m2 is retraced <u>61.8% or more</u> by m3, place ":L5" at the end of m1.

If m1 (plus one time unit) is completely retraced in the same amount of time (or less) that it took to form, a Trending Impulse pattern may have completed with m1; place ":L5" at the end of m1.

If m1 (plus one time unit) is completely retraced in the same amount of time (or less) that it took to form **and** m3 is shorter than m2 **and** m2 (or m2 through m4) approaches (or exceeds) the beginning of m(-3) within a time frame which is 50% (or less) of that taken by <u>m(-3) through m1</u> **and** m0 and m(-2) share some of the same price territory, <u>then</u> m1 may have completed a Terminal Impulse pattern; add "(:L3)" to the Structure list at the end of m1. If m3 is between 61.8-100% (exclusive) of m2 and ":L3" is used as the preferred Structure label, m2 is probably an x-wave OR the Terminal pattern which completed with m1 is within a larger Triangle; place "x:c3?" at the end of m2.

If m1 is completely retraced <u>slower</u> than it took to form **and** m2 does not exceed the end of m(-2) **and** m(-1) is at least 61.8% of m1 **and** m(-2) is shorter than m(-1), m1 may be wave-a of a Flat pattern which concludes a Complex Correction and m0 is the x-wave of the pattern; place ":F3" at the end of m1 and "x:c3?" at the end of m0.

Condition "b" -- m0 is at least 100%, but less than 161.8% of m1

**(if m3 is composed of more than three monowaves [or monowave groups])

If the first three monowaves of m3 **do not** retrace more than 61.8% of m2, a Complex Correction may be unfolding with the first or second monowave (immediately after the end of m2) moving in the opposite direction of m2 representing an x-wave OR m2 contains a "missing" in its center OR m2 is the 3rd-wave of a 5th-Failure Impulse pattern (Trending or Terminal); place ":5/:s5" at the end of m1. NOTE: for the "missing" wave scenario, circle the center of m2 and place ":5 ?" to the left of the circle and ":F3/x:c3?" to the right of the circle, the end of the Complex Correction will be confirmed complete when the market violently turns (in the opposite direction of m1) and exceeds the 61.8% retracement level of m2. The termination of the Complex Correction would be at the beginning of the violent turn-around.

If the first three monowaves of m3 retrace more than 61.8% of m2, m2 may have completed wave-a of a Flat with a complex b-wave OR m1 may have completed wave-3 of a 5th Failure Impulse pattern; place ":F3/:5" at the end of m1 to show these two possibilities respectively.

**(if m3 is composed of three monowaves [or monowave groups] or less)

If m1 is less than or equal to m0 in time <u>or</u> m1 is less than or equal to m2 in time **and** m(-2) is shorter than m(-1), m1 may be the x-wave of a Complex Correction; place "x:c3" at the end of m1.

If m1 takes the same amount of time (or more) as m0 or m1 takes the same amount of time (or more) as m2 and m0 is close to 161.8% of m1, m1 may be part of a Zigzag or Impulse pattern, place ":F3" at the end of m1.

If m1 (plus one time unit) is completely retraced by m2 in the same amount of time (or less) that m1 took to form and m2 is retraced less than 61.8% OR more than 100% in a period of time equal to (or less) than that taken by m2 and m(-1) is at least 61.8% of m1 in price and time and if m1 is a Compacted pattern, make sure no part of m1 moved beyond the beginning of m1 during m1's formation, then there is a chance m1 completed the C-wave of a Flat; place ":L5" at the end of m1 to show this possibility.

If m1 (plus one time unit) is completely retraced by m2 in the same amount of time m1 took to form (or less) and m2 is retraced less than 61.8% and m(-1) is at least 61.8% of m0 in price and time, there is a chance m1 completed a Contracting Triangle or one of several Flat variations [depending on the length of m(-1)]; place ":L3/:L5" at the end of m1 to show these two possibilities (respectively). If m1 is a polywave and part of m1 exceeds m1's beginning, drop ":L5" as a possibility.

If m1 is completely retraced slower than it took to form and m2 is composed of three or more monowaves and m2 is longer than both m(-1) and m0, then m1 may be one of the middle legs of a Triangle; place ":c3" at the end of m1.

If m2 (plus one time unit) is completely retraced in the same amount of time (or less) that it took to form and m0, when compared to m(-2) and m2, is not the shortest of the three and the market approaches (or exceeds) the beginning of m(-2) in a period of time 50% (or less) of that taken by m(-2) through m2, a Terminal pattern may have concluded with m2; add ":sL3" to any current Structure list present at the end of m1.

If m3 is between 101%-161.8% of m2, an Expanding Triangle may be forming; if there is an ":F3" in m1's Structure list, add brackets around it to indicate ":c3" is the better choice.

Condition "c" -- m0 is between 161.8% and 261.8% of m1 (inclusive)

Despite any particular circumstances, ":F3" is a good possibility; place ":F3" at the end of m1.

If m1 (plus one time unit) is completely retraced in the same amount of time (or less) that it took to form and m2 is retraced less than 61.8% and during a period equal to the time of m0, m2 exceeded the length of m0 and m(-1) is between 61.8-161.8% of m0 and m2's price coverage is larger and its movement more vertical than that exhibited by m0, m1 may have completed a Contracting Triangle or a severe C-Failure Flat; place ":L3/(:L5)" at the end of m1 to show these two possibilities (respectively).

If m2 (plus one time unit) is completely retraced in the same amount of time (or less) that it took to form and m(-1) and m1 share some of the same price range and m0, when compared to m(-2) and m2, is not the shortest of the three and the market approaches (or exceeds) the beginning of m(-2) in a period of time 50% (or less) of that taken by m(-2) through m2, then a Terminal pattern may have concluded with m2; add ":sL3" to the end of m1.

If m3 is between 101%-161.8% of m2, there is the unlikely possibility that an Expanding Triangle is forming; add "(:c3)" to the Structure list.

Condition "d" -- m0 is more than 261.8% of m1

If m0 (minus one time unit) is less than or equal to m1 in time OR if m2 (minus one time unit) is less than or equal to m1 in time, and as long as m1 is not simultaneously less in time than both m0 and m2, m1 is either the first phase of a larger correction or the completion of a correction within a Zigzag or Impulse pattern; place ":F3" at the end of m1.

If m2 is retraced less than 61.8% and the combined time of m2 through m4 is equal to or less than that consumed by m0 and m2-m4's combined price coverage is larger and their movement more vertical than that exhibited by m0, there is a small chance that m1 completed a Contracting Triangle or a severe C-Failure Flat; place "(:L3)/[:L5]" at the end of m1 to show these two possibilities (respectively).

If m2 (plus one time unit) is completely retraced in the same amount of time (or less) that it took to form **and** m(-1) and m1 share some of the same price range **and** m0, when compared to m(-2) and m2, is not the shortest of the three **and** the market approaches (or exceeds) the <u>beginning</u> of m(-2) in a period of time 50% (or less) of that taken by m(-2) through m2, <u>then</u> a Terminal pattern may have concluded with m2; add ":sL3" to the end of m1.

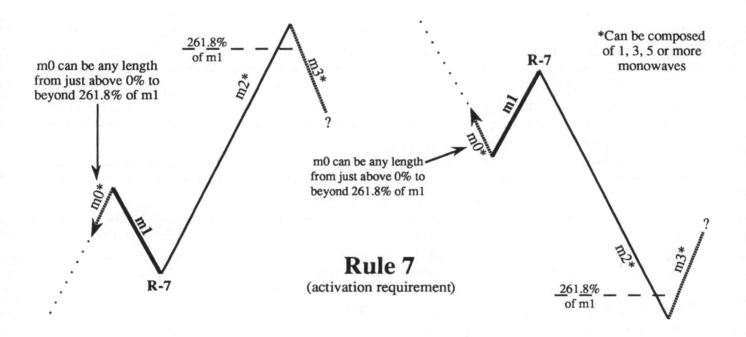

261.8% of m1

m0 can be any length from just above 0% to beyond 261.8% of m1

m0 can be any length from just above 0% to beyond 261.8% of m1

*Can be composed of 1, 3, 5 or more monowaves

R-7

R-7

261.8% of m1

Rule 7
(activation requirement)

<u>Rule 7</u> { any Structure possible, if no descriptions apply, use Position Indicator Sequences on page 3-61 for guidance }

Condition "a" -- m0 is less than 100% of m1

**(if m2 is composed of more than three monowaves [or monowave groups])

If the first three monowaves of m2 <u>**do not**</u> retrace more than 61.8% of m1, a Complex Correction may be unfolding with the first or second monowave (immediately after the end of m1), moving in the opposite direction of m1, representing the x-wave OR m1 contains a "missing" x-wave in its center OR m1 is the 3rd-wave of a 5th-Failure Impulse pattern (Trending or Terminal); place ":5/:s5" at the end of m1 and "x:c3?" at the end of m2. <u>NOTE</u>: if the "missing" x-wave is used, circle the center of m1 and place ":5 ?" to the left of the circle and ":F3 ?" to the right of the circle, the end of the Complex Correction will be confirmed immediately before the point when the market violently turns (in the opposite direction of m1) and exceeds the 61.8% retracement level of m1.

If the first three monowaves of m2 retrace more than 61.8% of m1, m1 may have completed wave-a of a Flat with a complex b-wave OR m1 may have completed wave-3 of a 5th Failure Impulse pattern; place ":F3/:5" at the end of m1 to show these two possibilities respectively.

**(if m2 is composed of three monowaves [or monowave groups] or less)

No matter what the surrounding evidence, an ":L5" is very likely; place it at the end of m1.

If m2 is retraced <u>less than 61.8%</u> **and** m(-2) is shorter than m(-1) **and** m(-2) and m0 share some similar price territory, there is the possibility a Terminal Impulse pattern completed with m1; add":L3)" to the existing Structure list.

Condition "b" -- m0 is at least 100%, but less than 161.8% of m1
**(if m3 is composed of more than three monowaves [or monowave groups])

If the first three monowaves of m3 **do not** retrace more than 61.8% of m2, a Complex Correction may be unfolding with the first or second monowave (immediately after the end of m2), moving in the same direction as m1, representing an x-wave - - this might also mean m2 is "missing" a b-wave in its center OR m2 is the 3rd-wave of a 5th-Failure Impulse pattern (Trending or Terminal); place ":F3/:c3/:L3/:L5" at the end of m1. If more than five monowaves are required to retrace more than 61.8% of m2, drop ":F3" and the "3rd-wave/5th-Failure" scenario as a possibility. NOTE: for the "missing" wave scenario, circle the center of m2 and place ":5 ?" to the left of the circle and "b:F3 / x:c3 ?" to the right of the circle. The end of the Complex Correction will be confirmed complete when the market violently turns (in the opposite direction of m1) and exceeds the 61.8% retracement level of m2. The termination of the Complex Correction would be at the beginning of the violent turn-around. To account for all of these possibilities, place "x:c3 ?" at the end of the first and second monowave (immediately after the end of m2) which are moving in the same direction as m1.

If the first three monowaves of m3 retrace more than 61.8% of m2, m2 may have completed wave-a of a Flat with a complex b-wave OR m1 may have completed wave-3 of a 5th Failure Impulse pattern; place ":F3/:5" at the end of m1 to show these two possibilities respectively.
**(if m3 is composed of three monowaves [or monowave groups] or less)

If m0 is at least 61.8% of m1 **and** m3 is between 100-261.8% of m2, m1 may be part of an Expanding Triangle; place ":c3" at the end of m1. If m4 is more than 61.8% of m3, add ":F3" to m1's Structure list.

If m1 is not much more than 61.8% of m0 **and** m2 is retraced less than 61.8% OR more than 100% in a period of time equal to or less than that taken by m2 **and** the time it took for m2 to equal the length of m0 was equal to or less than that consumed by m0 **and** m2's price coverage is more vertical than m0, there is a good chance m1 completed a Contracting Triangle or a C-Failure Flat; place ":L3/:L5" at the end of m1 to show these two possibilities (respectively).

If m2 is retraced at least 61.8%, but less than 100%, an Elongated Flat is the most likely pattern which concluded with m2; place ":c3" at the end of m1.

If m2 (plus one time unit) is completely retraced in the same amount of time (or less) that it took to form, a Trending Impulse pattern may have concluded with m2; place ":L5" at the end of m1's Structure list. If m(-1) is shorter than m0 **and** m0, when compared to m(-2) and m2, is not the shortest of the three **and** the market approaches (or exceeds) the beginning of m(-2) in a period of time 50% or less of that consumed by m(-2) through m2, a 5th Extension Terminal may have completed with m2; add ":sL3" to the list of possibilities at the end of m1.

Condition "c" -- m0 is between 161.8% and 261.8% of m1 (inclusive)

If m1 takes the same amount of time (or more) as m0 **OR** m1 takes the same amount of time (or more) as m2, despite any other circumstances, ":F3" is a good possibility; place ":F3" at the end of m1.

If the time it took for m2 to equal the length of m0 was equal to (or less) than that taken by m0 **and** m2's price coverage is larger and more vertical than m0 **and** m(-4) is longer than m(-2), then m1 may have completed a Contracting Triangle; place ":L3" at the end of m1.

If the time it took for m2 to equal the length of m0 was equal to (or less) than that taken by m0 **and** m2's price coverage is larger and more vertical than m0 **and** m(-2) is at least 161.8% of m0 **and** m(-2) is at least 61.8% of m2 **and** one of m(-1)'s Structure possibilities is an ":F3," m1 may have completed an Irregular Failure Flat; place ":L5" at the end of m1.

If m2 (plus one time unit) is completely retraced in the same amount of time (or less) that it took to form **and** the market approaches (or exceeds) the beginning of m(-2) in a period of time 50% (or less) of that consumed by m(-2) through m2 **and** m0 is longer than m(-2), an Expanding Terminal Impulse pattern may have concluded with m2; add ":sL3" to the end of m1.

If m1 (plus one time unit) is completely retraced in the same amount of time (or less) that it took to form **and** m2 is at least 161.8% of m0 **and** m1 breaks through a line drawn across the end of m(-3) and m(-1), a Running Correction could have completed with m1; place :L5 at the end of m1.

Condition "d" -- m0 is more than 261.8% of m1

If m0 (minus one time unit) is less than or equal in time to m1 <u>OR</u> if m2 (minus one time unit) is less than or equal in time to m1, **and** <u>as long as m1 is not simultaneously less in time than both m0 and m2</u>, m1 may be part of a Zigzag or Impulse pattern, place ":F3" at the end of m1.

If m1 takes the same amount of time (or less) as m0 **and/or** m1 takes the same amount of time (or less) as m2 **and** m(-2) is 161.8% or more of m(-1) **and** m(-1) is shorter than m0 **and** m1 is less than 61.8% of the distance from the beginning of m(-2) through the end of m0 **and** <u>if</u> m3 is longer than m2, <u>then</u> make sure m4 is shorter than m3 **and** <u>if</u> m3 is longer than m2, <u>then</u> make sure that 61.8% of m(-2) through m2 is retraced before the end of m2 is exceeded, THEN m1 may be the x-wave of a Double Zigzag or a Complex correction which begins with a Zigzag; place "x:c3" at the end of m1.

If m1 takes the same amount of time (or less) as m0 <u>OR</u> m1 takes the same amount of time (or less) as m2 **and** m0 is between 100-161.8% of m(-1) **and** m2 is not more than 161.8% of m0 **and** m4 is at least 38.2% of m2 **and** <u>if</u> m3 is longer than m2, <u>then</u> make sure m4 is shorter than m3, THEN m1 may be the x-wave of a Complex correction which starts with a Flat and ends with a Flat or Triangle; put "x:c3" at the end of m1.

If m1 is less than or equal to m0 in time **and/or** m1 is less than or equal to m2 in time, place ":c3" at the end of m1. If m(-1) and m1 are almost equal in price or time or both (or related in either case by 61.8%) **and** m(-1) is shorter than m0 **and** when comparing the price lengths of m(-2), m0 and m2 you find that m0 <u>is not</u> the shortest of the three and not one of the three is more than 161.8 of the next smaller, **then** m1 may be part of a Complex Double Zigzag (which will involve one or two x-waves); add an "x" in front of ":c3." If m0 is not the longest of the three mentioned above, the x-wave is probably at the end of m1, but if there are other Structure label possibilities for m1 other than ":c3," the x-wave could be at the end of m(-1) or m(3). If m0 is the longest of m(-2), m0 and m2, the x-wave may be "missing" in the center of m0, mark the center of m0 with a dot and put "x:c3?" to the right of the dot and place ":s5" to the left of the dot; m(-2), in this instance, would be the beginning of the pattern and m2 would be the end. If m(-2) through m2 constitute a Complex Correction with a "missing" x-wave, the market should retrace between 61.8-100% of it before the next wave group (of the same Degree as the Complex Correction) begins. If the "missing" x-wave Complex Correction is retraced less than 61.8% **and** then the market exceeds the end of the Complex Correction, either m(-2) through m2 does not make-up such a pattern or the Complex Correction is part of a Terminal Impulse pattern.

If m1 (plus one time unit) is completely retraced in the same amount of time (or less) that it took to form **and** m(-1) and m1 are equal in price <u>and</u> time (or related by 61.8% in either case) **and** m2 is at least 161.8% of m0 **and** m1 and m(-1) do not share any similar price range **and** m2 <u>is not</u> retraced faster than it took to form, then m1 may have completed a Running Correction; place ":L5" at the end of m1. If m(-2) is less than 161.8% of m0 **and** m2 is retraced less than 61.8% **and** the ":L5" is used, m1 simultaneously terminates more than one Elliott pattern, each of consecutively larger degree.

If m2 is retraced less than 61.8% **and** m2's price coverage is larger and more vertical than m0 **and** m(-1) is no more than 161.8% of m0 **and** m(-1) and m1 share some of the same price range **and** at least one of m0's Structure possibilities contains a ":3" (any variation), there is a small chance m1 completed a Contracting Triangle; place "(:L3)" at the end of m1. If m(-1) and m1 are equal (or related by 61.8%) in price or time or both **and** m(-1) and m1 share some of the same price range, m1 may have completed an Irregular **or** C-Failure Flat; place ":L5" at the end of m1.

If m2 is retraced less than 61.8% **and** m2's price coverage is between 61.8% and 161.8% of m0 **and** m(-1) is shorter than m0 and m(-1) is not more than 161.8% of m0, m1 may be an "x-wave" of a Complex Corrective pattern; make note of that on the chart next to m1 and add "x:c3" to m1's Structure list if it is not already present.

If m2 (plus one time unit) is completely retraced in the same amount of time (or less) that it took to form **and** m3 is not retraced more than 61.8% **and** m(-1) is shorter than m0 **and** part of m(-1) shares some of the same price range as m1 **and** m0, when compared to m(-2) and m2, is not the shortest of the three **and** m3, in a period of time 50% (or less) of that consumed m(-2) through m2, closely approaches (or exceeds) the beginning of m(-2), a Terminal pattern may have concluded with m2; add ":sL3" to the Structure list at the end of m1.

Implementation of Position Indicators
(Rules to reduce Structure lists to a single possibility)

Position Indicators are the alphabetic companions ("c, F, L, s or sL") which precede most Structure Labels (:3's and :5's). Position Indicators do just what their name implies, they describe - sometimes generally, sometimes specifically - the position of a Structure Label within the context of surrounding market action. Since two varieties of Structure labels exist (those that contain Position Indicators and those that do not), specific phrases representing each variety have to be coined to allow for accurate and intelligible coverage of this subject. The phrase, "base Structure label," describes a Structure label which is not preceded by a Position Indicator (i.e., solely a ":3" or a ":5" is present). A Structure label which is preceded by a Position Indicator (i.e., :F3, :c3, :s5, :L5, etc.) will be termed a "positioned Structure label." In the paragraphs to follow, if the Structure of a pattern is not important, is unknown or bares no significance to the discussion, the truncated phrase, "Structure Label," will be used to encompass both possibilities.

The Position Rules must be used whenever there are wave segments which contain multiple Structure label possibilities. The use of Position Indicators is the only way novice Elliotticians can properly piece together reliable Wave patterns. Surprisingly, this same approach is used by expert Elliott Wave analysts, but most have internalized the process (through practice) to the point that they do not realize they are applying these concepts; they probably do not even realize they are applying Elliott Wave techniques in a standardized fashion - the same approach presented in this book.

The implementation of Position Indicators begins to intensify the bonds between adjacent wave segments due to rules which guide Position Indicator coexistence. They begin to exert a degree of order to the seemingly random fluctuations of market action. Position Indicators are the the next clarifying step toward unraveling current market action.

If you find a wave on your chart which possesses multiple Structure label choices, the Position Indicators will be needed to simplify your choices. These Indicators will assist you in eliminating all but one Structure choice from m1's list. Once all waves on your chart possess only one Structure label, later in this chapter you will learn how to use these Structure labels to isolate important beginning and ending points of Elliott patterns. For those of you who have had difficulty in the past deciding where to start your analysis, this section will be the answer to your prayers.

Directions

By the time you reach this section, the internal Structure of numerous monowaves on your chart should have been established. Unfortunately, the "Pre-Constructive Rules of Logic" are not always capable of reducing the Structure choices of each wave to a single possibility. Therefore, it is likely several waves on your chart contain two or more Structure labels. Studying each Position Indicator's unique characteristics (described in this section) and logically integrating that label's implications with surrounding Structure should enable you to eliminate all but one label from each wave segment.

Each Structure label in the listing to follow is separated by either a dash ("-") or two plus signs ("++"). All Structure labels connected with a dash are part of the same, Standard Elliott Wave pattern. If separated by double plus signs, a completed Standard pattern occurred right before the double plus signs and another Standard pattern begins right after the second set of double plus signs. The action isolated between the double plus signs constitutes an "x-wave." Explained in detail later, "x-waves" connect multiple Standard Corrections to form one larger Complex Correction (a Non-Standard formation). For a detailed explanation of Standard and Non-Standard patterns, refer to Chapter 8.

Under the sub-heading, "Position Indicator Definitions and Sequences," are descriptions of all Positioned Structure Labels. Knowledge of the characteristics of each Position Indicator is valuable when attempting to reduce a monowave's listing to one possibility. Acquaint yourself with these descriptions so you can quickly spot and eliminate illogical Structure possibilities. **Do not** eliminate positioned Structure labels (which are connected to question marks "?") or base Structure labels (which are the result of compaction) unless they are **within** the confines of a larger Compacted pattern.

If a Position Indicator's definition does not allow you to reduce a monowave's Structure list to a single possibility, below each label is a list of acceptable "before and after" Structure labels which can be found in partnership with the current label being analyzed. Tthe left-hand sequences usually start with corrective variations (:3's). The right-hand sequences usually begin with Impulsive variations (:5's). The darkened, larger positioned Structure label toward the middle of each sequence represents m1. Taking note of the Structure choices surrounding m1 and comparing them with allowable "before and after" possibilities on your chart, you should be able to eliminate improper Structure labels for all current m1's, leaving just one Structure label per wave. Keep in mind, once a pattern is compacted, its base Structure label (:3 or :5) instead of its positioned Structure label (such as :c3, :sL3, :s5, etc.) may be needed to create Complex Elliott Corrections. This is another reason to retain <u>base</u> Structure labels on a compacted wave group even after <u>positioned</u> Structure labels have been assigned to the pattern.

Position Indicator Definitions and Sequences
":F3"

This Structure label is the abbreviation for "First three(**3**)." An ":F3" either starts a series, occurs after an "x:c3" **or** is found between two ":5's" (all variations). If you find two ":F3's" next to each other, a new pattern (of a smaller magnitude) begins with the second ":F3." Circle the start of both ":F3's," but do not attempt to connect the two until the second ":F3" can become part of a polywave pattern using waves which proceed it (a polywave is an Elliott pattern composed of three or more monowaves). Examples of where an ":F3" can be found under real-time conditions follow:

1. ? - **F3** - c3 - L5 (circle start of F3)
2. ? - **F3** - c3 - c3 (circle start of F3)
3. x:c3 ++ **F3** - c3 - L5 (circle start of F3)
4. x:c3 ++ **F3** - c3 - c3 (circle start of F3)

5. 5 - **F3** - 5 - F3 - L5 (second 5 at least 38.2% of first 5)
6. 5 - **F3** - 5 ++ x:c3 (second 5 at least 38.2% of first 5)
7. 5 - **F3** - s5 ++ x:c3 (s5 must be at least 38.2% of 5)
8. 5 - **F3** - L5 (L5 must be at least 38.2% of 5)
9. s5 - **F3** - L5 (L5 must be more than 100% of s5)

":c3"

This Structure label is the abbreviation for "center three(3)." A ":c3" can <u>never</u> start or end a sequence; as a result, a large violent move will virtually never occur after ":c3." If the first wave in a group contains more than one Structure label, and one of those happens to be a ":c3," you can eliminate it. If trying to complete a three or five segment pattern and the last leg contains ":c3" as a possibility, the ":c3" can be eliminated. If any sequence from 1 to 7 below is used, circle the beginning of the ":F3." Following are examples of where ":c3's" can be found under real-time circumstances and the conditions which must be present for their proper positioning:

1. F3 - **c3** - c3 - c3 (second or third c3 must be smallest or largest of all four)

2. F3 - **c3** - 5 (5 must be larger than c3; if 5 is 161.8% or more of F3, 5 should be retraced 61.8% or more)

3. F3 - **c3** - s5 ++ x:c3 (read sequence 2 guidelines)

4. F3 - **c3** - L5 (if L5 shorter than c3, start of F3 **or** end of c3 must be exceeded swiftly)

5. F3 - c3 - **c3** - c3 - L3 (last c3 **or** L3 should be smallest or largest of all five Structure labels)

6. F3 - c3 - **c3** - sl3 - L3 (l3 must be the smallest of all five Structure labels)

7. c3 - c3 - **c3** - L3 (L3 must be smallest with the last c3 the next smallest **OR** L3 or last c3 must be the longest of all four)

8. :<u>3</u> - x:**c3****- :<u>3</u> (x:c3 should be less than 61.8% or more than 161.8% of the first :3)

9. 5 ++ x:**c3*** ++ 5 - F3 (c3 must be smaller than 5)

10. 5 ++ x:**c3*** ++ F3 - c3 (c3 must be smaller than 5)

11. s5 ++ x:**c3*** ++ 5 - F3 (c3 must be smaller than s5)

12. s5 ++ x:**c3*** ++ F3 - c3 (c3 must be smaller than s5)

13. L5 ++ x:**c3*** ++ F3 - c3 (c3 must be larger than L5, F3 must be smaller than c3)

*In this position, the ":c3" would be considered an "x-wave" of a Complex Correction. If the Structure labels surrounding ":c3" agree with those in this series, you are allowed to add an "x" in front of ":c3." NOTE: A ":c3" can become an "x:c3," but the reverse is not allowed.

**In this position, the ":c3" would be considered an "x-wave" of a Complex Correction. If the Structure labels surrounding ":c3" agree with those in this series, you are allowed to add an "x" in front of ":c3." The underlined ":<u>3</u>'s" represent compacted polywave (or higher) corrections which <u>should not</u> contain Position Indicators. NOTE: A ":c3" can become an "x:c3," but the reverse is not allowed.

"x:c3"

This Structure label is the abbreviation for "center three(3) in the x-wave position." An "x:c3" can <u>never</u> start or end a sequence, as a result, if a large violent move (relative to surrounding activity) takes place after an "x:c3," it is almost always wave-a of a Non-Limiting Triangle. An "x:c3" occurs **between** Standard Elliott patterns as a connecting agent to help build single corrections into larger formations. If the first wave in a group contains more than one Structure label in its list of possibilities **and** one of those happens to be an "x:c3," you can eliminate it. If trying to complete a three or five segment pattern and the last leg contains an "x:c3" in its list of possibilities, obviously the "x:c3" can be eliminated.

When working with compacted corrections (whose base Structure is ":3") like those found in series 3, 4 and 9 below, the complexity of "x:c3" should never be higher than the complete Elliott pattern which occurs before and after it. The "x:c3" will usually be one lower complexity level than the two patterns which surround it, but on rare occasions (such as series 4), the middle ":3" may be two levels higher in complexity than one or both "x:c3's" (see Chapter 7 for information on Complexity and Compaction).

The following are examples of where ":c3's" can be found under real-time circumstances and the conditions which must be present for their proper positioning:

1. L3 ++ **x:c3** ++ F3 - c3

(**extremely rare**, c3 must be larger than L3 and F3)

2. L3 ++ **x:c3** ++ 5 - c3

(**virtually impossible**, c3 must be larger than L3 and 5)

3. :<u>3</u>* ++ **x:c3** ++ :<u>3</u>*

(last :<u>3</u> is termination of pattern)

4. :<u>3</u>* ++ **x:c3** ++ :<u>3</u>* ++ x:c3 ++ :<u>3</u>*

(last :<u>3</u> is termination of pattern)

5. 5 ++ **x:c3** ++ 5 - F3 (x:c3 must be smaller than both 5's)

6. 5 ++ **x:c3** ++ F3 - c3 (x:c3 must be smaller than 5 and F3)

7. s5 ++ **x:c3** ++ 5 - F3 (x:c3 must be smaller than s5 and 5)

8. s5 ++ **x:c3** ++ F3 - c3 (x:c3 must be smaller than s5 and F3)

9. s5 ++ **x:c3** ++ :<u>3</u>* (last :<u>3</u> terminates pattern OR rare Triple

Combination pattern forming - if so, after :<u>3</u> another x:c3 would occur)

10. L5 ++ **x:c3** ++ F3 - c3 (x:c3 must be larger than L5 and F3)

*Represents a compacted polywave (or higher) correction which <u>should not</u> contain an alphabetic Position Indicator

The Structure Series under list "3." and "4." above have underlines beneath their ":<u>3</u>'s." Those underlines denote ":<u>3</u>'s" which represent Compacted Elliott corrections of polywave Complexity or higher (if you are unfamiliar with Compaction and Complexity and have at least read through the end of Chapter 5, refer to Chapter 7 for a detailed discussion of these concepts). The corrections that these ":<u>3</u>'s" represent could be Zigzags, Flats or Triangles. If a Triangle is involved in one of these Complex formations it will almost always be positioned as the last ":<u>3</u>" in the list.

":sL3"

This Structure label is the abbreviation for "**second to Last three(3)**." An ":sL3" <u>never</u> begins or ends an Elliott Wave pattern. This is a conditional Structure Label that cannot be present without its partner, the ":L3" label. Whenever you contemplate using ":sL3" there must be an ":L3" immediately afterward. This grouping of labels always indicates one of two Elliott patterns is forming; a Terminal or a Triangle. Therefore, the ":sL3" (along with its partner the ":L3") must be one of five **consecutive** corrective (:3's) Structure labels with the ":sL3" the "second-to last" of the series. The following is the only position an ":sL3" can be found in during real-time circumstances:

1. c3 - **sL3** - L3 - ? (L3 should not be more than 61.8% of sl3 <u>OR</u> it should be more than 100% of sl3)

":L3"

This Structure label is the abbreviation for "**Last three(3)**." Unlike an ":sL3" (which must be proceeded by an ":L3"), an ":L3" label does not insist on a <u>preceding</u> ":sL3." If it is the smallest wave, an ":L3" requires its own price distance (plus one time unit) be completely retraced in the same amount of time <u>or less</u> that the ":L3" took to form. Similar to the ":sL3" label above, an ":L3" must be part of a Terminal or Triangle pattern and therefore requires a total of five ":3's" be found adjacent to one another with the ":L3" the last of the five ":3's." By applying the Essential Impulse Construction Rules (Chapter 5, page 2) to the wave group, you will be able to decide whether the market is forming a Terminal or a Triangle. If the market action obeys the Rules found on page 5-2, a Terminal is forming; if it does not, a Triangle is unfolding. The following are examples of Structure labels which may surround an ":L3" under real-time conditions (NOTE: none of the ":c3's" in Structure list "1" or "2" [below] may have an "x" in front of them):

1. F3 - c3 - c3 - c3 - **L3** - ? (L3 must be larger than second OR larger than third c3, circle end of L3)

2. F3 - c3 - c3 - sL3 - **L3** - ? (L3 must be the smallest wave, it must be violently retraced; circle end of L3)

3. sL3 - **L3** ++ x:c3 ++ F3 (sL3 and c3 must both be larger [or sL3 and c3 must both be smaller] than L3)

":5"

This Structure label is simply pronounced "five" and can represent any Impulsive wave which <u>does not</u> terminate an Elliott pattern. It can be the first leg of a Zigzag or Impulse pattern OR the middle leg of a Complex Correction or Impulse pattern. This is one of the more common Structure labels and can be found under a multitude of conditions. If you are examining the first ":5" of a series, the market cannot retrace more than 61.8% of it and the market must then exceed the end of the ":5." If you find conditions counter to this requisite, the ":5" (as long as it is not the only Structure label representing m1) should be eliminated. Below is a complete listing of where ":5" can occur along with a parenthetical description of its implications to surrounding price action:

1. ? - **5** - F3 - 5 (the "?" indicates the start of the current pattern OR there is no market action before the 5)
2. ? - **5** - F3 - s5 - F3 (s5 must be longer than 5 **and** both F3's are shorter than the previous Structure label)
2. ? - **5** - F3 - s5 ++ x:c3 (x:c3 must be shorter than s5 and F3 must be shorter than both 5 and s5)
3. F3 - **5** - F3 - L5 (the two F3's cannot share any similar price territory and L5 must be longer than 5)
4. F3 - **5** ++ x:c3 ++ F3 (x:c3 and first F3 must be smaller than 5; second F3 almost always longer than x:c3)
5. F3 - **5** ++ x:c3 ++ 5 (x:c3 and F3 must be smaller than first 5; second 5 must be larger than x:c3)
6. c3 - **5** ++ x:c3 ++ F3 (x:c3 is either 161.8% or more of 5 or less than 61.8% of 5; if x:c3 is larger than 5 then F3 must be smaller than x:c3; if x:c3 is smaller than 5, then F3 would almost always be larger than x:c3)

":s5"

This Structure label is the abbreviation for "special five (5)." It imposes specific limits on past, present and future market action. An ":s5" can function as if it were an ":L5," but without the reversal action normally required to confirm an ":L5." An ":s5" usually occurs in Complex Elliott formations, but (under very limited conditions) it can be the 3rd wave of a Trending Impulse pattern with a 5th wave Failure **or** with a 5th wave Extension. If you find a wave with only an ":s5" Structure label, two previous labels (going backward in time) must be connected to the ":s5." The two sequences which must precede the ":s5" are "5 - F3" or " F3 - c3." The following are examples of where ":s5's" can be found under real-time circumstances:

1. 5 - F3 - **s5** ++ x:c3 ++ F3 (x:c3 and the first F3 must be smaller than s5)
2. 5 - F3 - **s5** - F3 - L5 (both F3's must be smaller than s5 and L5 must be longer than s5)
3. F3 - c3 - **s5** ++ x:c3 ++ F3 (both c3 and x:c3 must be smaller than s5; F3 will usually be larger than x:c3)

":L5"

This Structure label is the abbreviation for "Last five (5)." Since an L5 is always the conclusion of a larger Elliott pattern, the presence of an ":L5" can be the simultaneous termination of larger Degree patterns. The minimum confirmation of an ":L5" requires that the trendline drawn across the end of <u>m0 and m(-2)</u> be broken in a period of time equal to that of the ":L5" (plus one time unit) or less. To connect an ":L5" to previous market action, the Structure label immediately before the ":L5" must be an ":F3" or a ":c3." If you find an Impulsive Structure label (:L5, :s5 or :5) immediately before the ":L5," circle the end of both waves. When you reach Chapter 4, search for a possible Structure Series which ends with the earlier ":L5, :s5 or :5." Once the earlier Impulsive Structure label has been integrated into an Elliott pattern, Compact the pattern to its base Structure (Compaction is discussed in Chapter 7), **dot** the start and finish of it, <u>then</u> attempt to connect the second Impulsive Structure with the Compacted pattern and earlier Structure labels. Examples of where ":L5's" occur under real-time conditions follows:

1. 5 - F3 - **L5** - ? <small>(F3 must be shorter than both 5 and L5; circle end of L5)</small>
2. s5 - F3 - **L5** - ? <small>(L5 must be longer than s5, preferably L5 will be 161.8% or more of s5; circle end of L5)</small>
3. F3 - c3 - **L5** - ? <small>(if c3 is 138.2% or more of F3, L5 will almost certainly be shorter than c3; circle end of L5)</small>
4. F3 - c3 - **L5** ++ x:c3 ++ F3 <small>(both c3 and x:c3 must be larger than L5, F3 must be smaller than x:c3)</small>

Pattern Isolation Procedures

If there are no ":L5's" or ":L3's" on your chart, keep updating the price action each day - applying the Retracement and Pre-Constructive Rules - until there is a wave with only an ":L5" or ":L3" (or both) attached. Once an ":L5" or ":L3" has been identified, the ***Pattern Isolation Procedures*** can be initiated. From the far left of your chart, begin a **forward** search for the first wave which contains only an ":L5" or an ":L3" or both; when you have found the first occurrence, circle the end of that wave (see Figure 3-31), it will probably be the end of an Elliott pattern. [*The wave just circled will usually be followed by a significant, maybe even violent, market move in a relatively short period of time. Be on the look out for such activity; it is a reliable indication of recent Elliott pattern termination.*] By applying the Position Rules and techniques discussed in this section, that circled wave will now assist you in identifying the **start** of an Elliott pattern.

From the circled point, move backward **three** Structure labels. If that new wave contains only one label **and** it is one of the following, ":F3, x:c3, :L3, :s5 or :L5," stop there; you may have found the beginning of an Elliott Wave pattern. If the third wave back possesses more than one Structure label,

Figure 3-31

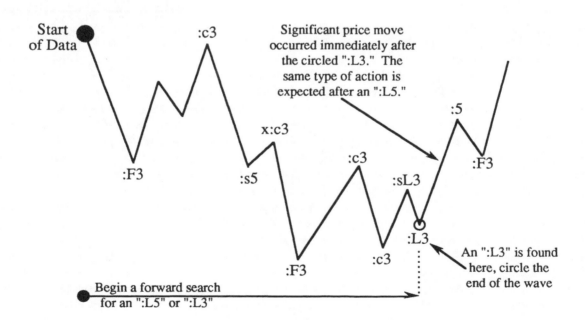

Figure 3-32

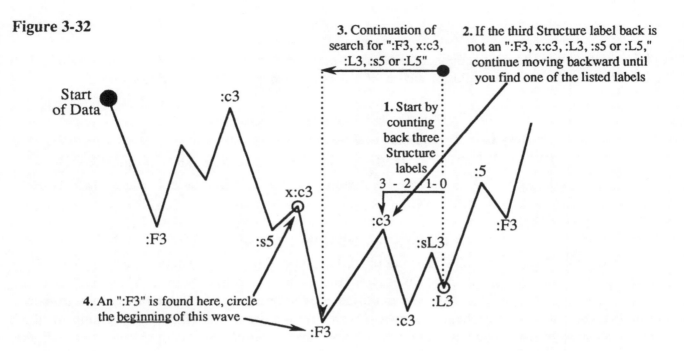

or the single Structure label which it does possess is not one of those listed above, continue moving backward one wave at a time until you find a wave with a single Structure possibility which is one of those listed above. When you finally find a wave which meets the above requirements **and** it is an ":F3," circle the beginning of that wave; if it is an "x:c3, :L3, :s5 or :L5," circle the end of that wave (refer to Figure 3-32 for a demonstration of this process). Later, after a pattern has been tested and proven to be legitimate, the circles encasing the pattern should be turned into dots (darkened). Since all Elliott patterns require an odd number of waves to complete, count the number of waves betweeen the two circled points. If an even number of waves is present, countinue moving backward in search of another ":F3, x:c3, :L3, :s5 or :L5." If an odd number of waves is present, you can proceed to the next page.

Figure 3-33

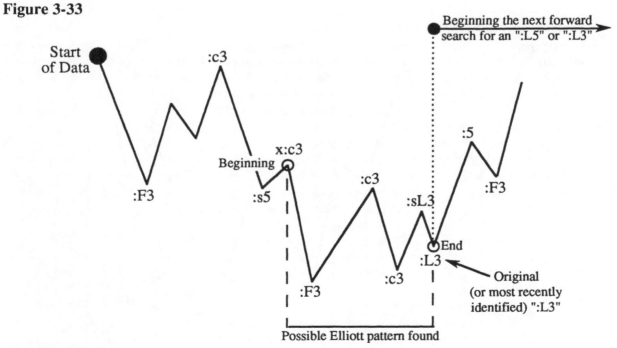

Figure 3-34

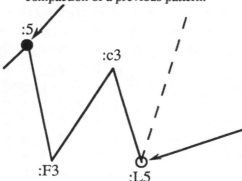

This high is a solid circle due to the compaction of a previous pattern.

:5

:c3

:F3

:L5

Moving backward from an identified ":L5," a previously circled high is found before an ":F3, x:c3, :L3, :s5 or :L5" is found. If the number of waves between the circled points is "three or more," an Elliott pattern may be unfolding. Make sure all waves between those two points possess only one Structure label.

After you have accomplished that task, return to the ":L5" or ":L3" you most recently circled; then, move **forward** in time until you get to the next wave which is labeled with only an ":L5" or ":L3" (or both) and repeat the above process (see Figure 3-33). If there is not another ":L5" or ":L3" to work with, proceed to Chapter 4 (and beyond) to verify and compact the patterns currently isolated.

When you repeat this process, frequently a high or low is found which has already been darkened (**dotted**) from previous analysis. That means the current pattern actually began at the previously dotted point (Figure 3-34) OR the previously dotted point is the end of a Compacted wave group which, as a whole, is the same Degree as the current ":L5" or ":L3." *[The previous statement would be true even if the ":L5" or ":L3" were one Complexity Level different from the Compacted pattern (see Chapter 7, page 4 for details on Compaction and Complexity Levels)].* The Compacted pattern will eventually become an integral part of a larger pattern terminating with the ":L5" or ":L3." If the current ":L5" or ":L3" occurs <u>immediately</u> after a dotted point (like in Figure 3-35), not enough waves occurred between the points to create an Elliott pattern. Therefore, the base Structure of the dotted point should be used for grouping waves when you proceed to Chapter 4. Usually, it is clear which of the two circumstances is unfolding. No matter what type of grouping occurs, the current ":L3" or ":L5" should conclude an Elliott pattern.

Figure 3-35

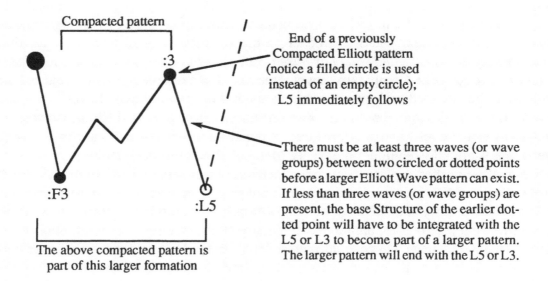

Compacted pattern

:3

:F3

:L5

The above compacted pattern is part of this larger formation

End of a previously Compacted Elliott pattern (notice a filled circle is used instead of an empty circle); L5 immediately follows

There must be at least three waves (or wave groups) between two circled or dotted points before a larger Elliott Wave pattern can exist. If less than three waves (or wave groups) are present, the base Structure of the earlier dotted point will have to be integrated with the L5 or L3 to become part of a larger pattern. The larger pattern will end with the L5 or L3.

Special Circumstances

When in the process of simplifying Structure lists, if you find a compacted pattern's price action exceeding its own beginning before it terminates (see Figure 3-36), the base Structure of the compacted pattern is unequivocally corrective in nature (a ":3") no matter what the Pre-constructive Rules suggest. Therefore, if the condition presented below in Figure 3-36 is found on your chart, do not attempt to apply the Pre-Constructive Rules of Logic during the **Reassessment** process; simply retain the compacted patterns base Structure instead. When trying to integrate these special compacted patterns with surrounding market action, use the Position Rules to help you decide if only the base Structure of the compacted pattern is required or if a Position Indicator is necessary to connect it to a larger Elliott pattern.

Figure 3-36

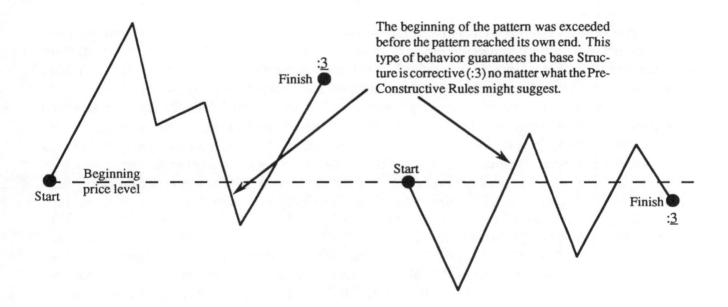

The beginning of the pattern was exceeded before the pattern reached its own end. This type of behavior guarantees the base Structure is corrective (:3) no matter what the Pre-Constructive Rules might suggest.

You may run into difficulty when piecing together multiple polywave (or higher) patterns if you do not understand Complexity and its impact on market behavior and pattern interaction. Make sure adjacent patterns *(represented solely by base [:3 or :5] or positioned Structure labels [:F3, :L5, etc.])* are never more than one Complexity Level removed. For example, if m0, m1 and m2 were previously identified as Standard Elliott Wave corrections and m0 were Complexity Level-1, m1 were Level-2 and m2 were Level-1, the three could be combined into a larger Standard Elliott Correction. If m0 were Level-1, m1 were Level-2 and m2 were Level-4, the difference between m1 and m2 is two (2) Complexity Levels; that is not allowed under the **Neely Method** of pattern formulation.

"Degree" is another important concept which must be understood and applied during the amalgamation of various wave segments. If you are unfamiliar with **Degree** and its application to price action, read the "Rule of Similarity and Balance" on page 4-4 before attempting to eliminate Structure labels from your chart. That section of Chapter 4 will assist you in detecting and combining waves of the same Degree. Chapter 7, page 11, goes into more detail on the Degree concept; as your understanding of Degree improves, you will want to become acquainted with that section, also.

Synopsis of Chapter 3

The complexity of Chapter 3 warrants a brief overview of all the processes covered. When beginning actual market analysis, first construct a chart of cash data using approximately 60 data points. Reconstruct a new chart (starting from an important top or bottom found on the first chart), but this time employ the Rule of Proportion. Using **dots**, identify the end of all monowaves and apply the Rule of Neutrality where necessary. Pick a recent monowave to analyze (do not go back more than 20 monowaves); implementing the Rules of Observation, place Rule, Condition and Category Identifiers at the end of each important monowave. Next, read through the appropriate section (dictated by the Rule and Condition [and sometimes, Category] Identifier) under the Pre-Constructive Rules of Logic and begin transforming the Rules into Structure labels. Locate any monowaves which contain more than one Structure label and, using the **Position Indicator Definitions and Sequences** section, eliminate all but one Structure label from the end of each m1 and then apply the "Pattern Isolation Procedures." Your chart will then be prepared for Chapter 4 where the grouping of Structure labels into an Elliott Series is required.

Intermediary Observations

4

The previous chapter, *Preliminary Analysis,* provided you with a strong foundation on which to continue your search for legitimate Elliott patterns. These Elliott patterns will be constructed from groups you isolated based on rules listed at the end of Chapter 3 (see "Pattern Isolation Procedures," page 3-65). To continue a wave group's qualification process in becoming an Elliott pattern, the Structure labels of the isolated group must adhere to a rigid sequencing scheme. In addition, certain tests will be applied (before proceeding to Chapter 5) to make sure all wave segments are of the same Degree.

Monowave Groups

As you already know, monowaves are the building blocks of the Elliott Wave Theory. Unfortunately, if studied individually, monowaves only provide you with a limited perspective on the future course of a market. A greater understanding of market possibilities can be derived by grouping monowaves which exhibit specific Structure label sequencing (which the word *polywave* was coined to describe). This specific sequencing is guided by the Position Indicators preceding each Structure label.

Just like monowaves, polywaves can either be Corrective or Impulsive in nature. It takes at least three monowaves to create a Corrective polywave. At least five monowaves are needed to create an Impulsive polywave. At the top of the next page, Figure 4-1 illustrates an idealized Corrective pattern while Figure 4-2 depicts an idealized Impulsive pattern. As can be ascertained from a study of Figures 4-1 & 4-2, Impulsive patterns occur when a market is gaining or losing ground (similar to *Directional action,* see Chapter 3) while Corrective patterns usually travel sideward (similar to *Non-Directional action*).

Figure 4-1

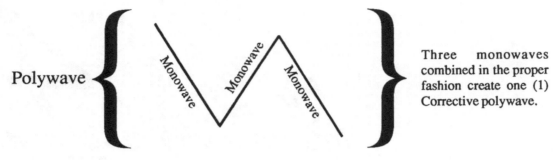

Polywave

Three monowaves combined in the proper fashion create one (1) Corrective polywave.

Figure 4-2

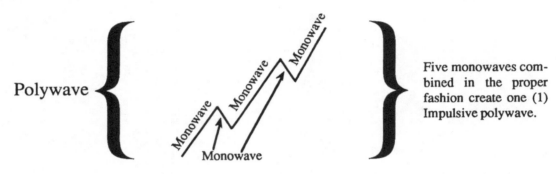

Polywave

Five monowaves combined in the proper fashion create one (1) Impulsive polywave.

Figure 4-3

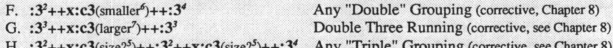

Standard Patterns (for all students)

A. **:5-:F3-:?5-:F3-:L5** Impulsion (trending nature, refer to page 5-2)

B. **:5-:F3-:?5¹** Zigzag (corrective nature, refer to page 5-19)

C. **:F3-:c3-:?5¹** Flat(corrective nature, refer to page 5-17)

D. **:F3-:c3-:c3-:?3-:?3¹** Triangle (corrective nature, refer to page 5-23)

E. **:F3-:c3-:c3-:?3-:L3** Terminal (conclusive nature, refer to page 5-2)

¹If the last Structure label of a Standard *Corrective* Series does not contain an "L" Position Indicator, the Series will need to be Compacted to a ":3" and made part of one of the Non-Standard Patterns listed below

Non-Standard Patterns (for advanced students)

F. **:3²++x:c3**(smaller⁶)**++:3⁴** Any "Double" Grouping (corrective, Chapter 8)

G. **:3³++x:c3**(larger⁷)**++:3³** Double Three Running (corrective, see Chapter 8)

H. **:3²++x:c3**(size?⁵)**++:3²++x:c3**(size?⁵)**++:3⁴** Any "Triple" Grouping (corrective, see Chapter 8)

²Could be any **Standard** Flat, Zigzag or Expanding Triangle (when the "x:c3" is larger than the previous ":3," a Zigzag should not occur before or after the "x:c3")

³Could be a **Standard** Flat or Triangle (if first ":3" is a Triangle, it should be Expanding)

⁴Could be any **Standard** *Corrective* pattern. ⁵The "x:c3" could be larger **or** smaller than the previous ":3"

⁶"x:c3" must be smaller than previous ":3"   ⁷"x:c3" must be larger than previous ":3"

With singular Structure "lists" inhabiting the end of numerous monowaves on your chart, locate those wave groups created during the "Pattern Isolation Procedures" at the end of Chapter 3. Of the groups isolated, always choose among those which contain only three or five monowaves, those three or five adjacent monowaves may form a Standard Elliott polywave pattern (as you gain experience, some or all of the three or five segments chosen could be compacted wave groups). IMPORTANT: Of those wave groupings chosen, work first on the group which consumes the least overall price and time.

The deciding factor in creating a polywave is the presence of a Structure *Series* found among the isolated string of monowaves. A Structure Series is an organized listing of properly Positioned Structure labels which combine to form an Elliott pattern (see top of Figure 4-3 on previous page). Through the process of Compaction, Structure Series' will allow you to amalgamate many simple Elliott Wave patterns into more complex patterns in a logical, natural fashion.

Figure 4-3 illustrates the Position Indicator sequencing and the exact number of ":3's" and ":5's" which *must* be present to create a Standard or Non-Standard Elliott pattern. If some of the waves you are dealing with are compacted and only possess their base Structure label, the compacted pattern's base Structure can represent any legal Positioned Structure label of the same Class (for a definition of "Class" refer to page 2-4). Any pattern above monowave development must conform to one of the <u>four</u> Standard Structure Series' **or** one of the various varieties of Non-Standard Series'. [*Notice, in the last sentence only <u>four</u> Structure Series' were mentioned. <u>Five</u> Standard Structure Series' are listed in Figure 4-3 , but only four of those Series' are actually unique. The last Standard Series, ":3-:3-:3-:3-:3," is repeated and can be used to create two separate patterns with vastly different implications. Depending on the shape of the waves grouped (and other conditions introduced in Chapter 5), you will be able to decide which of the two patterns is unfolding.*] Figure 4-3 illustrates and names these combinations allowing you to further define development on your chart. Notice, some Structure labels in Figure 4-3 do not contain Position Indicators. When Position Indicators **are not** present in this official listing, any Position Indicator "legally" allowed in front of the Structure label in question could be used.

When comparing the isolated wave groupings on your chart with those in Figure 4-3, and your grouping is composed strictly of monowaves, one of the "Standard Patterns" at the top portion of Figure 4-3 should be appropriate. If you have already Compacted some previous price action and an "x:c3" occurs anywhere in your isolated wave group, one of the Non-Standard Patterns will take effect.

No matter how large or time consuming, all grouped market action pegged for analysis *must* match one of the Structure Series listed in Figure 4-3. Your goal is to find a Structure Series listed which mimics the one on your chart. Ultimately, the identification of Structure Series' is paramount in your search for larger, reliable wave patterns. Once a series has been found, many additional tests (some covered immediately after this section and some covered in Chapters 5-12) must be conducted to elevate your monowave group to *polywave* status.

Rule of Similarity & Balance

Once an acceptable Structure Series has been identified, the **Rule of Similarity & Balance** must be considered. All Elliott patterns depend on combining <u>like with like</u>. Waves that are <u>not</u> "similar" cannot be combined to complete a larger Elliott formation.

Market action develops under two general categories - price and time. For two adjacent waves to be "similar" the relationship between the two waves should fall into a specific relational range based on price or based on time **or** a combination of the two. Similarities of price and time can simultaneously occur between adjacent waves but are not necessary to pass inspection under this Rule.

When comparing adjacent waves within an *Impulse* pattern, similarities of time are more common than price; in *Corrective* patterns, similarities of price are more common than time. To decide whether two adjacent waves are "similar" is a matter of simple mathematics. Following are instructions on how to detect similarities based on the two elements (price and time) which define market action.

Price

For there to exist a *price similarity* between two adjacent waves, the smaller of the two waves should be no less than one-third (1/3) in price of the larger wave.

Time

For *time similarity* to be present between two adjacent waves, the shorter time pattern should not be less than one third of the longer time pattern.

If you have been following along with your own chart and have one or more isolated groups which adhere to an Elliott Structure Series, **only investigate those Series where all adjacent waves adhere to one or both of the above Rules of Similarity and Balance.** If in any comparison, neither of the above Rules is obeyed, the likelihood two adjacent waves are of the same Degree is very slim. When two waves are not of the same Degree, they cannot be directly or immediately connected to form a larger Elliott pattern; Compaction of several or more smaller waves will first be required.

Figure 4-4

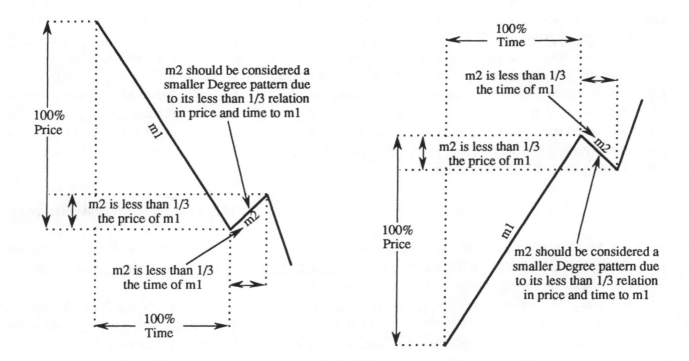

In Figure 4-4, the Rule of Similarity and Balance is not adhered to in either respect indicating m1 and m2 are not of the same Degree; therefore, m1 and m2 cannot be directly combined. To eventually include m1 and m2 within the same Elliott pattern, m2 has to be connected to waves of similar price and time. Those waves will have to be grouped and identified as an Elliott pattern, tested, confirmed and then reassessed. Afterward, if the waves grouped with m2 consume enough time or price to relate to m1 (based on the Rule of Similarity and Balance), m1 and m2's group can be combined with surrounding action to create a larger Elliott pattern. In Figure 4-5, both conditions of the Rule of Similarity and Balance are obeyed indicating the two waves **might** be of the same Degree. Consequently, if Structure labels allow, m1 and m2 may be directly attached as part of a larger Elliott pattern. NOTE: Just because two waves adhere to the Rule of Similarity and Balance does not mean they are of the same Degree, it just indicates they might be.

Figure 4-5

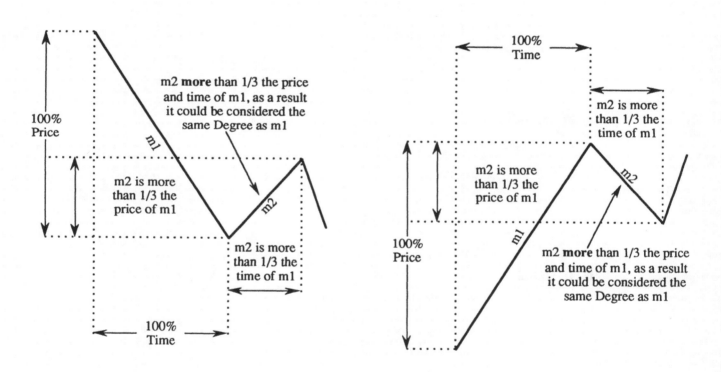

To illustrate how the identification of Structure Series' assists you in the wave building process over an extended period and how larger and larger Elliott patterns develop and are detected, the following **Rounds** have been created:

Round 1

Using the Rules and techniques described in Chapter 3, all monowaves in Figure 4-6 have been Structurally labeled. Isolated between the beginning of Chrono-2 and the end of Chrono-4, the Series "F3-c3-L5" occurs. Looking at the Table in Figure 4-3, that Series can be identified as a possible Flat pattern. Beginning with Chrono-9 and ending with Chrono-11, a Zigzag Series (5-F3-L5) is found. Chrono-12 through Chrono-16 form an Impulse Series.

Figure 4-6
(an updated version of Figure
3-18, taken from page 3-14)

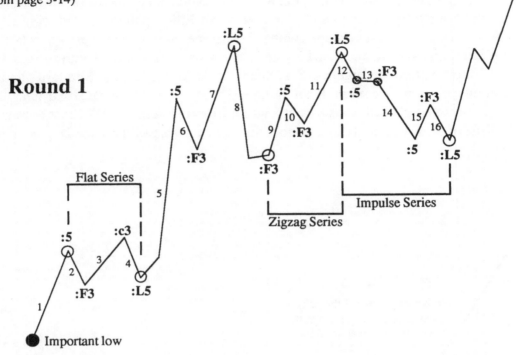

Round 1

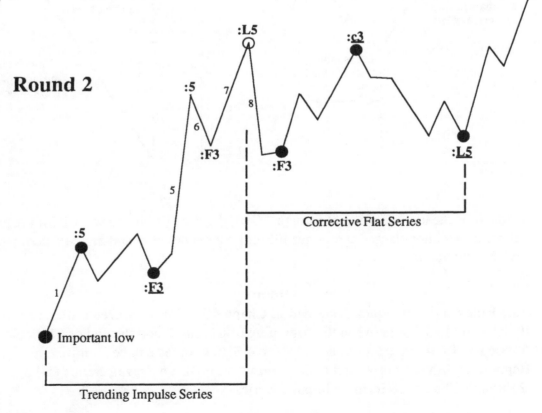

Figure 4-7

Round 2

Round 2

All wave groups (the Flat, Zigzag and Impulse) identified in **Round 1** have been thoroughly tested and have proven their validity. Therefore, they were Compacted (see Chapter 7 for details) to their base Structure of ":3" or ":5" (the underline represents Complexity level; refer to Chapter 7) and the circles encasing the Elliott wave pattern have been darkened. Figure 4-7 displays these transformations.

On your return to the Retracement Rules in Chapter 3, all Compacted patterns must be analyzed as if they were single waves containing one Structure label. Initially, the Compacted patterns must go through **reassessment** to detect for the possible existence of "missing" waves. Also, the two Structure labels adjacent to the Compacted pattern must be **reassessed** to decide if the changed environment (caused by the Compacted pattern) has had any influence on surrounding market Structure.

In Figure 4-7, all of these procedures have been carried out with the results exhibited therein. Notice, just as directed in Chapter 3, all labels and markings between the boundaries of the Compacted patterns have been eliminated.

Round 3

During Round 3, no Structure labels containing "L" Position Indicators are present. Until additional price action is plotted and a new ":L5" or ":L3" appears, no further progress can be made with pattern construction. What can be derived from the chart (Figure 4-8) as it currently stands. An uptrend began with Chrono-1 which, according to the table in Figure 4-3, cannot terminate until another Impulsive advance unfolds (very valuable information indeed). Under these circumstances, you know the proper strategy would be to stay long until another advance satisfied the required Impulsive characteristics.

Figure 4-8

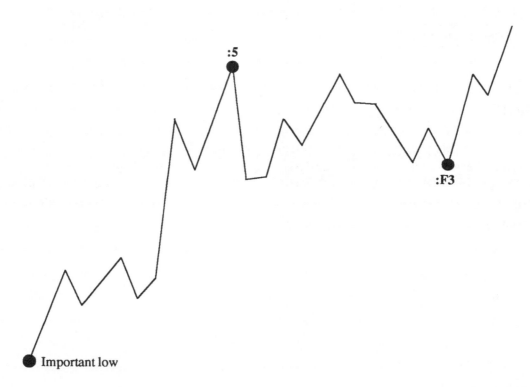

Zigzag "DETOUR" Test

Whenever you find an ":L5" which concludes a possible Zigzag pattern, always be aware that the Zigzag could really be the last three segments of an Impulse pattern. For example, in Figure 4-7 after the first ":F3" (moving forward from the low marked "Important Low") a ":5 - :F3 - :L5" Zigzag Series can be found. In isolation, the waves make a legitimate Zigzag pattern. On the other hand, if you connect market action preceding the first leg of the Zigzag, an Impulsive Series can be created beginning at the "Important Low."

Whenever you are working with Zigzags, always check the two Structure labels preceding to assure no Impulsive pattern is accidentally overlooked. If an Impulse Series can be assembled by connecting two previous Structure labels to the Structure Series (see Figure 4-9), always test the Impulsive pattern (applying the guidelines described in Chapters 5-12) before attempting to work with the Zigzag. If the Impulse pattern obeys typical construction Rules, use it. If the Impulse does not adhere to all the Rules, drop the first two Structure labels, return to the Zigzag scenario and test it for reliable construction. If the Zigzag is well constructed, use it. If the Zigzag does not obey the Rules, skip that section of the chart and work on another area until surrounding action clarifies the situation.

Figure 4-9

Zigzag Series at the end
of an Impulse pattern

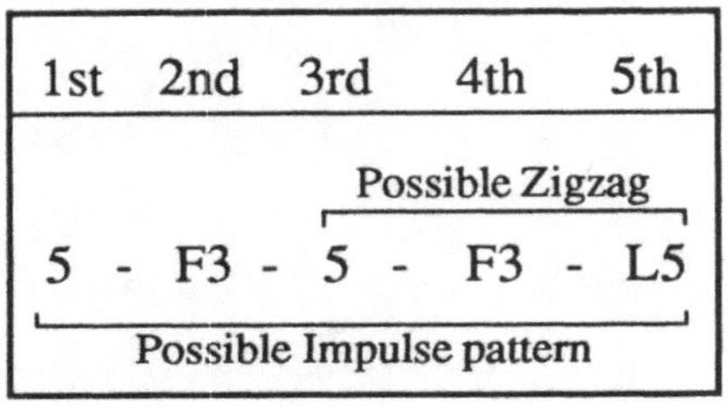

1st	2nd	3rd	4th	5th
		Possible Zigzag		
5 -	F3 -	5 -	F3 -	L5
Possible Impulse pattern				

Whenever working with a Zigzag, to guarantee an Impulse pattern is not overlooked, make sure the two Structure labels before the Zigzag cannot be used to make a larger Impulse pattern. If an Impulse Series is present, proper testing of the pattern must occur before considering the Zigzag possibility. If the Impulse idea is invalid, return to the Zigzag scenario.

What Next?

Once you reach this point, you have identified a possible Elliott pattern which will now need to be run through a battery of tests to see if it really is what it appears to be. If your wave group obeys one of the Standard Patterns Series' in Figure 4-3, move on to Chapter 5. If the wave group follows one of the Non-Standard Pattern Series' and you feel you have enough experience to tackle a Complex pattern, proceed to Chapter 8.

Until now, the main focus of the book has been on monowaves, a very basic concept. The more advanced stages of Wave Theory application require a "group" mind set, Monowave **combinations**. The "Structure Series" was a step in that direction. To continue the analytical process, more specific rules will need to be introduced to further differentiate between Impulsive and Corrective action. Exacting, *bottom line* rules will be presented for every <u>standard</u> Elliott pattern and its respective variations.

As you know from the previous section, each Structure Series represents a particular Elliott pattern. One of the more important aspects of pattern identification depends on appearance. Unfortunately, because of numerous *Impulsive* and *Corrective* pattern variations, there is no "standard" way of diagraming either category that reflects the <u>realistic</u> appearance of each variation. Elliott, in all of his writings, used illustrations similar to those found in Figure 5-1. The figures on the left were intended to represent Impulsive patterns. Those on the right were drawn to represent a certain type of Corrective pattern. The unrealistic nature of these diagrams tends to wrongly influence the perceptions and expectations of the beginning student on how wave patterns should really look. Avoiding this problem, which has continued up to the present, required this book contain hundreds of diagrams which <u>do</u> reflect realistic wave behavior. These diagrams will allow you to quickly become accustomed to the correct appearance of Elliott patterns, thus significantly diminishing the period between preliminary investigation and actual application of the Theory to *real-time* market action.

Construction of Polywaves

We have been working with "Structure Series" combinations to detect possible polywave development, but no rules have been presented to explain how Polywaves should look or behave based on their Class. Let's do that now.

Figure 5-1

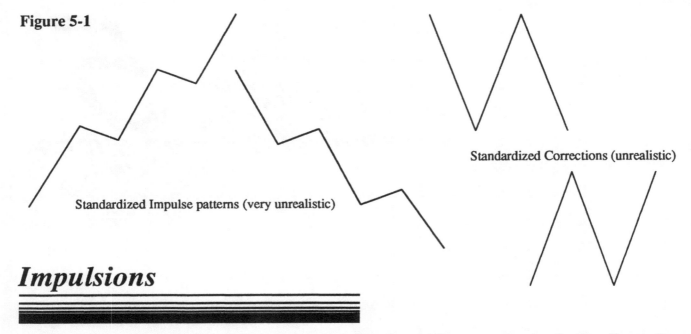

Standardized Corrections (unrealistic)

Standardized Impulse patterns (very unrealistic)

Impulsions

There are specific guidelines which apply to *Impulse* patterns that are not present in most *Corrective* patterns. Presented here will be <u>behavior confining</u> rules designed to solidify (or negate) groups of monowaves into patterns which are "*Impulsive in nature.*"

Essential Construction Rules

Market action *must* adhere to the following rules to be considered a <u>candidate</u> for *Impulsive* behavior:

1. **Five adjacent segments must be present (monowave or larger), which meet the Structure Series requirements of a Trending or Terminal pattern.**
2. **Three of those five segments *must* "thrust" in the same upward or downward direction.**
3. **Immediately after the first segment, a minor move in the opposite direction takes place (the second segment). This second segment can *never* retrace all of the first.**
4. **The third segment *must* be longer than the second.**
5. **Immediately after the third segment, a minor move in the opposite direction of the third (but the same direction as the second) takes place (segment four). The fourth segment must *never* retrace *all* of the third.**
6. **The fifth segment will *almost always* be longer than the fourth, but only <u>has to be</u> 38.2% of the fourth segment (price wise). When the fifth segment is shorter than the fourth, the fifth segment is termed a "failure."**
7. **When the vertical price distances covered by the first, third, and fifth segments are measured and compared, the third does *not* have to be the longest, but it can <u>never</u> be the shortest of the three segments.**

If a single one of the above rules is not adhered to, the market action being analyzed, by default, must be Corrective in nature, *not* Impulsive OR your wave group is incorrectly combined. If the pattern you are analyzing does not follow the above criteria, proceed to the **Correction** section (page 5-16).

Application to market action

We now have some general, broad application rules at our disposal for the detection of potential Impulse behavior. To make use of them, we need to work with five adjacent, "similar" monowaves. Figure 5-2 has multiple "five monowave" groups for us to study. Once we have gone through each pattern in the illustration, you should get out your own "wave chart" and practice the application of the "Essential Construction" rules on groups of real-time monowaves. These are the most critical rules of the entire Theory. You should not continue with your study of "Mastering Elliott Wave" until you are able to correctly apply these rules to real-time market action.

Making use of the "Essential Construction" rules, let's take a look at some monowave combinations and decide which patterns have Impulsive *potential* and which do not. Remember, if a pattern does not meet all of the rules, it *cannot* be Impulsive, it must automatically be Corrective. In Figure 5-2:

A fails the test since the 2nd segment retraces all of the first (see Rule 3, page 5-2).
B does not break any of the rules, so is possibly an Impulsive pattern.
C also obeys all of the aforementioned rules.
There are only four segments in **D**, five are required (see Rule 1).
In **E**, the 3rd segment is the shortest of the three advancing segments, an Impulsive pattern is impossible (see Rule 7).
F is up for consideration.
In **G**, the 4th segment retraces **all** of the 3rd, that is not allowable (see Rule 5).
There are no rules broken in **H, I**, nor **J** (in J, the 5th segment is shorter than the 4th, creating a "failure").
K is acceptable in **all** respects except for one, the third segment is not longer than segment two (see Rule 4)

Figure 5-2

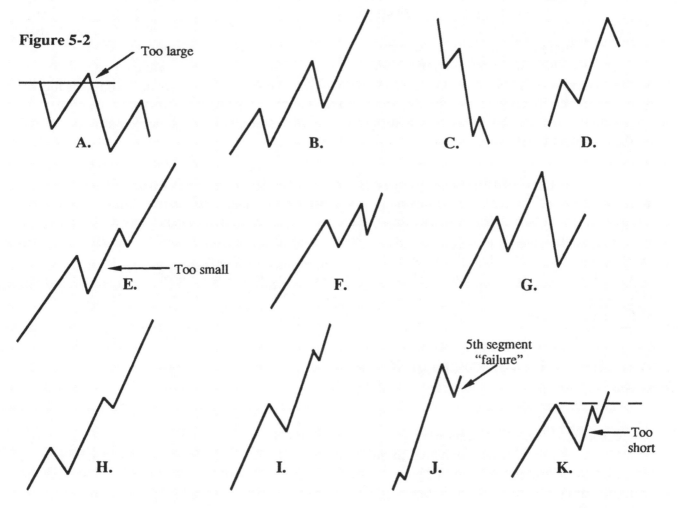

Extension Rule - *(the Litmus Test)*

An ***Extension*** is an <u>essential</u> element of any reliable Impulse pattern and is exclusively associated with them. The term **Extension** is used to describe *"the longest wave"* in an Impulsive grouping. The presence (or lack) of an Extended wave is the analytical "fork in the road," delineating Impulsive from Corrective behavior. This test will virtually always separate *real* Impulse patterns from imitations (Corrections).

By following all of the rules up to this point, you should be working with a group of five monowaves (or higher, if your abilities have increased). Among this group, one wave should be noticeably longer in price than all the others. The longest wave will be the Extension <u>candidate</u>. **To actually qualify as an Extension, the longest wave should be at least 161.8% of the next longest wave (price-wise).**

If a pattern has followed *all* of the Impulse rules up to the present, including the Extension Rule, move on to the next section in this chapter to verify whether the wave group meets further Impulsive requirements. **[NOTE: If this test is passed, but you still feel the pattern is corrective** (based on severe retracement of the move by later market action or other more subtle observations, which will be explained later), **a wave hidden in the price action may be the reason for the odd Corrective construction** (see *Missing Waves,* page 12-34).]

On the other hand, if any of the rules up to this point were not followed, it would virtually guarantee the pattern is corrective with two important, but <u>rare</u> exceptions:

1. When the <u>first</u> segment in a pattern is the longest, it *might* be slightly shorter than 161.8% of the third segment, but the third <u>will not</u> conclude higher than 61.8% <u>beyond</u> the end of the first.
2. When the third segment is the longest, but less than 161.8% of the first, and the fifth segment is shorter than the third, there is the remote possibility the market is forming an extremely rare *Terminal Impulse* pattern variation (see Terminal Impulse, Wave-3 Extended, page 11-6).

When the **Extension Rule** is broken, and neither of the two above exceptions is unfolding, the market is <u>absolutely</u> experiencing a Correction; proceed to the "Corrections" section. If one of the above two situations does appear to be unfolding, continue with your assumption the pattern is Impulsive and proceed to the next heading.

Introduction of *Progress Labels* to the Wave Group

If your pattern has passed the Extension test, you are finally ready for the placement of *Progress Labels* to your price charts. *Progress Labels* serve the purpose of testing market activity for the presence of very specific attributes. The testing will provide you with enough "ammunition" to allow for confident conclusions and the eventual solidification of the proposed wave pattern.

Here is how the *Progress Labels* should be placed: The first segment of your pattern will be considered wave-1. The second segment will be wave-2, the third wave-3, the fourth wave-4 and the fifth wave-5. Place these labels in order on your chart at the termination of each wave.

Conditional Construction Rules
▐▌▌▌▌▌▌▌▌▌▌▌▌▌▌▌▌▌▌▌▌▌▌▌▌▌▌▌▌▌▌▌▌▌ ▌ ▌ ▌ ▌

There is a great deal of difference between market action which does not break any of the *Essential Impulse Construction* rules and an actual impulse pattern. The Essential Impulse Construction rules, discussed a few headings back, are for general application to <u>all</u> Impulse wave patterns. They are the first rules to employ on market action <u>believed</u> to be Impulsive in nature. If a single "Essential" rule is broken, there is no need to proceed to the next stage of Impulse examination. Automatically assume the pattern is Corrective, or you have started your analysis from an improper point in the count, or all of the monowaves involved in the sequence are not of the same degree (for details, see "Degree," page 5-12). If you are following along with your own real-time price chart, all the rules after this point will be dependent on the Progress Labels you recently placed on your wave pattern.

Rule of Alternation

The concept of *Alternation* is one of the most important, guiding influences on market action and is at the very foundation of the Wave Principle. Without it, the Theory would be basically useless. It applies to virtually all aspects of application (for additional information on Alternation, see page 8-19).

The Rule states that: *when adjacent or alternate waves of the same Degree are compared, they should be distinctive and unique in as many ways as possible.* The major, deciding factor in the full manifestation of the rule is time. The greater the time period covered by each formation, the more complete, in all respects, the Alternation between the two formations will become. There are many ways Alternation can occur. Regarding Impulse patterns, the most important application of the Rule of Alternation is to its "counter trend" phases (i.e., waves 2 & 4). If and when you find the market is unfolding in a Corrective manner, the Rule best applies to waves A & B. The possible types of Alternation, which should always be considered in every formation, are as follows:

A. Price (the distance covered in vertical units)

B. Time (the distance covered in horizontal units)

C. Severity (the percentage retracement of the preceding wave)
   (applicable only to waves 2 & 4 of an Impulse pattern [all Impulse variations])
D. Intricacy (the number of subdivisions present in a pattern)

E. Construction (one pattern may be a Flat, the other a Zigzag, etc.)

To continue with the assumption that your wave group is Impulsive, Alternation must be present between waves 2 & 4 in *at least one* of the ways mentioned above. *[At this stage in the analysis process, you may or may not be working with more complex polywave patterns, so illustrations of simple and complex polywaves are presented on the next two pages for completeness.]*

Figure 5-3a shows the area of an Impulse pattern where Alternation is most critical and reliable.
A. Diagram A shows Alternation on a price, time, and severity basis; wave-3 is the extended wave.
B. In Diagram B, again the 2 & 4 wave Alternate in similar ways as shown in Figure A, but the extended wave has changed to wave-1.
C. Diagram C changes some of Figure B's conditions; wave-4 is more severe, time consuming and larger in price than wave-2 and the extended wave is now wave-5.
D. The only alternation in Diagram D is severity. The two corrections appear so similar as to make the pattern very suspect. Other possibilities would need to be considered and kept in mind until further market action clarifies the situation. If the entire Impulse pattern gets retraced more than

Figure 5-3a

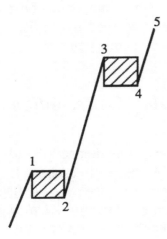

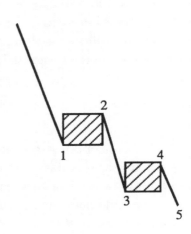

In order for waves 2 and 4 to interact correctly, there must be some type of Alternation between them in at least one (preferably more) of the ways listed under the Rule of Alternation on the previous page. On shorter term charts, you may <u>not</u> get any Alternation other than retracement severity, but it is important at least one factor is present. The larger and more time consuming the patterns being compared, the more important it is for adjacent and local (same Degree) market action to Alternate.

If this rule is not satisfied, the move is most likely a <u>Complex Correction</u> with a "Missing" x-wave in the center of what you are assuming is wave-3 (for details on Complex Corrections, see Chapter 8). If you recently began your study of Elliott Wave, avoid confusion by returning to the Preliminary Observations chapter and study a new group of monowaves.

Figure 5-3b

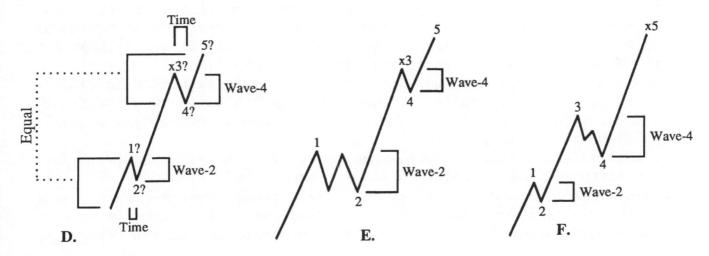

Figure 5-3b continued

G.

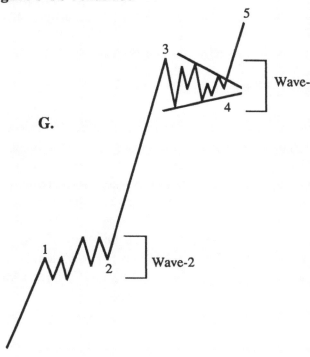

Wave-4

Wave-2

(Continued from page 5-5)

61.8%, but less than 100%, by the preceding market action, it would almost assure the pattern is a corrective Double Zigzag with a "missing x-wave" toward its center (see "Missing Waves," Chapter 12). If that is the situation on your real-time chart, go to Chapter 8 ("Construction of Complex Polywaves...") to continue your analysis. If you do not feel ready for more complex market discussions, move to a completely new group of monowaves and restart the analysis process. If the pattern is retraced <u>less</u> than 61.8%, and later its end is exceeded, the Impulse scenario would be reinforced.

E. Diagram E is an example of all facets of Alternation (price, time, severity, intricacy and construction) between waves 2 & 4.

F. In Diagram F, the reverse situation occurs from that of figure E, it is still an excellent example of Alternation.

G. The similarities of time and price between waves-2 and 4 in Diagram G make it necessary for Alternation to occur in a different way. In this case they are Alternating in construction (wave-2 is a Double Three Running correction and wave-4 is a triangle [covered later]). Time Similarities are most common between Waves 2 & 4 when the 3rd wave extends.

Note: Alternation is a relative phenomenon. A simple or complex correction is not dependent so much on what Elliott pattern is created as much as how the two Alternating patterns relate and "look" to each other. For example, a Double Three correction for wave-2 could be considered the simple pattern if wave-4 were a much larger and more time consuming Horizontal Triangle (see Figure 5-4).

Figure 5-4

(3)

(5)

(4) "Complex" Horizontal Triangle

(1)

(2) "Simple" Double Three

NOTE: This diagram is strictly for the illustration of the Rule of Alternation. The time consumption of waves-(2) & (4) is excessive in relation to waves-(1), (3) & (5).

From this diagram it should be obvious that the name given to a pattern does not automatically indicate its nature is simple or complex. Simple and Complex are relative terms which make sense only in a comparative environment. (Triangles will be covered in the Corrections section of this chapter and Complex patterns are covered in Chapter 8 ["**Construction of Complex Polywaves, Multiwaves, etc.**"]).

Rule of Equality

In any Impulse pattern, as discussed under the Extension Rule, one of the waves must be significantly *longer* than any other wave. Once you have identified the longest wave, the Rule of Equality needs to be considered. This Rule applies to only two of the following three waves: 1, 3, & 5. Whichever wave extended in the pattern, the Rule refers to the other two. The combinations would be:

1. If wave-1 extended, the rule applies to waves 3 & 5.
2. If wave-3 extended, the rule applies to waves 1 & 5.
3. If wave-5 extended, the rule applies to waves 1 & 3.

The Rule of Equality states that the two <u>unextended</u> waves should tend toward equality in price and/or time **OR** relate by a Fibonacci ratio (usually 61.8%) under either parameter or both. Price is the far more important consideration when looking for equality of two waves. In addition, the Rule exerts the most influence in Impulse patterns when the 3rd wave extends. The Rule is especially useful when the 5th wave fails <u>after</u> a 3rd wave extension. The rule is least useful when a pattern contains a 1st wave extension <u>or</u> is a Terminal Impulse.

Refer back to Figure 5-3b for some examples of how this takes place. In Diagram D, wave-3 extended, so waves 1 & 5 are approximately equal in price and time. Diagram B shows a 1st wave extension, in which waves 3 & 5 relate by 61.8% in price and 61.8% in time. In Diagram C, the 5th wave extended, so waves 1 & 3 relate to each other by 61.8% in price and 100% in time.

Overlap Rule

The *Overlap Rule* can be applied in two different ways, depending on whether you are analyzing a Trending or Terminal Impulse wave. Both categories are listed below.

Trending Impulse (5-3-5-3-5)

In a <u>Trending</u> Impulsive polywave (or higher), no part of wave-4 can fall into the price range covered by wave-2 (Figure 5-5a). This Rule creates one of the most obvious, visual differences between Trending Impulse patterns **and** Terminal Impulse or Corrective patterns.

Terminal Impulse (3-3-3-3-3)

Contrary to Trending Impulse patterns, it is <u>required</u> in a Terminal pattern that the price action zone of wave-2 be partially violated by the price action zone of wave-4 (Figure 5-5b).

Figure 5-5a

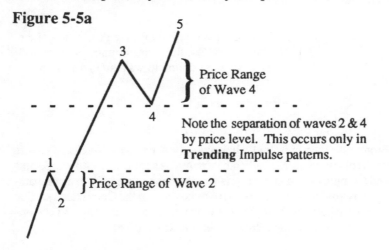

Price Range of Wave 4

Note the separation of waves 2 & 4 by price level. This occurs only in **Trending** Impulse patterns.

Price Range of Wave 2

Figure 5-5b

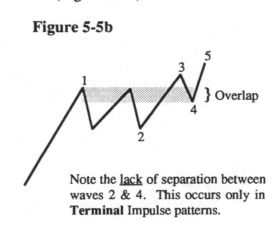

} Overlap

Note the <u>lack</u> of separation between waves 2 & 4. This occurs only in **Terminal** Impulse patterns.

Review

If the price action you have been working with has passed *all* the tests from the Preliminary section up to the Overlap Rule, the action is all but guaranteed to be Impulsive. If it has not passed all these rules, the probabilities are extremely high the move is Corrective. Anytime one of the rules, up to this point, is not followed by market action (with the few exceptions listed taken into account), move to the *Corrections* section and go through that market behavior check list to decide what type of Correction is unfolding. The wide variance of Corrective possibilities and flexibility allowed such phases makes the process more time-consuming and tedious than that required to decipher Impulsive behavior.

Breaking Point - Impulsions

From this point forward (limited to the Impulsions section), additional Impulse formation rules develop in one of two directions:

1. They become more subtle, conditional and difficult to work with requiring experience and confidence to properly apply, **OR**

2. They are easy to apply and not very subtle. They serve more as supplemental evidence of an interpretation's validity than as concrete proof.

Not all of the rules which follow (until the end of the "Impulsions" section) need to be obeyed by market action, but usually they are (see Exception Rule). If you are working with a pattern of Complex Polywave or Multiwave development, and are having trouble with solidifying your formation into an Impulse wave, side track to the "Progress Label and Logic Rule" chapters. Subtle Impulse behavior requirements are covered wave-by-wave. Once you are satisfied with the conclusions, return here, or if you are not satisfied, you may want to go to the Corrections section in this chapter.

Channeling

Channeling is a very important part of the analysis process. It is crucial in deciding when an Impulse pattern has completed and is fundamental in the search for the terminal points of waves 2 & 4.

There are two types of base channel lines when working with an Impulse pattern. They are the 0-2 trendline and the 2-4 trendline. Each serves a different purpose and provides important clues as to what type of Impulse pattern is forming, along with the type of Correction the market will create for waves 2 & 4. Channeling is most useful in Impulsions when wave-2 or wave-4 is a Polywave or higher.

Figure 5-6 shows how base channels are drawn. Illustrated are various combinations of development for 2nd waves. During the early stages (diagram A), the 0-2 trendline is employed and is essential in deciding when and where the 2nd wave completes. In diagram A, the first retracement of wave-(1) was originally believed to be the end of wave-(2). This was a legitimate assumption. Unfortunately, the minute the 0-2 trendline was broken *and* the break occurred with action still in wave (2)'s price zone, the 2nd wave was confirmed incomplete and the original trendline touch point becomes wave-a of wave-(2).

If after moving away from the revised 0-2 trendline, the market breaks the trendline again (before any significant advance occurs **or** if price again falls into the 2nd wave price zone), it is still likely wave (2) is incomplete. The trendline will have to be revised once again and the movement previously thought

Figure 5-6

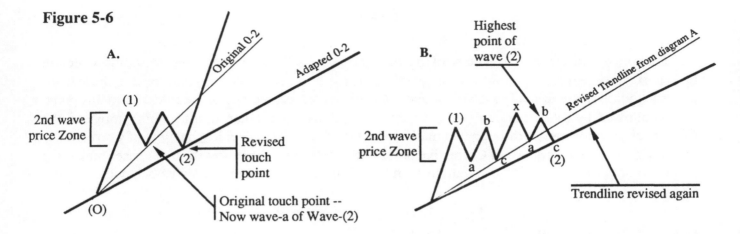

In diagram A, the first retracement of Wave (1) was originally considered the end of Wave (2). This was a correct assumption at the time. However, the instant the 0-2 trendline got broken (diagram A) and that break occurred with price action within the 2nd wave zone, price action indicates that wave (2) is not over and the original touch point becomes wave-a of wave (2). If the revised 0-2 trendline is broken again (diagram B) before a significant advance occurs **or** the price action falls into the 2nd wave zone again, it is likely Wave (2) is not complete and the movement to the "revised trendline" is only the first a-b-c of a Complex Correction which will form Wave (2). After a sequence like the one diagramed in Figure B, the correction will probably conclude, and Wave (3) will commence. If you witness a significant move (relative to the x-wave) followed by a Correction which terminates outside of the price range of Wave (2), the 3rd wave should be in progress, or perhaps even complete. After a pattern such as the one in Figure B, the wave-3 should be much longer than wave-1 and it should definitely be the extended wave.

> **Important: If the 0-2 trendline is _real_, no part of wave 1 _or_ 3 should break the line.**

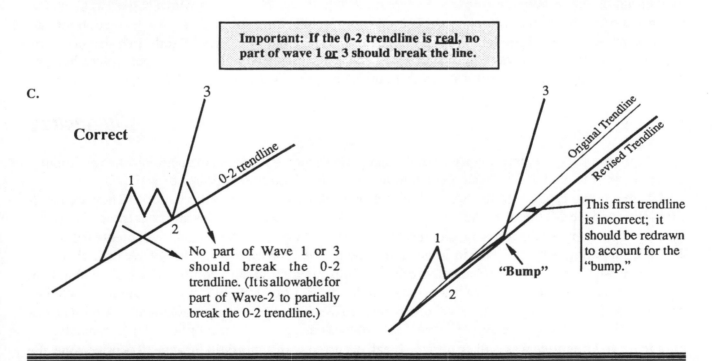

to be the end of wave-(2) will likely be only the a-b-c of a complex 2-wave [i.e. wave-2 may be a Double or Triple Three, (for details on Complex patterns such as Double and Triple Threes, refer to Chapter 8)].

After a sequence like Diagram B, the Correction will probably conclude and wave-(3) can get underway. Once you witness a significant advance (more than 161.8% of wave-1) followed by a

Figure 5-6 continued

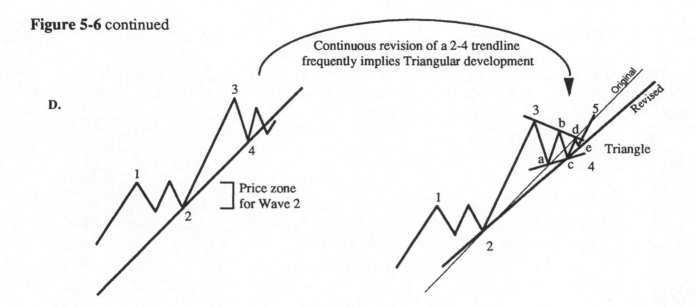

Continuous revision of a 2-4 trendline frequently implies Triangular development

D.

Price zone for Wave 2

Original
Revised
Triangle

correction which stays above the highest price range of wave-1, the 3rd wave is probably in progress (or may have even completed). **Important:** If the 0-2 trendline (of any degree) is "real," no part of wave-1 or wave-3 should break it (see the left-hand side of diagram C).

Once it is decided wave-3 has terminated, you will need to begin a similar channeling process to detect the conclusion of wave-4. Diagram D (above) shows the steps necessary to arrive at a viable conclusion. In the diagram, it is assumed wave-3 is finished since it is much longer than wave-1 and because a correction following wave-3 stayed above the price zone of wave 2. To decide whether wave-4 has finished, you need to draw a trendline across the lowest point of the correction after the supposed 3rd wave high. If the market runs to new highs shortly after without breaking **or** touching the 2-4 trendline, the fifth wave is probably in progress. If, before making a new high, the market breaks the 2-4 trendline, wave-4 is probably still forming and the first corrective low is only wave-a of wave-4. *[Note: More than one revision of the 2-4 trendline might be necessary. All of the above procedures and techniques also apply to downward Impulse movement.]*

Fibonacci Relationships

Fibonacci relationships are most prevalent when the 5th wave Extends in an Impulse pattern. Relationships are least common in a 3rd wave Extension pattern. More specific relationship discussions take place in *Advanced Fibonacci Relationships* (Chapter 12). For now the discussion will be limited to general observations based on which wave Extended in the Impulse sequence.

1st Wave Extension

The most common arrangement is for wave-3 to relate to wave-1 by 61.8% and wave-5 to relate to wave-3 by 38.2%. The next most likely combination is the reverse of the above; wave-3 will be 38.2% of wave-1 and wave-5 will be 61.8% of wave-3.

3rd Wave Extension

This situation allows for a limited number of relationships, which usually occur between waves-1 & 5. Wave-1, if not equal to wave-5, will normally be 61.8% or 161.8% of wave-5. Wave-3 *must be more* than 161.8% of wave-1 <u>if</u> it is the Extended wave.

5th Wave Extension

It is during 5th wave Extensions that wave-3 will normally be related to wave-1 by 161.8%. The 5th wave most commonly will relate to the entire move from the beginning of wave-1 to the end of wave-3 by 161.8%, added to the end of wave-3 or wave-4.

Degree

In order to keep track of a market's varying levels of development, Elliott devised stratified wave Degree Titles. Disappointingly, to some, **Degree** cannot be described in absolute terms such as days, weeks, dollars and cents. It is a relative concept describing how one pattern interacts with another. The smallest **Degree** Title that Elliott coined was *Sub-Minuette*. Since the monowaves we have been working with are the simplest wave patterns on a chart, it is logical to adopt the *Sub-Minuette* Title for those monowaves being analyzed.

Adoption of a **Degree** Title also requires the utilization of a specific Symbol. Looking over your chart, if you are still working with monowaves, replace the plain Impulsive Symbols you have been using on your pattern (1, 2, 3, 4, 5) into the specific Impulse Symbols which represent *Sub-Minuette* Degree. Those symbols are as follows:

> **i - for wave-1**
> **ii - for wave-2**
> **iii - for wave-3**
> **iv - for wave-4**
> **v - for wave-5**

As your skills improve, you will need to Title and Symbolize larger **Degree** patterns. If you are working with polywave (or higher) patterns at this point, the steps involved in deciding **Degree** are described in the "More on Degree" section (Chapter 7).

NOTE: **Degree** is a very difficult concept. It is something the beginning student <u>should not</u> be very concerned with, just aware of. Whether something is of Supercycle or Minor degree is purely subjective and not at all important in the early stages of analysis. What is not subjective is the relationship between each wave. For example, if you label one move *Minor* and it completes a larger Elliott sequence, the larger sequence will have to be of Intermediate Degree.

Realistic Representations - (Impulsions)

On the next two pages there are diagrams depicting **realistic** Impulse patterns just as they might occur on a plot of real-time market action. Based on the author's discoveries of standardized pattern design, the diagrams illustrate the typical shape an Impulse pattern will exhibit as a direct result of which wave segment Extended. The extended wave of an Impulsive Elliott pattern is the most important factor in deciding its appearance. Furthermore, each diagram lists Progress Labels on the left of the colon (:) and Structure Labels on the right. The "X" in front of some Progress Labels' (only used in front of Progress Labels' "1, 3 or 5") indicates which wave eXtended.

If you plot data in the proper fashion (clearly described and illustrated in Chapter 2 & 3 of "Mastering Elliott Wave"), real-time market action will closely, and sometimes exactly, resemble the diagrams which follow. If you use bar charts, hourly closing figures, or other types of incorrectly plotted or calculated data, the action <u>sometimes will</u> and <u>sometimes will not</u> look like the diagrams to follow.

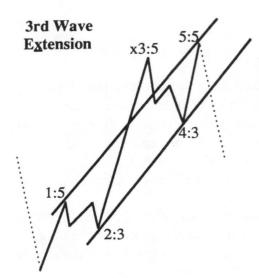

3rd Wave Extension

x3:5
5:5
4:3
1:5
2:3

If the 5th wave is shorter than wave-4, it creates a situation called a 5th-wave Failure. A 5th-wave Failure is only possible in Impulse patterns with 1st or 3rd-wave Extensions (5th-wave Failures in Impulse moves with 3rd-wave Extensions being the **far more common** of the two). Of all 3rd-wave Extension variations, the 5th-wave is most likely to Fail when wave-1 is "microscopic" in relation to wave-3 (see diagram on right).

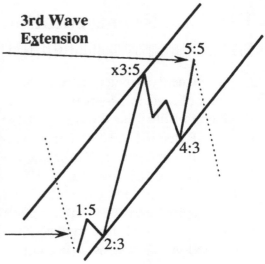

3rd Wave Extension

x3:5
5:5
4:3
1:5
2:3

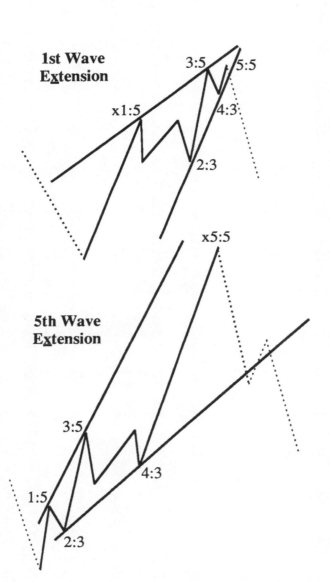

1st Wave Extension

3:5
5:5
x1:5
4:3
2:3

5th Wave Extension

x5:5
3:5
4:3
1:5
2:3

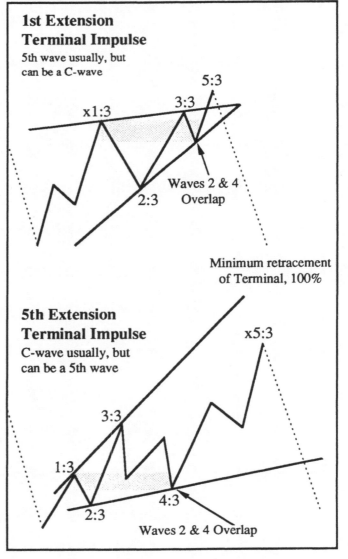

All Impulsions <u>outside</u> of this rectangle are Trending

1st Extension Terminal Impulse
5th wave usually, but can be a C-wave

x1:3
3:3
5:3
2:3
Waves 2 & 4 Overlap

Minimum retracement of Terminal, 100%

5th Extension Terminal Impulse
C-wave usually, but can be a 5th wave

x5:3
3:3
1:3
4:3
2:3
Waves 2 & 4 Overlap

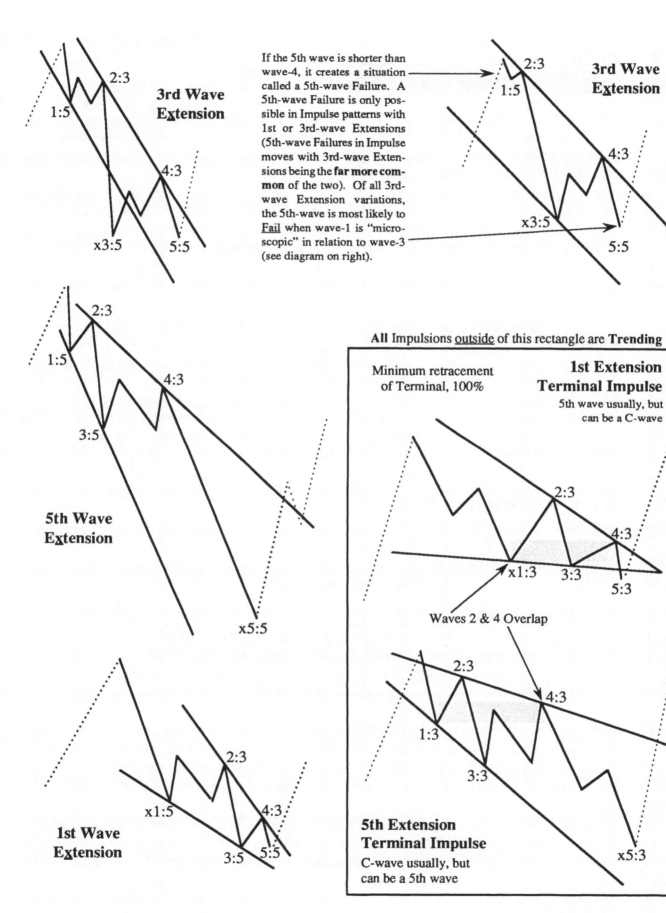

3rd Wave Extension

2:3
1:5
4:3
x3:5
5:5

If the 5th wave is shorter than wave-4, it creates a situation called a 5th-wave Failure. A 5th-wave Failure is only possible in Impulse patterns with 1st or 3rd-wave Extensions (5th-wave Failures in Impulse moves with 3rd-wave Extensions being the **far more common** of the two). Of all 3rd-wave Extension variations, the 5th-wave is most likely to <u>Fail</u> when wave-1 is "microscopic" in relation to wave-3 (see diagram on right).

3rd Wave Extension

2:3
1:5
4:3
x3:5
5:5

5th Wave Extension

2:3
1:5
4:3
3:5
x5:5

1st Wave Extension

2:3
x1:5
4:3
3:5
5:5

All Impulsions <u>outside</u> of this rectangle are **Trending**

Minimum retracement of Terminal, 100%

1st Extension Terminal Impulse
5th wave usually, but can be a C-wave

2:3
4:3
x1:3
3:3
5:3

Waves 2 & 4 Overlap

5th Extension Terminal Impulse
C-wave usually, but can be a 5th wave

2:3
1:3
4:3
3:3
x5:3

Corrections

Corrections are patterns which occur <u>between</u> Impulse waves. As you already know, Corrections are normally made up of three monowave (or larger) segments. How to relate those segments and form a **Standard**, reliable Elliott pattern is the purpose of this chapter.

In general, Corrective phases are more difficult to interpret than Impulsive phases due to a larger pool of variation possibilities. Deciphering Corrective phases often requires a complex understanding of market behavior and, sometimes, lots of patience. Do not get discouraged when unable to figure out a wave count during corrective activity, that is typical; simply give the market additional time to develop. Corrections (and Impulsions) only become "crystal clear" when they have <u>completed</u> or are near completion. At termination of a pattern it is usually obvious which variation is occurring.

Introduction of **Progress Labels** *to the Wave Group*

Unlike Impulsions, which require a multitude of specific criteria exist before Progress Labels can be placed on the price action, Corrections are less demanding. When you reach this stage of the analysis process, you are not trying to decide whether the pattern is Corrective or Impulsive. You "know" it is Corrective simply because it is **not** Impulsive. This allows for the <u>immediate</u> placement of Progress Labels to the price action. The first Structure Label in the group should be marked wave-a. The second is marked wave-b, the third wave-c. If still working strictly with monowaves, and there is a fourth and fifth monowave, mark them waves-d and wave- e, respectively.

Essential Construction Rules

Since there are many different types of Corrections, general rules covering all eventualities cannot be devised. The only way to describe **all corrective patterns** is indirectly:

> If market action <u>does not</u> follow all necessary Impulse rules from the "Preliminary Observations" chapter up to, but not including, the Breaking Point discussion under the Impulsions section, by*default* the market action is Corrective.

The following rules are essential in the construction of specific Corrective polywaves. Even though larger patterns should almost always abide by the same parameters, these guidelines are designed to help you form Corrective polywaves from only three or five adjacent monowaves. As a result of the numerous patterns (and variations) falling under the Corrective umbrella, the parameters for each will have to be listed separately.

First, all patterns which fall under the category of *Flats* (B Failure, C Failure, Common, Double Failure, Elongated, Irregular, Irregular Failure, Running) will be covered. *ZigZag* patterns will be discussed second; and finally, the most difficult and important Corrective pattern, the *Triangle*, will be introduced last.

To continue with your analysis, return to the monowave group you have been working on. Match the "Series" possibility you are currently testing with one of those listed below and branch to the heading (in this section) indicated by that series.

ZigZag 5-3-5
Flat 3-3-5
Triangle 3-3-3-3-3

[If you are not following along with your own chart **or** this is your first time reviewing this section, just read straight through.]

Flats (3-3-5)

Let's start off by defining the minimum retracement requirements for each wave in a Flat to make sure the "Series" is properly constructed. If the requirements below are not met, go back to your chart, pick a new wave group isolated in Chapter 3 and start the process over beginning from the **Intermediary Observations** chapter.

The three monowaves (or higher patterns, if you have developed your skills to that level) under observation must exhibit the following criteria before they can fall under the **Flat** category:

1. Wave-b **must** retrace at least 61.8% of wave-a. (see Figure 5-19)

Figure 5-19

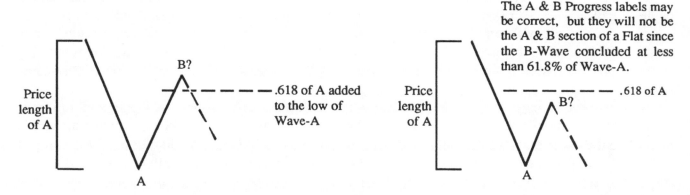

Price
length
of A
B?
A
.618 of A added
to the low of
Wave-A
The A & B Progress labels may be correct, but they will not be the A & B section of a Flat since the B-Wave concluded at less than 61.8% of Wave-A.
Price
length
of A
.618 of A
B?
A

2. Wave-c **must** be at least 38.2% of wave-a (see Figure 5-20).

Figure 5-20

B
.382
C
Price
length
of A
A

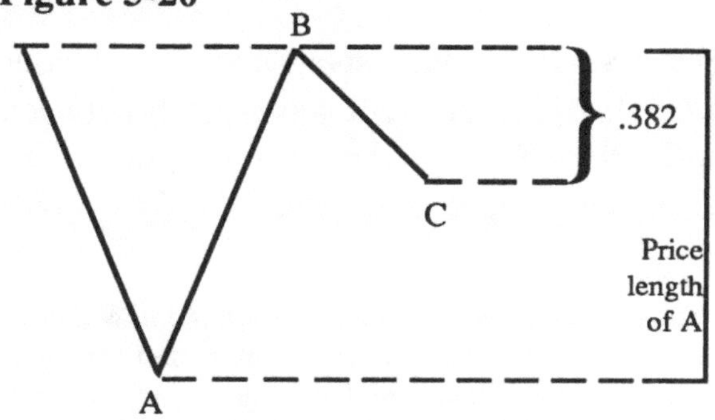

This illustrates the minimum percentage retracement of wave-A which wave-C must attain to even consider wave-C has completed. For this arrangement of monowaves to complete a Flat, other factors must also be present (see Channeling).

Figure 5-21

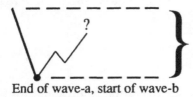

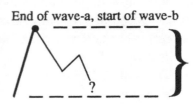

End of wave-a, start of wave-b

Horizontal parallel lines off the highest and lowest price of the first monowave in the group

End of wave-a, start of wave-b

There are more variations on Flat patterns than any other Elliott formation. To get a general idea of what type is taking place, the following technique should be implemented. Draw two horizontal, parallel lines, one off the highest point of the first monowave and the other off the lowest point of the first monowave (see Figure 5-21). This will provide an accurate measuring tool to detect the differences between one *Flat* variation and another.

Beginning the observations, if wave-b breaks the horizontal line opposite its starting point, the market action indicates a "stronger than normal" *Flat* is forming (refer to Strong B-Wave, below). If the b-wave retraces between 81-100% of wave-a, refer to Normal B-Waves. If the retracement of wave-a is between 61.8-80%, refer to Weak B-Waves.

Strong B-wave

Depending on the size of the b-wave in relation to wave-a, the c-wave may or may not exceed the beginning of wave-b. If the b-wave falls between 101-123.6% of wave-a, there still exists a relatively good chance wave-c will completely retrace wave-b. If the b-wave falls within the above range, the c-wave is 100% or more of wave-b and wave-c is not more than 161.8% of wave-a, the market is forming an *Irregular* correction. If the c-wave is more than 161.8% of wave-a, designate the pattern an Elongated Flat.

If the b-wave is more than 123.6% of wave-a, there is little chance wave-c will retrace all of wave-b. If it does, it will still be an *Irregular* pattern. When the b-wave exceeds 138.2% of wave-a, there is no chance the c-wave will retrace all of wave-b (the c-wave of a Triangle might, but not the c-wave of a Flat). As long as part of wave-c falls <u>within</u> the range of the horizontal parallel lines, but does not completely retrace wave-b, the pattern should be considered an *Irregular Failure*. If the c-wave does <u>not</u> fall within the horizontal, parallel lines, the pattern should be considered a ***Running Correction***.

Once the specific Flat variation taking place is known*, proceed to the "Conditional Rules" section (page 5-34) for further testing of the pattern.

Normal B-wave

The b-wave, to be considered "normal," should fall between 81-100% (inclusive) of wave-a. Under these conditions, the c-wave will most likely retrace all of wave-b. If the length of wave-c falls between 100-138.2% of wave-b, the pattern should be considered a *Common Flat*. If the c-wave is more than 138.2% of wave-b, the market is forming an *Elongated Flat*. If the c-wave is less than 100% of wave-b, the pattern is a C-Failure.*

* Knowing what type of Flat pattern is forming during your early stages of study will not be that helpful, but will be of great assistance and application at a later date. If you feel you have assimilated everything up to this point and are ready for some new "formation solidifying" guidelines, refer to Chapters 10 & 11 at this juncture and then return to Chapter 5.

Weak B-wave

Weak b-waves are characterized by their <u>less than normal</u> retracement of wave-a. To classify as a weak b-wave, it must retrace between 61.8% - 80% (inclusive) of wave-a. If the c-wave is less than 100% of wave-b, the pattern should be called a *Double Failure*. If wave-c is between 100% - 138.2% of wave-b, the pattern should be called a *B-Failure*. If wave-c is longer than 138.2% of wave-b, it will once again fall under the *Elongated Flat* category.*

ZigZags (5-3-5)

There is a limited number of variations on *ZigZag* patterns. ZigZags and their complex combinations (see Chapter 8) are the only corrective patterns which can temporarily "resemble" Impulsive activity. To avoid misinterpretation, very specific limits must be placed on Zigzag behavior. Below are the minimum requirements which allow a pattern to be categorized as a ZigZag.

1. Wave-a should not retrace more than 61.8% of the previous Impulse wave (if present) *of one larger degree* (see Figure 5-22a).

Figure 5-22a

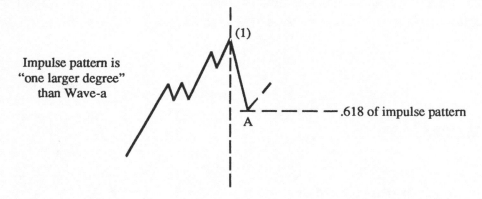

2. Wave-b should retrace at least 1% of wave-a (Figure 5-22b).

Figure 5-22b

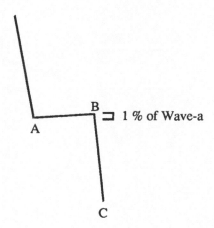

3. Wave-c <u>must</u> move, even if only slightly, beyond the end of wave-a (Figure 5-23).

Figure 5-23

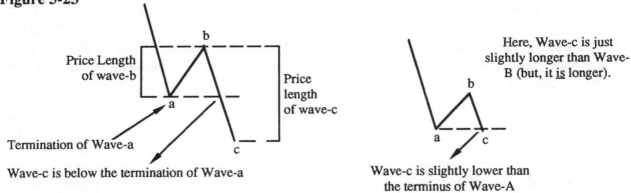

Price Length of wave-b

Price length of wave-c

b

a

c

Termination of Wave-a

Wave-c is below the termination of Wave-a

Here, Wave-c is just slightly longer than Wave-B (but, it <u>is</u> longer).

b

a   c

Wave-c is slightly lower than the terminus of Wave-A

If your wave group has followed these three minimum parameters, it is time to check for adherence to the <u>maximum limits</u> imposed on wave-b of a ZigZag.

1. No part of wave-b will normally retrace more than 61.8% of wave-a.
2. If part of wave-b retraces more than 61.8% of wave-a, that part will not be the end of wave-b. It will only be the first segment of a more complex correction for wave-b. The termination of wave-b will complete at 61.8% of wave-a <u>or less</u> (see Figure 5-24).

Figure 5-24

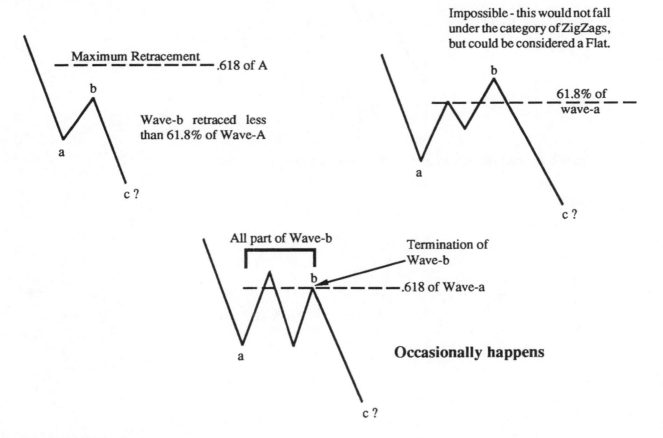

Maximum Retracement ———— .618 of A

b

a

c ?

Wave-b retraced less than 61.8% of Wave-A

Impossible - this would not fall under the category of ZigZags, but could be considered a Flat.

b

61.8% of wave-a

a

c ?

All part of Wave-b

Termination of Wave-b

b

.618 of Wave-a

a

Occasionally happens

c ?

The length of wave-c is the deciding factor for categorizing a ZigZag formation. It's length says many things about the current and future market action. If the c-wave of the Zigzag you are working with is less than 61.8% of wave-a, refer to **Truncated** Zigzag. If the c-wave completes at a point which is more than 161.8% of wave-a added to the end of wave-a, move to the *"Elongated Zigzag"* section (be careful with this pattern, it could be part of an Impulse wave). In any other situation, refer to *"Normal Zigzags"*.

Normal

In a ***Normal*** Zigzag, the c-wave can be anywhere from 61.8% to 161.8% (Internally and Externally, for details, refer to page 12-22) of wave-a. The listing below describes and illustrates the conditions necessary for ***Normal*** Zigzag formation.

1. Wave-b (when measured from its terminus) ***must not*** retrace more than 61.8% of wave-a. Figure 5-24 (previous page) shows several situations which allow for, or negate, Zigzag formation.

Figure 5-25a

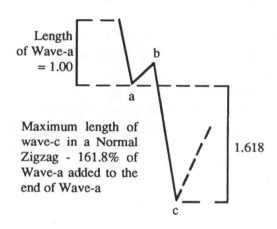

Length of Wave-a = 1.00

Maximum length of wave-c in a Normal Zigzag - 161.8% of Wave-a added to the end of Wave-a

1.618

Figure 5-25b

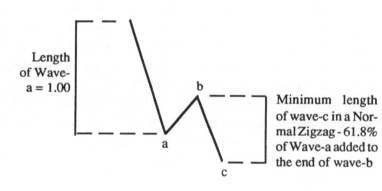

Length of Wave-a = 1.00

Minimum length of wave-c in a Normal Zigzag - 61.8% of Wave-a added to the end of wave-b

2. Wave-c should not exceed 161.8% of the length of wave-a added to the end of wave-a (Figure 5-25a), but should be at least 61.8% of wave-a, Internally (see Figure 5-25b).

When a move, which you think is a c-wave, exceeds the limits described above, refer to *"Elongated Zigzag"* (page 5-22) and/or *"Impulsions"* (page 5-2).

Truncated

This is the rarest Zigzag variation, to be justified it must meet the following criteria:

1. Wave-c cannot be shorter than 38.2% of wave-a, but should be less than 61.8% of wave-a.
2. After completion of the Zigzag, the market must retrace at least 81% of the entire Zigzag, and preferably, it should retrace 100% or more (Figure 5-26, next page).
This is essential due to the counter trend strength indicated by the extremely short c-wave.
3. The pattern will most likely be found as <u>one</u> of the five legs of a Triangle **or** <u>as a segment</u> of one of the legs of a Triangle.

Figure 5-26

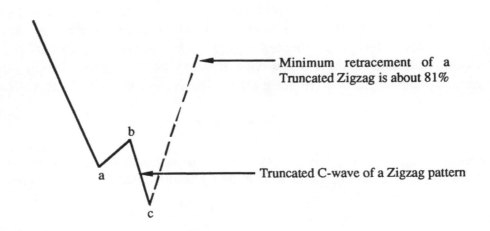

Minimum retracement of a Truncated Zigzag is about 81%

Truncated C-wave of a Zigzag pattern

Elongated

An *Elongated Zigzag* is characterized by an <u>oversized</u> c-wave. As stated earlier, Zigzags temporarily resemble Impulsive activity. Of the Zigzag variations, the Elongated is the best imitator of Impulsive behavior. This makes them very difficult to recognize and dangerous to assume while they are unfolding. Usually, they can only be confirmed after the fact. Their redeeming quality is that they only take place in the early stages of Complex Polywave (or higher) Contracting Triangles (see Chapter 8) or the late stages of Complex Polywave (or higher) Expanding Triangles.

Whenever a c-wave is more than 161.8% of an a-wave, the probabilities greatly favor the assumed a-b-c is actually the 1-2-3 of a five segment Impulse pattern. The criteria which helps decide between the two different patterns is **retracement.** After an Elongated Zigzag, the market should reverse and retrace more than 61.8% of the c-wave <u>before</u> the end of wave-c is violated. If those conditions are met, assume the pattern is an *Elongated Zigzag*. If those conditions are not met, it likely the pattern forming is an Impulse wave. Return to the **Preliminary Analysis** chapter and tack on some additional monowaves to the wave group you are working with and see if you can come up with a possible Impulse Series (also refer to the Zigzag "DETOUR" Test on page 4-8). If not, go on to a new group of monowaves and start from the beginning. Eventually, through all the techniques described in this book, the pattern you could not figure out will become clear once the surrounding price action is correctly deciphered.

Figure 5-27

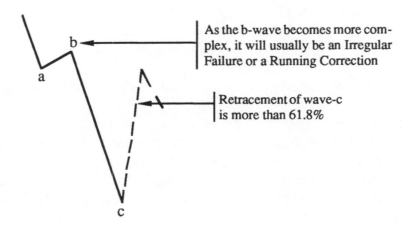

As the b-wave becomes more complex, it will usually be an Irregular Failure or a Running Correction

Retracement of wave-c is more than 61.8%

Some of the most difficult patterns to work with under the Wave Theory are variations of *Triangles*. They have no specific time limits for completion. Absolute certainty as to the direction of post-Triangular action is often impossible. One of their redeeming qualities is that after they complete, they provide a significant amount of information on current market position and offer numerous clues on how post-triangular price action should behave for extended periods of time.

Despite their difficulties, Triangles are some of the most common Elliott patterns; therefore, a thorough understanding of them is essential. Learning to identify Triangles **early** in their formation can save you from many hours of from frustration and from unnecessary trading losses (especially in Options). Following are the most important Rules and characteristics of Triangular formations. The majority of these Rules (which are essential) are presented here for the first time, so pay close attention.

Following is a list of *minimum* requirements in the formation of all Triangle variations:

1. Elliott said there are five segments to a Triangle, no more, no less. This rule applies no matter how simple or complex each segment is. In order of occurrence, each segment of the Triangle is given a letter of the alphabet: (a,b,c,d,e).
    (see Figure 5-29)
2. Each Segment of the Triangle *is* (or represents, in the case of monowaves) a Complete Corrective phase (a ":3").

Figure 5-29

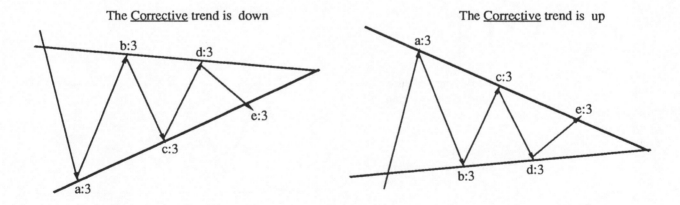

As in an Impulse sequence, three of the Corrective patterns go
with the corrective trend and two go against the corrective trend.

3. Unlike an Impulse pattern, which tends to trend up or down, the five segments of a Triangle will oscillate over and over in the same price territory (overlap) with a slightly Expanding or Contracting bias (Figure 5-30, next page).
4. The Triangle can drift slightly upward or downward without affecting these general guidelines (see Figure 5-31, next page).
5. The length of wave-b must fall between 38.2-261.8% of wave-a. NOTE: there is a strong tendency for wave-b to avoid a 100% price relation to wave-a.

Figure 5-30

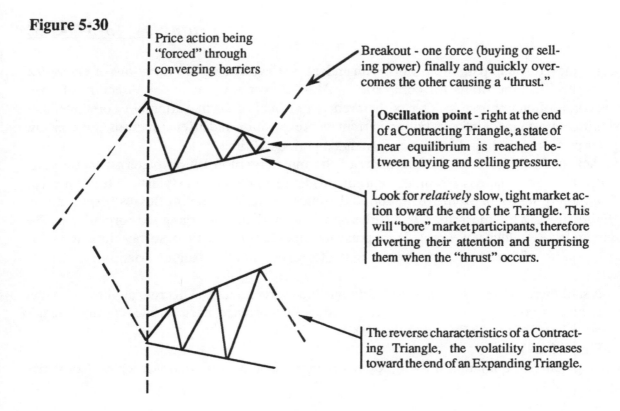

Price action being "forced" through converging barriers

Breakout - one force (buying or selling power) finally and quickly overcomes the other causing a "thrust."

Oscillation point - right at the end of a Contracting Triangle, a state of near equilibrium is reached between buying and selling pressure.

Look for *relatively* slow, tight market action toward the end of the Triangle. This will "bore" market participants, therefore diverting their attention and surprising them when the "thrust" occurs.

The reverse characteristics of a Contracting Triangle, the volatility increases toward the end of an Expanding Triangle.

Figure 5-31

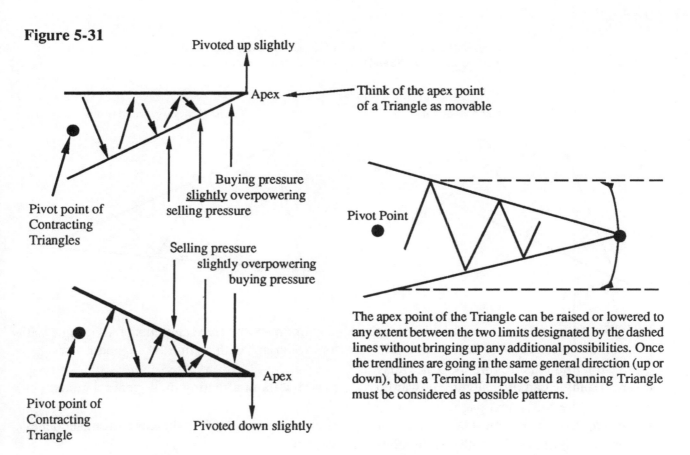

Pivoted up slightly

Apex

Think of the apex point of a Triangle as movable

Buying pressure slightly overpowering selling pressure

Pivot point of Contracting Triangles

Selling pressure slightly overpowering buying pressure

Pivot Point

Pivot point of Contracting Triangle

Apex

Pivoted down slightly

The apex point of the Triangle can be raised or lowered to any extent between the two limits designated by the dashed lines without bringing up any additional possibilities. Once the trendlines are going in the same general direction (up or down), both a Terminal Impulse and a Running Triangle must be considered as possible patterns.

6. Of the five segments in a Triangle, four retrace a previous segment. The retracing segments are waves b,c,d & e. Of those four, <u>three segments must retrace at least 50% of the previous wave</u> (see Figure 5-32). During a rare Running Triangle, this parameter <u>may not</u> be completely met.

Figure 5-32

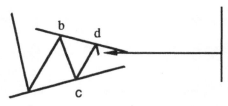

The e-wave does not need to retrace a lot of wave-d in this case since three of the other retracements were substantial enough. If the e-wave had retraced at least 50% of wave-d, then one of the others <u>would not</u> have had to retrace 50%, but probably would have anyway.

Three 50%+
Retracements
{
1. <u>Wave-b</u> retraces more than 50% of Wave-a
2. <u>Wave-c</u> retraces more than 50% of Wave-b
3. <u>Wave-d</u> retraces more than 50% of Wave-c
}
This sequence meets minimum retracement criteria .

7. When dealing with the important reference points in a Triangle, there are six you need to be concerned with. They are all of the same degree:

   i.   The beginning of wave-a, which is called point "0" (zero)
   ii. End of wave-a, point "a"
   iii. End of wave-b, "b"
   iv. End of wave-c, "c"
   v. End of wave-d, "d"
   vi. End of wave-e, "e"

Only four of the terminal points (of the same degree) in a Triangle should be channeled between contracting trendlines. NOTE: Non-Limiting Triangles usually follow this rule, but there are occasional exceptions which may create a 5th touch point (see Figure 5-33 for an illustration of this Rule).

Figure 5-33 (continued on next page)

Each dark dot represents a possible touch point which you can employ to draw converging trendlines which encase the triangle. Only four (of the six) turning points can <u>simultaneously</u> touch the converging trendlines.

Too many Touchpoints on this side

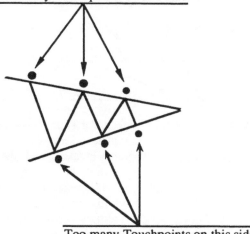

This pattern would not qualify as a proper triangle. If market action did channel like this, the action could be <u>part of a Triangle</u>, but it almost certainly would not be the completion of a Triangle at the last point on the lower right.

Too many Touchpoints on this side

Figure 5-33 continued

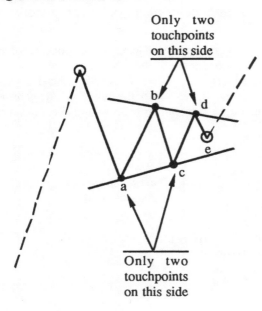

Only two
touchpoints
on this side

Only two
touchpoints
on this side

This is how channeling should occur in
a <u>normal</u> Horizontal Contracting Triangle.

Note that there are two points *of the same degree* which do not
touch the trendlines. The beginning of wave-a (where the Triangle began) and the end of wave-e (the end of the Triangle).

8. The channel line crossing waves B & D in a Triangle should be thought of as the ***Base*** line. Its
function is similar to a 2-4 trendline in an Impulse wave. ***As a general rule,*** the B-D trendline
should not be broken by any part of wave C or E in the Triangle (Figure 5-34a).   In other words,
there should be a clear path from wave B to D and from wave D until the end of wave E. Figure
5-34b illustrates behavior around the Triangle's B-D trendline which is not acceptable.

Figure 5-34a

Correct

This break is fine
since it occurred after
wave-e completed

End of Wave-e

No part of waves
c or e touched the
b-d trendline

Most Typical
Arrangement

Figure 5-34b

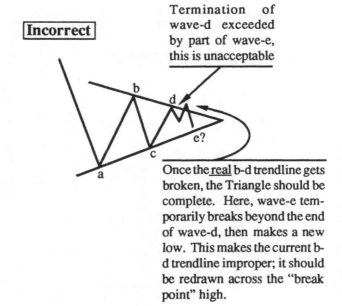

Incorrect

Termination of
wave-d exceeded
by part of wave-e,
this is unacceptable

Once the <u>real</u> b-d trendline gets
broken, the Triangle should be
complete. Here, wave-e temporarily breaks beyond the end
of wave-d, then makes a new
low. This makes the current b-
d trendline improper; it should
be redrawn across the "break
point" high.

Contracting Triangles (general)**

Contracting Triangles are, by far, the most common type of triangle. Following is a list of necessary elements in the formation of a Contracting Triangle.

Minimum Requirements (all Contracting Triangles):

1. After a Contracting Triangle completes, there occurs a "thrust" that <u>must</u> be at least 75% of the widest segment of the Triangle and under "normal" circumstances will not exceed 125% of the widest segment (Figure 5-35).

Figure 5-35

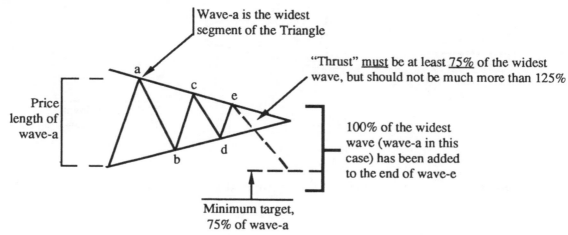

Wave-a is the widest segment of the Triangle

"Thrust" <u>must</u> be at least <u>75%</u> of the widest wave, but should not be much more than 125%

Price length of wave-a

100% of the widest wave (wave-a in this case) has been added to the end of wave-e

Minimum target, 75% of wave-a

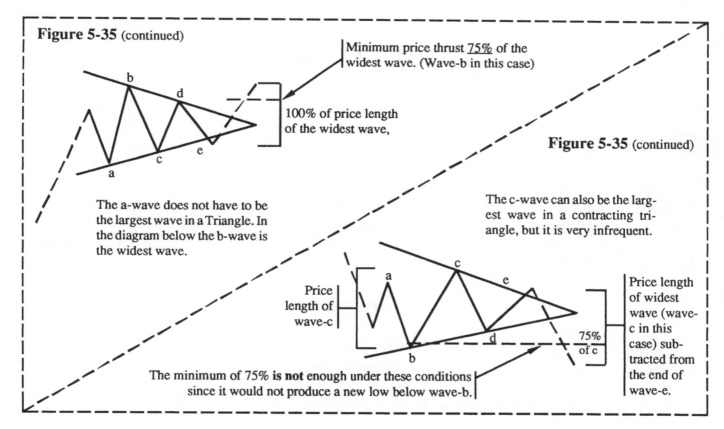

Figure 5-35 (continued)

Minimum price thrust <u>75%</u> of the widest wave. (Wave-b in this case)

100% of price length of the widest wave,

Figure 5-35 (continued)

The a-wave does not have to be the largest wave in a Triangle. In the diagram below the b-wave is the widest wave.

The c-wave can also be the largest wave in a contracting triangle, but it is very infrequent.

Price length of wave-c

75% of c

Price length of widest wave (wave-c in this case) subtracted from the end of wave-e.

The minimum of 75% **is not** enough under these conditions since it would not produce a new low below wave-b.

2. In a Contracting Triangle, the thrust *must* exceed the highest or lowest price (depending on the direction of the thrust) achieved during the formation of the triangle (Figure 5-36).

Figure 5-36

When a Triangle is about to complete, draw two parallel horizontal lines. One should be drawn across the highest price level achieved during the formation of the Triangle; the other across the lowest price level. Following the guidelines below, those parallel lines will help you predict what to expect after the Triangle completes. The only exception to this Rule would occur when the Triangle noticeably drifts in the opposite direction of its thrust.

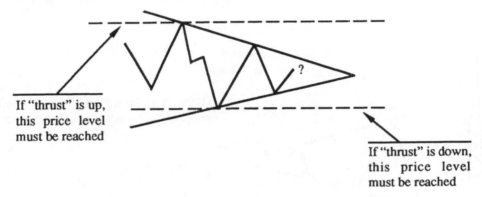

If "thrust" is up, this price level must be reached

If "thrust" is down, this price level must be reached

3. The e-wave *must* be the smallest wave in the Triangle (based on price, not time), see Figure 5-37.

Figure 5-37

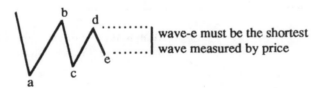

wave-e must be the shortest wave measured by price

I. Limiting (specific)

"Limiting" are the type of Triangles that Elliott discovered and talked about in his original writings. They occur in 4th-waves and b-waves. The post triangular action of a Limiting Triangle is constricted within very specific parameters, hence the name *Limiting*. The termination of wave-e in these patterns should occur from 20-40% before the apex point of the Triangle. Below are all three variations of *Limiting* Triangles and the specific formation rules which make them unique

a. Horizontal Variation

Of the Contracting Triangles, the **Horizontal** is the most common. When the market obeys the list of rules below, it indicates a **Horizontal** triangle is forming.

1. The trendlines of the Triangle *must* travel in opposite price directions (see Figure 5-38).

Figure 5-38

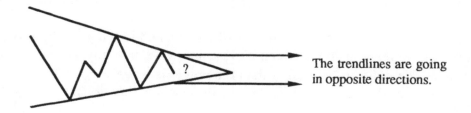

The trendlines are going in opposite directions.

2. The Apex point of the Triangle *must* fall within a range 61.8% of the longest segment of the Triangle, centered in the middle of the longest segment (see Figure 5-39).

Figure 5-39

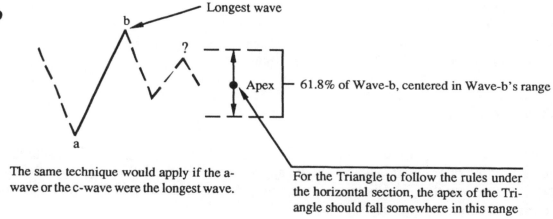

Longest wave

Apex — 61.8% of Wave-b, centered in Wave-b's range

The same technique would apply if the a-wave or the c-wave were the longest wave.

For the Triangle to follow the rules under the horizontal section, the apex of the Triangle should fall somewhere in this range

3. Wave-d *must* be smaller than wave-c (see Figure 5-40).

Figure 5-40

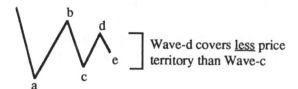

Wave-d covers <u>less</u> price territory than Wave-c

4. Wave-e *must* be smaller than wave-d (see Figure 5-37, previous page).

b. Irregular Variation

This type of Triangle implies slightly greater "thrust" and velocity potential than a Horizontal Triangle. The key element in the formation of an *Irregular Triangle* is the b-wave. It must be longer than wave-a. The parameters below should be obeyed by market action to indicate an *Irregular Triangle* has formed:

1. Wave-b should not be more than 261.8% of wave-a and will usually be less than 161.8%. An exact Fibonacci relationship between wave-a and wave-b of any Triangle is very unusual.
2. Waves c, d, & e must all be smaller than the previous wave.
3. The trendlines of the Triangle should move in opposite directions.

c. Running Variation

This is among the most misinterpreted patterns of the Wave Theory. It compares with the confusion associated with Double Three Running corrections. The important features of a Running Triangle are:

1. Wave-b is longer than wave-a and is the longest wave of the Triangle.
2. Wave-c is smaller than wave-b.
3. Wave-d is larger than wave-c.
4. Wave-e is smaller than wave-d.
5. Both trendlines will slope upward or downward.
6. The "thrust" after the Triangle will be much larger than the widest leg of the Triangle, sometimes as much as 261.8%, but no more.

II. Non-Limiting (specific)

There is very little difference between <u>Contracting Non-Limiting</u> and <u>Contracting Limiting</u> Triangles. All of the rules covered above should apply, except for the following subtle differences that set the two patterns apart. This listing is completely original, developed by the author after years of observing and carefully quantifying triangular behavior.

1. Channeling - a Non-Limiting Triangle will behave slightly different around the converging trendlines than a Limiting Triangle. The behavior occurs in one of three variations, all of which are noticeably different from Limiting action:

    a. The most common and distinguishing Non-Limiting action is congestion right into, or very near the apex point of the converging trendlines. "Right into, or very near" is quantified by the following calculation. Measure the time consumed by the triangle from its beginning to the end of <u>wave-e</u>. If the converging trendlines occur before 20% of that time (added to the end of wave-e) has elapsed, that defines the above quoted statement.

    b. Measuring the time distance from the beginning of the Triangle to the end of wave-e, if the apex point of the converging trendlines occurs after 40% of that time has elapsed, it should, again, be considered a Non-Limiting Triangle. This is harder to anticipate since the Triangle formation is not quite as obvious.

    c. The last way a *Non-Limiting Triangle* can signal its existence is by a <u>post-thrust</u> correction <u>into</u> the apex's time zone of the converging trendlines. In a Limiting Triangle, the *time zone* occupied by the apex point is where the "thrust" generally concludes* (unless the thrust forms into a *Terminal* pattern). The most common way for a Correction to retrace into the apex point of the recent Triangle is for the thrust, out of the Triangle, to be <u>very</u> violent, reaching its initial price length (based on the widest wave of the Triangle) well before the apex time period occurs. This allows the market to correct back into the time period of the apex point until time "runs out" <u>during</u> the apex time zone.

If any one of the above three situations arises toward the end of, or immediately after, the formation of a Triangle, the Triangle should be considered Non-Limiting in nature.

Post-Triangular Thrust

The "thrust" distance out of a Non-Limiting Triangle is not confined to any specific amount. It may (and usually does) temporarily react when it has achieved a price length approximately equal to the widest leg of the Triangle, but that is generally short-lived. The move will then usually resume in the direction of the original thrust and go for a distance which can only be determined by examining the larger patterns under formation.

Expanding Triangles (general)**

Expanding Triangles are <u>most common</u> during very large Complex Corrections. They are created when five Corrective phases occur in a row with <u>most</u> or all of the segments, from left to right, covering more price territory than the previous. The word "most," underlined in the last sentence, was used based on the observation that frequently one (and possibly two during *Running* Expanding Triangles) of the segments in an Expanding Triangle will be smaller than the previous (Figure 5-41, next page).

The general rules which apply to all Expanding Triangle formations are as follows:
1. The a-wave or the b-wave will always be the smallest segment of the Triangle.
2. Wave-e will almost always be the largest wave of the pattern.
3. Expanding Triangles cannot occur as b-waves in Zigzags or <u>b,c or d-waves</u> of a larger Triangle.
4. The e-wave will usually be the most time consuming and complex segment of the Triangle. The most typical construction of the e-wave (if you can see any subdivisions), would be a Zigzag (in small Expanding Triangles), or a Complex Combination of corrections in larger patterns.
5. The e-wave will almost always break beyond the trendline drawn across the top of wave-a and wave-c.
6. The b-d trendline should function the same as it would in any Contracting Triangle.
7. The "thrust" out of an Expanding Triangle should be *less* than the widest wave of the Triangle (which, in this case, is wave-e) unless it <u>concludes</u> a powerful, larger Correction.
8. Backward from wave-e, three of the **previous** waves must be at least 50% of the wave to the right.

Just like Contracting Triangles, Expanding Triangles divide into two distinct categories; Limiting and Non-Limiting . The same category names were used to keep things as simple as possible, but unlike their Contracting counterparts, the two terms do not have any significant post-triangular implications. [As stated above in Rule #7, the "breakout" of an Expanding Triangle is *less* than the widest segment of the Triangle.] The terms Limiting and Non-Limiting under this formation merely indicate whether the Triangle is in a <u>Standard</u> wave position or part of a more elaborate Correction involving multiple formations connected end to end.

* This concept was first presented in "Elliott Wave Principle, Key to Stock Market Profits," by Frost and Prechter (New Classics Library, Gainesville, GA)

Figure 5-41

Irregular Most common	**Running** Less Common	**Horizontal** Least common
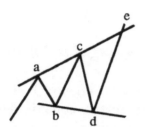	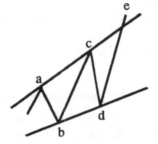	
Wave-b is shorter than Wave-a	Wave-d is shorter than Wave-c	Every succeeding wave is longer than previous

I. Limiting *(specific)*

The term *Limiting* always refers to 4th-wave or b-wave Triangles. Personally, I have never seen a "4th-wave Expanding Triangle." It must logically be assumed they exist, but must be considered rare. Of the few b-wave Expanding, Limiting Triangles witnessed, the following are the only descriptions which seem to regularly apply:

1. A b-wave Expanding, Limiting Triangle appears to be possible only in an Irregular Failure or C-Failure Flat pattern.
2. The "thrust" out of the Triangle is minimal, retracing approximately 61.8% of the Triangle from highest to lowest point.

a. Horizontal Variation

A perfectly Horizontal Expanding Triangle is probably the least common way for an Expanding formation to manifest itself. Why? It defies the natural tendency of a market to accumulate or distribute. If you continually make new highs and new lows in a market, no trend is established. The market would have to be in a state of <u>Fundamental</u> (as opposed to Technical) limbo. No large economic force pushing the market one direction or the other would be at work. That is a very unlikely position for a market to be in, especially for long periods of time. As a personal assessment, the longer the period of time covered, the more improbable the formation of a Horizontal Expanding Triangle becomes.

The parameters to create an Expanding Horizontal Triangle would be as such:
   (1). Wave-a *must* be the smallest wave in the pattern.
   (2). Wave-b, c, d, and e *must* each exceed the previous segment's <u>termination</u> point (which may or may not be the highest or lowest price).
   (3). The e-wave will probably break beyond the trendline drawn across wave-a & c.

b. Irregular Variation

The Irregular Expanding Triangle is a little more common than the Horizontal type and is characterized by several factors:

(1). Either wave-b is smaller than wave-a and all the rest of the waves are larger than the previous, *or* wave-d is smaller than wave-c and all the other waves are larger than the previous.

(2). The longer the period of time covered by the pattern, the more likely the channeling of the pattern will be upwardly or downwardly skewed.

c. Running Variation

The Running Triangle has a trending appearance which is created by both wave-b being slightly larger than wave-a and wave-d being slightly shorter than wave-c. Instead of the trendlines going in opposite directions, they both go in the same direction, but still diverge. The e-wave in this type of pattern can be quite violent. Another variation occurs when all waves are larger than the previous except wave-c (which is shorter than wave-b).

II. Non-Limiting *(specific)*

The ***Non-Limiting Expanding*** refers to Triangles which form within more complex formations. For example, the first or last phase of a Correction involving one or more x-waves could have as one of its components a *Non-Limiting Expanding* Triangle. This is due to Rule #7 (page 5-31), which generally states that the "thrust" out of an *Expanding* Triangle does not follow the same guidelines required of Contracting Triangles. Since the "breakout" should not be as large as the width of the *Expanding* Triangle (wave-e), obviously the "thrust" cannot start a major new advance or decline. This is an ideal situation for an x-wave to develop. The "thrust" out of the *Non-Limiting Expanding* Triangle will usually be an x-wave, but it could be the 5th wave of a Terminal or the second x-wave of a Triple Three or Triple Combination.

Non-Limiting Expanding Triangles will be formed in the same manner as *Limiting* Expanding Triangles with one exception regarding the apex point (which is backward in time). Measure the time consumed by the entire Triangle then take 40% of that amount and subtract it (i.e. go backward in time) from the beginning of wave-a. The apex will occur before the 40% time frame is reached. Usually it will be reached within 20% of the larger time frame.

Conditional Construction Rules - Corrections

Alternation

Price

The *price relations* in a Correction are one of the least useful applications of the Rule of Alternation. Why? The vast majority of waves in Corrections will be close to the same price length. When the Rule is applied to Corrections, it has its greatest usefulness in ZigZags. The a-wave and b-wave of a ZigZag should alternate in price. Wave-b will be 61.8% or less of wave-a. That is about the extent of *Price Alternation* in Corrective patterns. The Intricacy and Construction aspects of Alternation (presented on page 5-5) should also be considered at this time if your correction is composed of monowaves and polywaves (or higher).

Time

The Rule of Time Alternation assumes full force in Corrective patterns. To apply correctly, three adjacent corrective phases must be compared. The first two patterns will normally be very different in time duration. The first pattern may take "n" number of time units while the second one takes n(1.618 or greater) or n(.618 or less) number of time units. The third segment of the group will either be equal to one of the previous two segments, relate to one of them by 61.8% or 161.8% OR equal the total time consumed by both previous patterns combined. For more detail on this subject, refer to the "Time Rule" section. NOTE: When dealing with polywaves that consume only three time units, Alternation of time is impossible. Lack of time Alternation is a warning there may be a better way to arrange the monowaves , but if it appears correct, it can be used as a legitimate pattern.

Breaking Point - Corrections

From this point onward, the rules covered in this section concerning Corrections develop in one of two directions:
1. The rules get more subtle, conditional and difficult to apply, requiring lots of experience;
   or
2. They become less reliable, where they serve more as a reinforcement for an interpretation than as a critical deciding factor.

As your analytical skills increase, there are additional considerations which will improve your real-time accuracy and forecasting ability. It is recommended that the beginning student stop at this point and reread the earlier sections of the book, study, practice and commit to memory *all* of the rules and techniques covered up to this point. The rest of the book is dedicated to more complex discussions of market behavior. There is no need to go through the more difficult sections until you are comfortable with your knowledge of the basics.

Channeling

In **channeling** Corrections, the important point of reference is the b-wave. For Zigzags and Flats, always draw a trendline from the start of wave-a to the end of wave-b (called the 0-B trendline). A parallel line should be drawn across the termination of wave-a. If the pattern you are working with is a ZigZag, the c-wave may break beyond or stay away from the parallel trendline, but it should not touch it. If it does, it indicates the ZigZag will be part of a more complex correction, such as a Double, Triple Combination or Double, Triple ZigZag pattern (for details on this concept, refer to page 9-3). Once the 0-B trendline is broken, the c-wave (and the larger pattern) is <u>probably</u> over. For Triangles, the trendline is drawn across the end of wave-b and wave-d. When the B-D trendline is broken, the Triangle is probably over. More advanced Channeling techniques are covered in Chapter 12.

Fibonacci Relationships

Fibonacci relationships are among an array of "final touch" formation tests which help to solidify a proposal. *Almost* all Elliott patterns have their own unique combination of Fibonacci Relationships. This is one of the more difficult aspects of the Theory since there are a multitude of relationship possibilities. First, relationships are dependent on which general Correction category the market is in. Then, each Corrective variation has its own subtle differences (advanced Fibonacci concepts are discussed in Chapter 12).

Below is a listing of each "standard" Corrective category. Under each heading is a description of which waves generally relate in a particular pattern. After each heading, a more specific breakdown is given of the unique relationships usually associated with each Corrective variation.

a. <u>Flats</u> (3-3-5)

Of all Corrective patterns, *Flats* are least likely to display Fibonacci relationships since each wave is approximately equal to the previous one. When the b-wave of a Flat is much smaller or much larger than wave-a, then relationships can start to appear. Below is a basic listing of what to expect. [For a more detailed rundown of Fibonacci relationships, refer to "Advanced Fibonacci Relationships" Chapter 12.]

Strong b-wave
When the b-wave exceeds the end of wave-a, it would basically be limited to a 138.2% (not a true Fibonacci ratio, but a combination of two separate ones - 1.00 & .382) or 161.8% relationship, but <u>neither</u> may be reached. Usually, if the b-wave is longer than wave-a, especially if it is substantially longer, wave-a & c will be nearly equal in price. If the c-wave did relate by a Fibonacci ratio to wave-a, the ratio would be 161.8% or 61.8%.

Normal b-wave
This is the *Flat* pattern least likely to exhibit Fibonacci relationships. The only possibility for relationships would occur if the c-wave failed or elongated. If it failed, it is highly probable the c-wave would relate to wave-a by 61.8%. On very rare occasions, wave-c can relate to wave-a by 38.2%, but that is the <u>minimum</u> requirement.

If the c-wave Elongates, it is not likely there will be any relationship between wave-a & c. If there is, it would have to be 161.8% or 261.8%.

Weak b-wave

This situation allows for the greatest possible number of Fibonacci relationships. If wave-a and wave-b related, it would be by 61.8%. Waves-a & c could relate by the same amount (Internally or Externally, refer to page 12-22 for details). The c-wave could also be 61.8% of wave-b.

b. Zigzag (5-3-5)

Since there are not many ZigZag variations (compared to those possible in Flats and Triangles), there are only a few relational possibilities.

Normal

Fibonacci relationships between adjacent waves of a pattern cannot be strictly depended upon. If waves-a & b do relate, it will be by 61.8% or 38.2%. The more reliable relationships will occur between waves-a & c. Wave-a will be either 61.8%, 100% or 161.8% of wave-c (either Internally or Externally, see "Advanced Fibonacci Relationships," page 12-22 for details).

Elongated

When a ZigZag is given the title, "Elongated," it means the c-wave is excessively long in relation to wave-a. Usually an Elongated c-wave will have no relation to wave-a; if it does, it will be by 261.8%.

Truncated

The only relationship likely in a Truncated Zigzag is between wave-a and wave-c. Wave-c would be 38.2% of wave-a.

c. Triangles (3-3-3-3-3)

Triangles are composed of more wave segments than any other "standard" Elliott Correction. As a result, they have a high probability of exhibiting multiple Fibonacci relationships. As a matter of fact, a Contracting Triangle without Fibonacci relationships should be considered impossible. The normal way Fibonacci relations manifest themselves in Triangles is the same way they usually happen in most other patterns, between *alternate* waves. The most common setup is for waves-a, c, & e to relate by 61.8% or 38.2% and the b & d-waves to relate by 61.8%. The only two adjacent waves which will consistently relate to each other by a Fibonacci ratio (usually .618) are waves-d & e.

Important Note: If wave-b is 61.8% of wave-a, you are probably **not** in a Triangle.

Looking back at your chart, if you are working with monowaves, transform the plain Corrective symbols you have been using on your chart pattern into the specific symbol which represents Sub-Minuette Degree (a-b-c-d-e-x). With all Monowave patterns, start off by labeling each segment Sub-Minuette degree. If you are beyond the monowave development stage, the Degree Symbols used for the current pattern will be dictated by combining the smaller patterns (which have already been carefully analyzed and labeled) into larger Elliott patterns. How is this done? Generally, three patterns of one degree form to make one Corrective pattern of one larger degree (Triangles being the exception since they require five segments to complete the Correction of one larger Degree). Taking three Minor Degree patterns and combining them into one larger Elliott pattern raises the Degree of that pattern one level. The three Minor labels will be replaced with one Intermediate label. *[For a more complete discussion of Degree, refer to "More on Degree" on page 7-11 of Chapter 7.]*

Realistic Representations - (Corrections)

The diagrams on the next three pages show the most common arrangement of wave segments on a *Simple* and *Complex* Polywave scale. Each category variation is diagramed separately. When reviewing the section on Triangles, keep in mind the "Pivotal" discussion at the bottom of page 5-24.

If you plot data in the manner described earlier in this book (see pages 2-11 and 3-2), real-time market action will closely, and sometimes exactly, resemble the diagrams which follow. If you use bar charts, hourly closing figures, futures data or other types of incorrectly plotted or calculated data, the action sometimes will and sometimes will not look like the diagrams to follow and price behavior will not always obey the rules in this book.

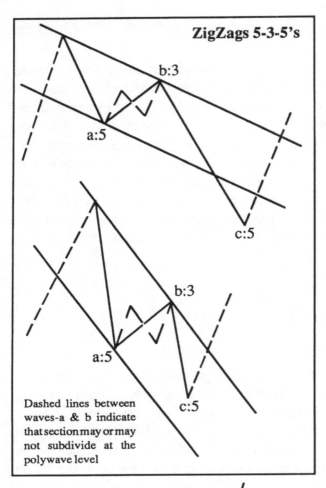

ZigZags 5-3-5's

b:3

a:5

c:5

b:3

a:5

c:5

Dashed lines between waves-a & b indicate that section may or may not subdivide at the polywave level

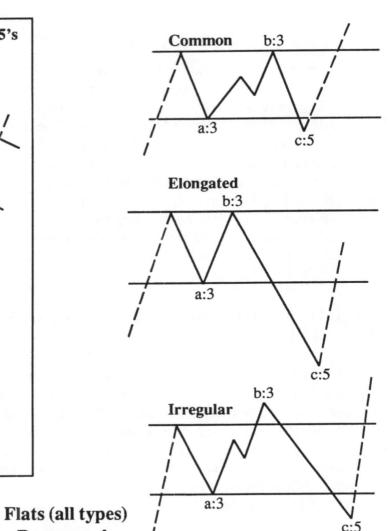

Common

b:3

a:3

c:5

Elongated

b:3

a:3

c:5

Irregular

b:3

a:3

c:5

**Flats (all types)
Downward
Corrections
3-3-5's**

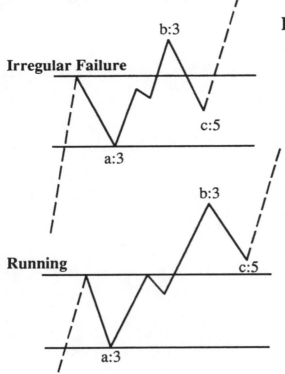

Irregular Failure

b:3

a:3

c:5

Running

b:3

a:3

c:5

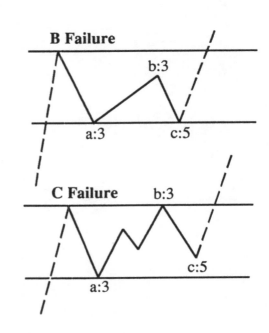

B Failure

b:3

a:3

c:5

C Failure

b:3

a:3

c:5

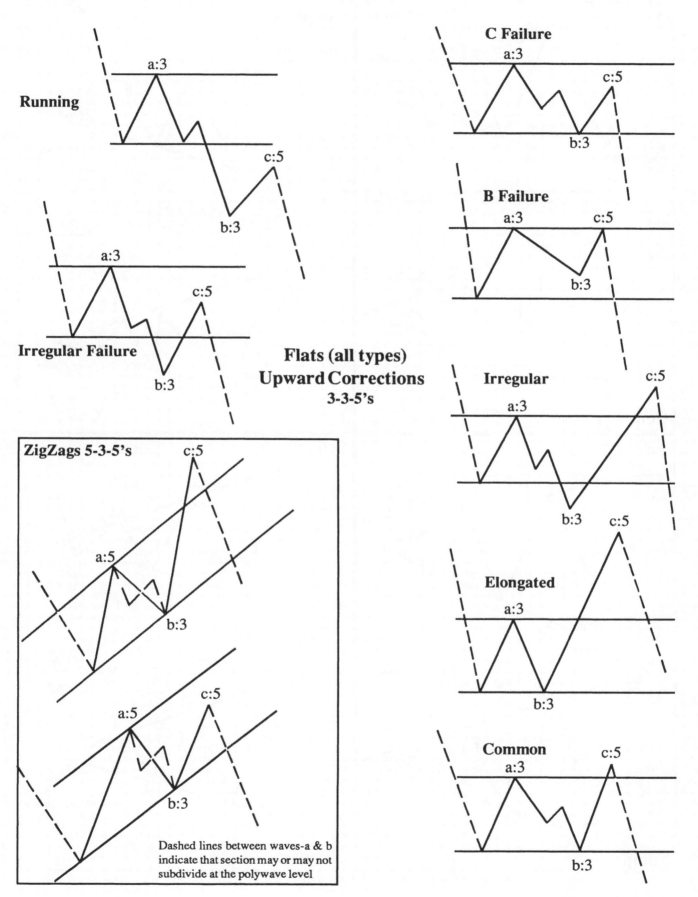

Running

a:3

b:3

c:5

Irregular Failure

a:3

c:5

b:3

Flats (all types)
Upward Corrections
3-3-5's

C Failure

a:3

c:5

b:3

B Failure

a:3   c:5

b:3

Irregular

a:3

c:5

b:3   c:5

Elongated

a:3

b:3

Common

a:3   c:5

b:3

ZigZags 5-3-5's

c:5

a:5

b:3

a:5   c:5

b:3

Dashed lines between waves-a & b
indicate that section may or may not
subdivide at the polywave level

Contracting Limiting Triangles

Simple

a:3

c:3

Wave-4 or B

e:3

d:3

b:3

Wave-3 or A

Correcting Down Trend

Wave-5 or C

Simple

Wave-5 or C

Wave-3 or A

b:3

d:3

e:3

a:3

c:3

Wave-4 or B

Correcting Up Trend

Complex

a:3

c:3

Wave-4 or B

e:3

d:3

b:3

Wave-3 or A

Wave-5 or C

Complex

Wave-5 or C

Wave-3 or A

b:3

d:3

e:3

a:3

c:3

Wave-4 or B

Simple

a:3

c:3

End of larger combination pattern

e:3

d:3

b:3

x-wave

Simple

x-wave

b:3

d:3

c:3

a:3

e:3

End of larger combination pattern

Complex

a:3

c:3

End of larger combination pattern

e:3

d:3

b:3

x-wave

Complex

x-wave

b:3

d:3

c:3

a:3

e:3

End of larger combination pattern

Contracting Non-Limiting Triangles

Expanding Limiting Triangles

Simple

Wave-4 or B

e:3

c:3

a:3

Wave-3 or A

Wave-5 or C

b:3

d:3

Simple

Wave-3 or A

b:3

d:3

Wave-5 or C

a:3

c:3

e:3

Wave-4 or B

Wave-4 or B

e:3

c:3

a:3

Wave-3 or A

b:3

d:3

Wave-5 or C

Complex

Wave-3 or A

b:3

d:3

a:3

c:3

Wave-5 or C

e:3

Wave-4 or B

Complex

Simple

e:3

c:3

a:3

*

b:3

d:3

*

*For Progress Label possibilities before and after this pattern, check the section on Expanding Triangles (page 5-31) and the Advanced Logic Rules (page 10-9).

Simple

*

b:3

d:3

*

a:3

c:3

e:3

*For Progress Label possibilities before and after this pattern, check the section on Expanding Triangles (page 5-31) and the Advanced Logic Rules (page 10-9).

e:3

c:3

a:3

*

b:3

d:3

*

Complex

*

b:3

d:3

*

a:3

c:3

e:3

Complex

Expanding Non-Limiting Triangles

S ince the beginning of this book, the main focus has been on explaining the step-by-step process of proper monowave detection and the assemblage of monowave groups into *Standard* Elliott polywave patterns. Each chapter and heading was organized in a very logical way to allow objective, methodical application of the Rules to real-time price action. As your understanding of the Wave Theory expands, the implementation of more sophisticated Rules and techniques is necessary to correctly manage larger wave formations.

The last seven chapters of this book have been broadly titled **The Neely Extensions**. This is not to be confused with the term "extension" which is the longest wave of an Impulse pattern. In this context, the term "extensions" is used to indicate the presentation of new, but essential techniques and concepts developed by the author which you should use when constructing wave patterns. Not every item mentioned over the next seven chapters is completely new. A few ideas which have previously been presented publicly are noted appropriately. Some previously known concepts have been greatly expanded upon. Most of the ideas to follow, however, are completely new and should be very helpful as you gain experience as an Elliott Wave analyst. Due to the prevalence of information now available on this subject (books, newsletters, etc.), it is possible that concepts presented here may have, at sometime, been presented before. If so, the previous presentation of that information is completely unknown to the author.

From this point forward, the concepts and techniques for analyzing market behavior get more involved and subtle. Many of the Rules, observations and tests require some knowledge of what the market did before the current pattern and what it has done since the pattern completed. Until you are comfortable with the construction of polywaves (and associated Rules), it is suggested you review the previous sections and practice the Rules presented on real-time market action.

To give you a "taste" of what is to follow, some of the important topics covered in the rest of this book are: discussions on the finalization and confirmation of polywave patterns; the reduction of polywaves to simpler, more manageable forms; the ranking of polywaves based on Complexity level; the grouping of polywaves (using Progress Labels) into larger, Standard or Non-Standard Elliott formations; the Integration of several polywaves to form a multiwave (or Non-Standard polywave) pattern, and much, much more. If you learned a lot from the first five Chapters of "Mastering Elliott Wave," the next seven are even more informative.

Post-Constructive Rules of Logic

Chapters 6 through 12 are almost exclusively the

Neely Extensions
of Elliott Wave Theory

6

Post-Constructive Rules of Logic

An important test of the validity of an interpretation involves the integration of **Logic Rules**. The **Logic Rules** have evolved from years of diligent market observation by the author. The Rules require all market action conform to specific behavior based on the implications of the pattern which preceded it. The expected behavior *must* be present or there is a flaw in your <u>current</u> interpretation *(these flaws can sometimes be explained if the pattern or market environment allows for exceptions; see Exception Rule, page 9-7)*. In other words, all action *must* be accounted for, it *must* logically integrate with its surroundings and post-pattern behavior *must* meet specific requirements.

If you have been following along with your own price chart, at this point the **Post-Constructive Rules of Logic** will need to be employed (more advanced rules are discussed under the **"Advanced Logic Rules"** Chapter 10). Deciding which Rules to apply is a result of the type of pattern developing, Impulsive or Corrective. Refer to the appropriate heading now if you are analyzing your own chart, otherwise just read straight through.

Impulsions

Two Stages of Pattern Confirmation

Stage 1 - Break 2-4 Trendline

The first point to apply the **Logic Rules** is immediately after an Impulse pattern finishes. Draw a trendline across the end of waves (2) and (4). To confirm whether the Impulse pattern you discovered is authentic, post-Impulsive market action *must* break the 2-4 trendline in the same amount of time consumed by the 5th wave <u>or less</u> (see Figure 6-1). If it takes more time, the 5th wave is developing into a Terminal **or** wave-4 is not complete **or** your Impulse interpretation is wrong.

Figure 6-1

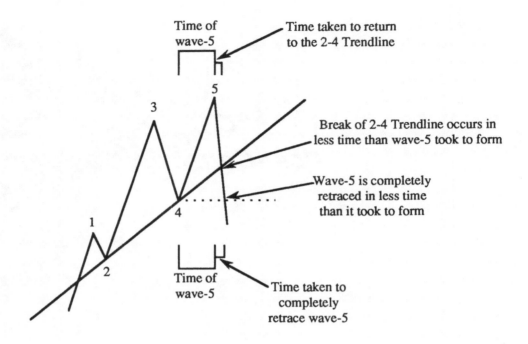

Time of
wave-5

Time taken to return
to the 2-4 Trendline

Break of 2-4 Trendline occurs in
less time than wave-5 took to form

Wave-5 is completely
retraced in less time
than it took to form

Time of
wave-5

Time taken to
completely
retrace wave-5

Stage 2 - 5th Wave Retracement Requirements

Next, check to see which wave of the Impulse pattern Extended. Depending on which wave is the Extension, price action will return to a support/resistance level defined by the price range of wave-2 or wave-4 of the Impulse pattern. Soon to follow is a list of <u>minimum</u> and <u>maximum</u> retracement expectations based on which Impulse variation took place.

1st Wave Extension

The retracement to follow this Impulse variation *must* return to the termination of wave-4. Elliott said it <u>usually</u> will return to the price zone of wave-2. This is the case when the 1st Extension is wave-1 or wave-5 of a larger Impulsive pattern. If the 1st Extension is wave-3, price action (depending on conditions) may not be able to return to the 2nd wave zone. If the market moves beyond the termination of wave-2, the whole Impulse pattern ended an even larger Impulsion or Correction.

3rd Wave Extension

Price action *must* return to the 4th wave zone of the completed Impulse pattern and will usually end near the termination of wave-4. If the move retraces more than 61.8% of the entire pattern, the 3rd Extension Impulse also completed an Impulse wave of *one higher degree*.

5th Wave Extension

The correction to follow a 5th extension pattern **must** retrace at least 61.8% of the 5th wave, but must not retrace all of the 5th wave <u>if</u> the trend is to remain in effect. If the 5th extension <u>does</u> get completely retraced, it indicates the extension terminated a larger trend. Here are the different ways that can happen:

1. The 5th Extension pattern was part of a larger Impulse which was also a 5th Extension; **or**
2. The 5th Extension was the c-wave of a Flat or Zigzag pattern.

5th Wave Failure

A 5th wave Failure occurs when wave-5 is shorter than wave-4. This situation implies a great deal of counter trend power. If the 5th wave failed in your wave group, the move to follow should completely retrace the entire Impulse wave. If the Impulse was moving upward (downward), there should be **no** new highs (lows) in the market until the Impulse is completely retraced.

If the price action <u>after</u> the Impulse pattern did not conform to the above, applicable behavior requirements, something is wrong with your interpretation. Changes in the *Progress Labels* used will probably need to occur. Possibly, reinstituting the entire wave identification and Structural labeling process of the waves on your chart will be required (if so, return to Chapter 3).

A Word of Warning: If market action has, up to this point, followed all the necessary rules for Impulse formation, do not be quick to throw out your entire count when <u>one factor</u> does not develop as expected. Generally, a <u>simple</u> change in the Progress Labels will put the entire count back on track. The most frequent reason the market will not follow through after breaking the 2-4 trendline is because the 4th wave is not actually complete (under **Channeling** in Chapter 5, see "Wave-4" for an explanation of this process and **Localized Progress Label Changes** on page 12-45 for a more complete description of how to make <u>alterations</u> of a scenario without destroying the whole interpretation).

Corrections

Requirements for Pattern Confirmation

Unlike Impulse patterns, the confirmation of a Correction ***does not*** require a <u>sequential</u> unfolding of specific retracement events. The order of the confirmation stages depends on whether wave-b is shorter or longer than wave-a in the Correction. Since Complex Corrections terminate with Standard Elliott patterns, the only requirement to confirm that a Complex Correction has ended is to confirm the Standard Correction which terminates the Complex Correction. Complete confirmation of a pattern always involves two Stages. If the confirmation of both Stages occurs, there is virtually no doubt of the authenticity of the Correction on your chart. If only one of the Stages passes inspection the pattern can still be legitimate, but consider it a warning and be on the look out for more appropriate possibilities. See the description (below) for the Corrective category you are currently studying.

Flats and Zigzags

Wave-b Shorter than Wave-a

Under these conditions, draw a trendline across the <u>beginning</u> of wave-a and the <u>end</u> of wave-b. To satisfy Stage 1 and confirm the authenticity of the Correction, post-Corrective market action ***must*** break the "0-B" trendline in the same amount of time (or less) that wave-c took to form. If it takes more time, the c-wave is developing into a Terminal **or** wave-4 (of wave-c) is not complete **or** your Corrective interpretation is incorrect. If Stage 1 confirmation is met, Stage 2 requires wave-c be completely retraced in the same amount of time (or less) that wave-c took to form. In Figure 6-2, the b-wave was intentionally drawn so that it related to wave-a by 61.8%; this allows the one diagram to dually function as an illustration of how Flat confirmation and Zigzag confirmation occurs.

Figure 6-2

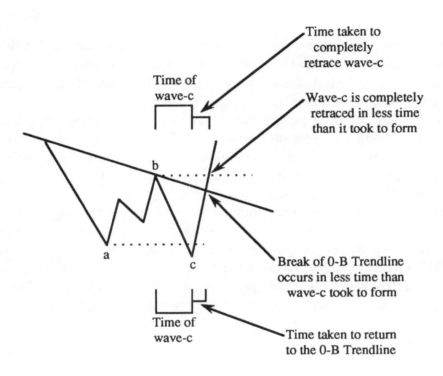

Wave-b Longer than Wave-a

In this situation, Stage 1 Confirmation is met if wave-c is completely retraced in the same amount of time (or less) that it took to form. Stage 2 Confirmation is satisfied if the market breaks the 0-B trendline in the same amount of time (or less) that wave-c took to form (see Figure 6-3). The larger the b-wave, the more the market will have difficulty attaining this second Stage of Confirmation. Therefore, be lenient on the time factor for Stage 2 during Running Corrections and Irregular Failures with very large b-waves.

Triangles

There are two Triangular categories, "Expanding" and "Contracting." Contracting Triangles can easily be confirmed based on post-Triangular market behavior. Expanding Triangles confirm their legitimacy through "non-confirmation." In other words, after the e-wave of an Expanding Triangle, the market should *either* not retrace the e-wave completely (obviously simultaneously eliminating the chance to break the B-D trendline) *or* it will take more time to completely retrace wave-e than wave-e took to form. Below are listed the two, sequential Stages of confirmation for a Contracting Triangle.

Stage 1

In Triangles, a B-D trendline is implemented instead of a 0-B. To obtain confirmation, the market must break the B-D trendline in the same amount of time (or less) as that consumed by wave-e.

Figure 6-3

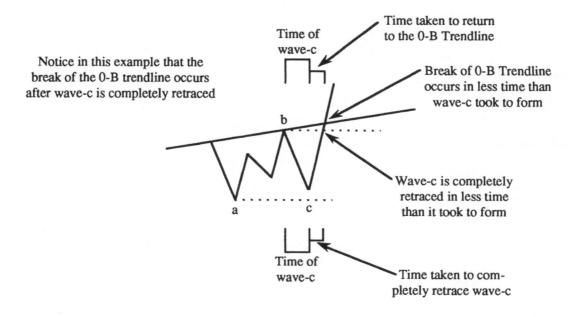

Notice in this example that the break of the 0-B trendline occurs after wave-c is completely retraced

Time of wave-c

Time taken to return to the 0-B Trendline

Break of 0-B Trendline occurs in less time than wave-c took to form

Wave-c is completely retraced in less time than it took to form

Time of wave-c

Time taken to completely retrace wave-c

Stage 2

The "thrust" after wave-e of a Triangle should exceed the highest or lowest price level achieved during the Triangle. *[This concept has already been discussed in Chapter 5, page 5-28.]* The Triangle's "thrust" must terminate in a time frame which is less than 50% of that taken by the Triangle (Non-Limiting Triangles are not bond to the 50% time rule) added to the end of wave-e.

All Elliott Wave patterns imply certain levels of strength or weakness (see **Pattern Implications** under the **Advanced Rules of Logic** heading, Chapter 10). When a pattern completes, the market *must* react in a fashion which is consistent with the *implications* of that pattern. For example, if a **Running** Correction just completed, the move afterward needs to be very explosive (upward or downward) and achieve a price distance equal to or greater than 161.8% of the wave (or wave group) preceding the **Running** Correction. To view diagrams of Running Corrections, see **Realistic Representations**, pages 5-38 and 5-39.

How will you know what to expect from other types of patterns? Multiple Corrective categories and numerous variations of each category preclude a simple discussion on this subject in Chapter 6. If you are at a more advanced level and feel ready to learn some new concepts, study the **Advanced Rules of Logic** in Chapter 10.

If you do not feel ready to tackle a *new* section, simply continue applying the more simplistic approaches outlined in this and previous chapters, but keep trying to form your wave pattern's with the *logical* concept in mind. Put another way, do not allow a weak Corrective pattern to precede a strong move or a strong Correction to occur before a weak move. That condition would produce a contradiction in the pattern's implications and virtually guarantee the interpretation is incorrect. With practice and keen observation, the concept will become easier to apply.

Conclusions

By meticulously following all the steps from **Preliminary Analysis** up to the **Central Considerations** section and carefully monitoring the effects of the **Post-Constructive Rules of Logic**, you should know whether the combination of waves you have been testing (since the beginning of Chapter 3) are Impulsive or Corrective in nature. It should also be relatively evident which Elliott pattern variation is unfolding. As a result of this testing, there should be little doubt as to the validity of the formation. Now that you have identified a legitimate Elliott pattern, most of the pattern management procedures are behind you. There are only a few more things to cover:

1. How is your completed pattern simplified to make it easier to work with in the future?
   and
2. How is the Degree **Title** and **Symbol** of that simplified pattern decided upon?

To answer the first question, the **Compaction Process** is used. Compaction allows you to <u>simplify</u> a pattern's *Structure Series* into its *base* Structure (simply either a ":3" or a ":5"). Using the **Complexity Rule**, a technique developed by the author, the *base* Structure has to be "stratified" contingent on the intricacies of the segments which made up the compacted pattern. This technique will greatly assist in the proper combining of large scale and complex formations as your skills develop. To answer the second question, the "<u>More On Degree</u>" heading at the end of this Chapter will enable you to deduce the proper Degree **Title** and **Symbol** of each pattern compacted.

Compaction Procedures

"Compaction" is a term used to describe an essential, heretofore undefined, Elliott process. It is a technique employed to reduce a completed series of Mono-, Poly-, Multi-, or Macrowaves into a single **Impulsive** or **Corrective** structure (":3" or ":5"). Due to the dynamic nature of this concept; any completed Elliott pattern, no matter how large or small, can be handled as and labeled as a single Corrective (:3) or Impulsive (:5) event. This process is necessary to continually simplify what would eventually become an unmanageable web of price action. The technique <u>cannot</u> be employed until after *all* previously discussed criteria has been considered (i.e., Structure, Series identification, Essential and Conditional Construction Rules, Channeling, Fibonacci ratios, Post-Constructive Rules of Logic, etc.). Once all basic Rules and procedures have been followed, the Compaction process is implemented.

Below is a listing of how to reduce *all* Elliott formations based on their Structure Series.

A. 5-3-5-3-5 = Trending Impulse= ":5"
B. 5-3-5 = Zigzag = ":3"
C. 3-3-5 = Flat = ":3"
D. 3-3-3-3-3 = Triangle= ":3"
E. 3-3-3-3-3 = Terminal Impulse = ":5"

F. All patterns which contain "x-waves" (immediately before Compaction is performed) can be reduced to a ":3."

Upon completion of a correctly formed Elliott pattern, the Structure Series which makes up the pattern can be *compacted* into the single number (the *base* Structure label) on the right of each Series. If you are diligent in the practice of **Compaction**, the larger and longer-term patterns can be put together with the same, and sometimes greater, ease as the shorter-term patterns.

Below is an example of a **Standard** wave pattern being Compacted into a single Structure label (see Figures 7-1 and 7-2).

Figure 7-1

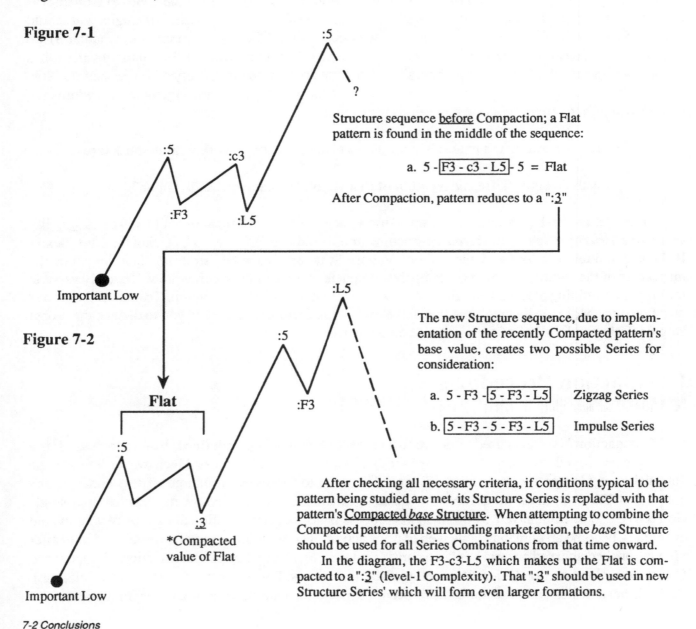

Structure sequence <u>before</u> Compaction; a Flat pattern is found in the middle of the sequence:

a. 5 - $\boxed{\text{F3 - c3 - L5}}$ - 5 = Flat

After Compaction, pattern reduces to a ":<u>3</u>"

Figure 7-2

The new Structure sequence, due to implementation of the recently Compacted pattern's base value, creates two possible Series for consideration:

a. 5 - F3 - $\boxed{5 - F3 - L5}$ Zigzag Series

b. $\boxed{5 - F3 - 5 - F3 - L5}$ Impulse Series

After checking all necessary criteria, if conditions typical to the pattern being studied are met, its Structure Series is replaced with that pattern's <u>Compacted *base* Structure</u>. When attempting to combine the Compacted pattern with surrounding market action, the *base* Structure should be used for all Series Combinations from that time onward.

In the diagram, the F3-c3-L5 which makes up the Flat is compacted to a ":<u>3</u>" (level-1 Complexity). That ":<u>3</u>" should be used in new Structure Series' which will form even larger formations.

Note: After Compaction, **reassessment** of that pattern's base Structure is required. Return to Chapter 3 and pretend the Compacted pattern is a monowave. Apply the necessary Rules to detect the groups internal Structure. If this produces a *base* Structure contrary to the pattern's nature (Class), a "missing" wave may exist. Note that, then **reassess** the two monowaves (or wave groups) on either side of the Compacted pattern. Afterward, proceed to Chapter 4 and resume normal analysis protocol. (***Warning****: do not reassess patterns which exceed their own beginning; for details see page 3-68*).

Regrouping

After **Compaction**, the wave group's *base* Structure (with the assistance of Chapter 4) will be used to form larger Standard or Non-Standard Structure Series'.

For example: The group of monowaves in Figure 7-1 has already been Structurally labeled. After making the proper observations and applying the appropriate tests, you decide the Flat pattern uncovered in Figure 7-1 is legitimate; therefore, it is <u>compacted</u> to produce a *base* Structure of ":3." When you return to Chapter 3, larger patterns may be formed by taking the *base* Structure of that pattern and combining it with Structure labels of surrounding patterns (which, themselves, may or may not have been through the Compaction process). Before searching for a new Structure Series, make sure there are at least five Structurally labeled wave segments to analyze. After Compacting the Flat in Figure 7-1, only three Structure Labels remained. Therefore, two additional monowaves were added to Figure 7-2. In Figure 7-2, moving backward from the ":L5" at the top of the wave group, only the compacted value of the recently found Flat should be used as part of this *new* Structure Series.

Integration

Integration is a term used to describe the transference of shorter-term, *compacted* wave Structure to longer-term charts allowing larger formations to slowly be pieced together. For example, if every time a polywave pattern completed on a shorter-term chart you were to place that pattern and its compacted Structure label on a slightly longer-term chart, eventually numerous polywaves would be present, each with its own *base* Structure identified. Combining those polywaves into larger formations would involve <u>exactly</u> the same procedures described for monowaves. Only a few additional Structure Series possibilities, which are covered in the next chapter (Construction of Complex Polywaves, Multiwaves, or higher), would need to be considered.

Another application of the **Integration** concept would include the cross-referencing of information (*i.e., comparing price/time projections of long and short-term charts to arrive at a consensus projection*) between charts to arrive at the most logical and accurate assessment of the future course of a market. Through a process of elimination and through the cross-referencing of target zones and time projections you can usually reduce the number of interpretation possibilities to a single choice. At the very least, you should be able to decrease the number choices and keep your trading with the trend of the market.

Progress Labels - *revisited*

In the process of transferring Structural information from a shorter to longer-term chart, do not transfer the Progress Labels associated with the pattern. Why not ? Progress labels serve a very specific and short-lived purpose. They solidify and confirm (or invalidate) your grouping of several adjacent mono-, poly-, multi-, or macrowaves by providing a list of necessary attributes which must be adhered to by the market action. They are an essential part of the procedures leading up to the **Compaction** process. After a pattern meets all necessary attributes and has been properly Compacted, the Progress Labels (1-2-3-4-5 or a-b-c, etc.), which organized the market action into an Elliott formation, no longer serve an important function to the pattern. The *base* Structure (:3 or :5) is now the <u>most important</u> consideration. It is the *base* Structure which properly guides you in the formation of larger Complex Polywave or Multiwave patterns.

Complexity Rule

This Rule, another tool developed by the author, provides a standard for the categorization of subdivisions within a pattern. It assists in combining <u>large scale</u> patterns and in deciding the *relative* Degree **Title** of a segment. As patterns get larger and more time consuming, it can be difficult to know which patterns can and cannot be combined with others to form larger ones (a function of Degree). Basically, all analysis begins with the combining of monowaves into polywaves, those polywaves into multiwaves, etc. As you progress, each pattern (visually and structurally) gets more difficult to manage *if* you do not keep track of the Complexity level of each pattern, *before* and *after* Compaction.

Complexity is not an important concept to understand when you are just beginning your studies of the Elliott Wave Theory, but it will become increasingly important as you begin to chart and keep track of long-term wave patterns. Complexity identification plays an important role in the detection of large scale patterns of the <u>same</u> **Degree**. Generally, patterns only of the same or of an adjacent Complexity Level can be considered the same **Degree** (Degree is explained in greater detail later in this Chapter).

In the early stages of market development, the **Complexity Level** of a pattern is relatively easy to ascertain. It is a direct product of the number of subdivisions visible in a price move. A single monowave has a Complexity Level of "0" (see Figure 7-3). Arranging three <u>or</u> five monowaves into a polywave constitutes the development of a pattern from its base level of Complexity to Level-1 (see Figure 7-5a). Level-2 is the point where one of the **Trending** patterns in a polywave noticeably divides into a smaller Impulse sequence, turning the polywave into a multiwave.

The detection of Complexity Level-3 (and higher) patterns is not quite so obvious. A larger pattern's Complexity Level is almost solely dependent upon the Complexity of smaller Impulse segments *within* the pattern. The "Complexity identification" techniques will keep you on track when attempting to combine patterns of extended price and time characteristics. For example, you cannot expect the straight line in Figure 7-4a (below left) to be of the same degree as the pattern in Figure 7-4b (below right). That is, of course, an obvious example. When a pattern covers several months or several years, the Complexity Level can be very difficult to discover without implementing the techniques to follow. The next few pages describe how Complexity Level is arrived at on varying scales of market activity.

Figure 7-4a   **Figure 7-4b**

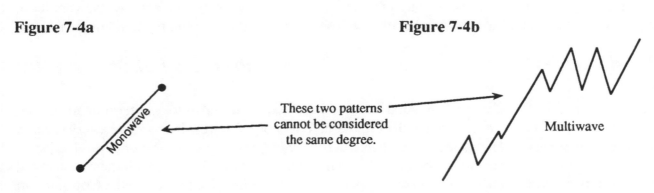

Monowave

These two patterns cannot be considered the same degree.

Multiwave

The Complexity Level of a monowave is easy to determine. With no visible subdivisions, the Complexity Level of a monowave is always "0." When it comes to integrate monowaves in the Complexity evaluation process, assign all monowaves a mathematical value of "0."

Figure 7-3

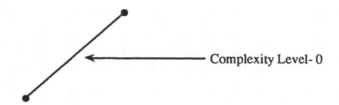

Complexity Level- 0

Anytime a market completes an Elliott pattern which has *visible* subdivisions and conforms to all applicable rules, the Complexity Level **must** <u>automatically</u> be Level-1 or greater. In other words, anything greater in Complexity than a monowave is of Complexity Level-1 or higher. The Complexity Level of a pattern is designated by the presence (or absence) of <u>underlines</u> to existing Structure Labels (the greater the number of underlines, the higher the Complexity Level of the pattern completed). No underlines indicate Level-0. One underline stands for Complexity Level-1, two underlines, Complexity Level-2, etc.

A *simple* Polywave is composed of only three or five monowaves. In Figure 7-5a, two *simple* polywaves (one Corrective, one Impulsive) are diagramed which fall under typical Elliott guidelines. Since both patterns subdivide, they must immediately be given a Complexity rating of Level-1. To confirm whether the pattern **is** or **is not** of greater Complexity, it is necessary to investigate the Impulsive

Figure 7-5a

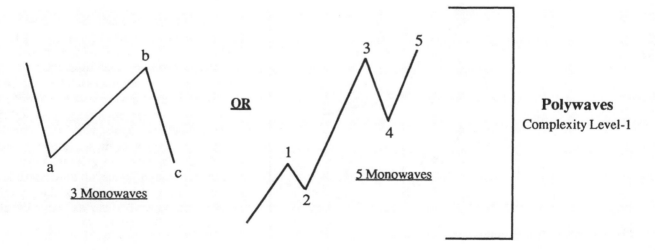

OR

Polywaves
Complexity Level-1

segments of the move and pick the segment of greatest Complexity. In Figure 7-5a, all the Impulse patterns (:5's) are monowaves. Monowaves, as mentioned earlier, are Level-0. This value would be added to the <u>assumed</u> Level-1 rating. The final result; both patterns in Figure 7-5a are Level-1.

Figure 7-5a
(Continued)

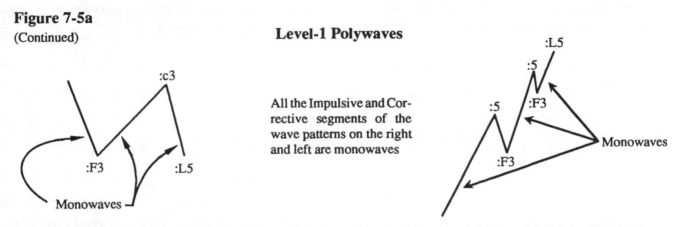

Level-1 Polywaves

All the Impulsive and Corrective segments of the wave patterns on the right and left are monowaves

Strictly due to visible subdivision, both of the above patterns automatically must be of Complexity Level **1** or higher.

To make sure there is no misunderstanding, several additional examples are needed. In Figure 7-5b (the bottom of this page and the top of the next), both patterns are of Complexity Level-1. Even though Corrective segments within each pattern subdivided (wave-b in the Flat and waves 2 & 4 in the Impulse pattern), the Complexity Level of the **Impulse** segments of those patterns remained at Level-0. This means the Complexity of the entire move is still Level-1.

Figure 7-5b

Complex Corrective Polywave

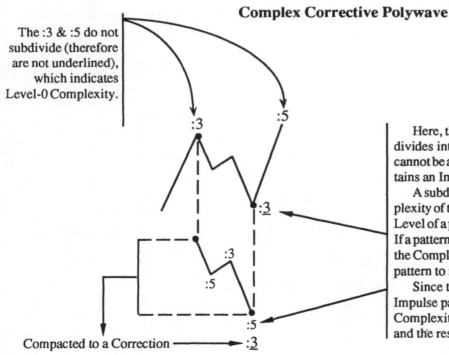

The :3 & :5 do not subdivide (therefore are not underlined), which indicates Level-0 Complexity.

Compacted to a Correction ⟶ :<u>3</u>

Here, the ":<u>3</u>" is underlined once because it subdivides into three segments. Its Complexity Level cannot be any greater than Level-1 since no part contains an Impulse pattern which subdivides.

A subdivided Correction does not raise the Complexity of the larger pattern. To raise the Complexity Level of a pattern always requires a <u>subdivided</u> ":5." If a pattern contains an Impulse segment, simply add the Complexity Level of the **most complex** Impulse pattern to its *automatic* Level-1 Complexity.

Since the formation on the left does contains an Impulse pattern, and that segment exhibits Level-0 Complexity, add Level-0 to the *automatic* Level-1 and the result is still Level-1.

Figure 7-5b

(continued from last page)

Complex Impulsive Polywave

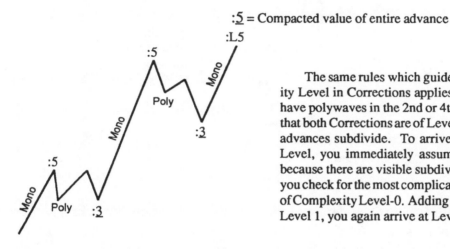

:5 = Compacted value of entire advance

The same rules which guide the discovery of Complexity Level in Corrections applies to Impulse patterns which have polywaves in the 2nd or 4th-wave position. Note here that both Corrections are of Level-1, but none of the Impulse advances subdivide. To arrive at the correct Complexity Level, you immediately assume a Level-1 rating simply because there are visible subdivisions in the pattern. When you check for the most complicated Impulse segment, all are of Complexity Level-0. Adding that amount to the automatic Level 1, you again arrive at Level 1 for the entire move.

Multiwaves

All **Multiwaves** are Complexity Level-2 patterns. What is the major difference between a Polywave and a Multiwave? In a Multiwave, at least one (and usually, only one) of the ":5's" *within* the multiwave subdivides into its own **Impulsive** polywave (see Figure 7-6a). On rare occasions, and under specific circumstances, more than one Impulsive polywave *might* occur in a multiwave. A Corrective Multiwave is diagramed in Figure 7-6b (next page).

Figure 7-6a

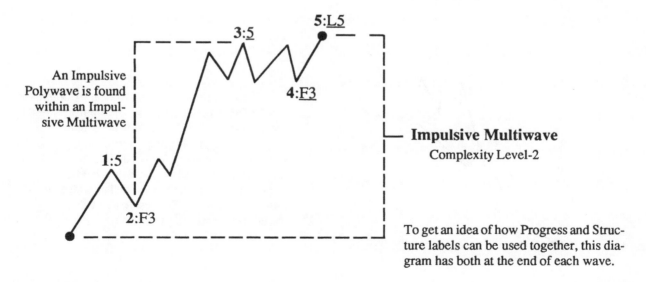

An Impulsive Polywave is found within an Impulsive Multiwave

Impulsive Multiwave
Complexity Level-2

To get an idea of how Progress and Structure labels can be used together, this diagram has both at the end of each wave.

Figure 7-6b

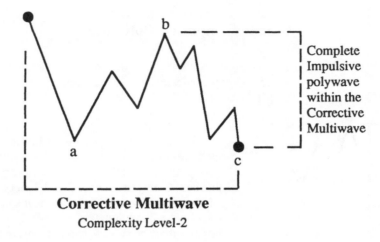

Complete
Impulsive
polywave
within the
Corrective
Multiwave

Corrective Multiwave
Complexity Level-2

Referring to Figure 7-6c, this is how Complexity of that pattern would be decided. First, observe if the pattern subdivides or not. It does, therefore it automatically qualifies for the minimum Complexity rating of Level-1. Next, examine each individual Impulse segment and note the Complexity Level of each. Out of the three Impulse patterns *of the same Degree*, choose the one with the highest Complexity rating. The most complex segment in this pattern is the middle advance (wave-3), which is Level-1 Complexity. Add "1" to the automatic Level-1 and you get Level-2. That is the Complexity of the diagram in Figure 7-6c. Figure 7-6d describes the Complexity delineation process for a Corrective Multiwave.

Figure 7-6c

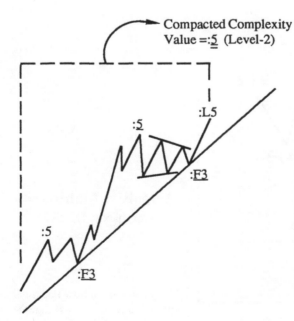

Compacted Complexity
Value =:5 (Level-2)

The same principles used to decide Complexity Level of a Correction are used for Impulse waves. In the diagram, the visible subdivisions again indicate a minimum Level-1 rating. Noting the Complexity of each individual segment of the Impulse wave, the most complex Impulse wave is wave-3 (it is a ":5"). This Level-1 pattern would be added to the *automatic* Level-1. The result, a Level-2 rating for the entire compacted pattern.

Figure 7-6d

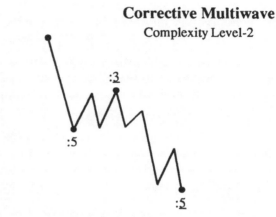

Corrective Multiwave
Complexity Level-2

:3

:5

:5

Compacted value of Zig Zag = :3

The Complexity Level is assumed to be at least Level-1 since there are visible subdivisions. Next, note that the second <u>Trending</u> pattern is the highest level of the two Impulsive patterns. The Level-1 Complexity of the final Impulse wave is added to the *automatic* Level-1. Consequently, the Zigzag is Complexity Level-2. *Note: each underline denotes an additional level of Complexity.*

Figure 7-6e is a discussion on how to derive the Complexity Level of a Complex Multiwave. Pay close attention; these are a little tricky since the Complexity Level is <u>less</u> than it would appear.

Figure 7-6e

Complex Multiwave Correction
Complex Corrections assume the Complexity Level of the most complex <u>Standard</u> Correction.

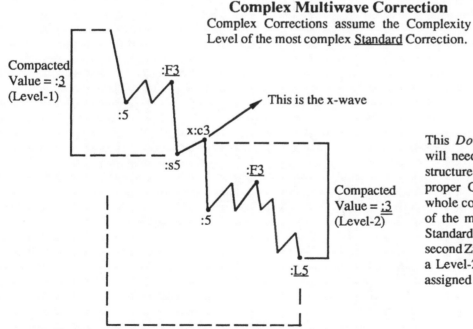

Compacted Value = :3 (Level-1)

:F3

:5

This is the x-wave

x:c3

:s5

:F3

:5

Compacted Value = :3 (Level-2)

:L5

This *Double Zigzag*, when complete, will need to be compacted to its base structure (which is ":3"). To maintain the proper Complexity Level, assign the whole correction the Complexity Level of the most complicated, independent Standard Elliott pattern. In this case, the second Zigzag is the most complex and is a Level-2 pattern. That would be the assigned value of the whole formation.

The entire formation would compact to the Complexity Level of the most complex **standard** Elliott pattern. The second Zigzag is the most complex **standard** pattern, so the rating of that Zigzag correction (Level-2) would be the rating of the <u>entire</u> larger Complex Correction.

Macrowaves

"**Macrowave**" is a less precise term than the previous three descriptions of market Complexity. As greater periods of time elapse, a pattern gets more and more complicated. Visually, there is an undefined point where it becomes very difficult to distinguish one Complexity Level from another. For this reason, there was no need to continue naming moves of greater and greater Complexity. Therefore, the term "Macrowave" will be used to describe any complex pattern above the Multiwave stage.

The minimum requirement for a pattern to fall into the **Macrowave** category is that it contains *at least* one Multiwave and one Polywave (usually there will be two polywaves, see Figure 7-7). To derive the Complexity Level of Figure 7-7, first use the "automatic" rule. The pattern should immediately be considered of at least Level-1 Complexity since subdivisions are visible. Examine each Impulsive segment (of the same **Degree**) *within* the Macrowave. Choose the one with the highest rating and add that to the automatic Level-1 rating. In this case, the final advance is a Level-2 pattern. Add that to the *automatic* Level-1 and you get Level-3. Since this is the minimum Construction of a Macrowave, all Macrowaves must have a Level-3 rating or higher.

Figure 7-7

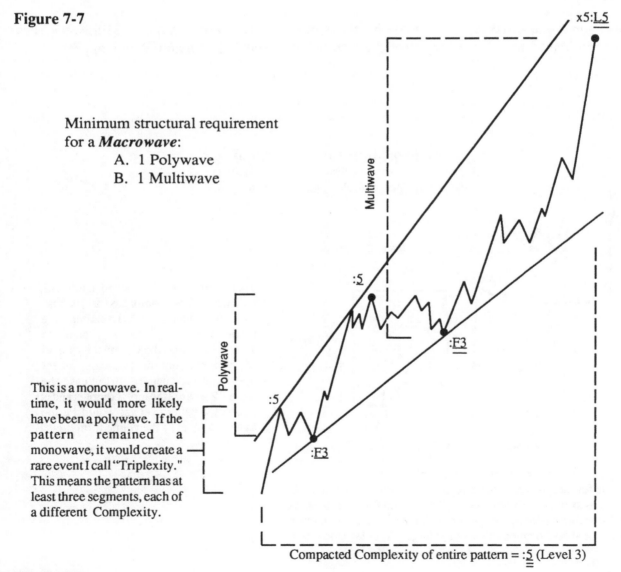

Minimum structural requirement
for a *Macrowave*:
> A. 1 Polywave
> B. 1 Multiwave

This is a monowave. In real-time, it would more likely have been a polywave. If the pattern remained a monowave, it would create a rare event I call "Triplexity." This means the pattern has at least three segments, each of a different Complexity.

Compacted Complexity of entire pattern = :5 (Level 3)

More on Degree

If you were to ask technicians their definition of **Degree** as it relates to the financial or agricultural markets, most would probably use non-specific descriptions such as *short-term*, *intermediate-term*, and *long-term*. Vaguely, they may describe how much time each Degree consumed (of course, from their own perspective). What constitutes a long period of time for some may be a short period of time for the more patient. This general description of **Degree** leaves a lot to be desired by the demanding technician and serious market student.

Precise rules are necessary to discern every aspect of market behavior if you expect to make accurate predictions. Having specific criteria for the evaluation of different **Degree** levels also aids in the discussion of a market, allowing you to speak about various types of price action with others from a common vantage point. Knowing the *relative* Degree of a pattern is essential to the proper application of numerous rules, the integration of information from a shorter to longer-term chart and the **Compaction** (discussed earlier) of a completed pattern into its *base* Structure (:3 or :5).

Degree has been purposely skimmed over (until now) to prevent any confusion. A good foundation of the more concrete concepts of the Elliott Wave Theory is required to properly understand Degree. Degree is important if you are employing a combination of short, intermediate and long-term charts, not when you are still learning to interpret and combine monowaves into polywaves.

Degree Titling

Degree is a concept which must be considered when you combine several (or more) Mono-, Poly- Multi, or Macrowaves to form a larger Impulsive or Corrective pattern. The realization of this process creates a higher Degree **Title** for the combined group of waves as a singular entity. In other words, when three or five segments are combined into one larger, legitimate Elliott pattern, a pattern of one *higher* Degree is created. Therefore, any visible subdivisions in a pattern are always at least of one lower Degree than the pattern as a whole.

The Degree concept cannot take root until you finally **Title** a specific move on your chart. The Title of a particular move is basically up to you, but some suggestions were given earlier to designate the original monowaves on your *first and smallest time frame* chart as ***Sub-Minuette*** in Degree. Once a segment is Titled, there is a frame of reference from which to compare all other patterns.

On the next page is a chronologically progressive listing of the **Titles** and **Symbols** which make up the various wave **Degrees**. These Title listings are the same as those R. N. Elliott originally devised (with two additions by the author).

TITLES:

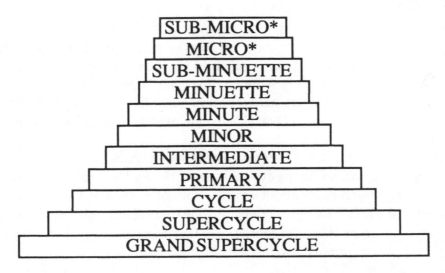

*Additions to Elliott's original titles

The Titles listed above represent progressively larger Degrees from the top downward. Beginning at the top, the Titles are encased in larger and larger rectangles to show their increasing size. By stacking the rectangles in pyramid fashion, it should provide you with a visual understanding of how each smaller Degree pattern builds the foundation for the next larger Degree.

When is it necessary to use the <u>additional</u> Degree **Titles** at the top of the list? The Complexity of a wave pattern expands and contracts as the time element allowed for its formation expands and contracts. The *Sub-Minuette* Degree designation used on the monowaves of your original chart may eventually represent patterns of increased complexity as time passes. The *Sub-Minuette* Title, which previously denoted only monowaves, may eventually represent polywaves, also. This takes place when the market expands the time element allowed to complete a formation of a specific Degree. The division of the *Sub-Minuette* pattern will require assigning each segment with a *lower* Degree **Title**. For each division, you would employ the *Micro* symbols. If the *Micro* waves subdivide, use the *Sub-Micro* **Titles** and **Symbols**.

Degree Symbolizing

A *Degree Symbol* is the simultaneous representation of a Progress Label (depicting pattern position), and a Degree Title (which, in a *relative* manner, loosely describes a pattern's price and time parameters and its intricacy *relative* to patterns of one larger and one smaller Degree).

On the next page is a repeat listing of all Degree Titles from the above diagram. This time, specific Progress Label *Symbols* have been added to show which Degree *Title* the *Symbols* represent. The *Symbols* in Figure 7-8 are not exactly like those Elliott used originally. A more logical labeling system has been devised by the author to make memorization easier.

Figure 7-8

		SYMBOLS		
	Impulsive		**Corrective**	
GSC — Grand Supercycle	[i̲]	— [v̲]	[a̲]	— [c̲]
SC — Supercycle	[1]	— [5]	[A]	— [C]
C — Cycle	[i]	— [v]	[a]	— [c]
P — Primary	①	— ⑤	Ⓐ	— Ⓒ
In — Intermediate	ⓘ	— ⓥ	ⓐ	— ⓒ
Mnr — Minor	(1)	— (5)	(A)	— (C)
Mnt — Minute	(i)	— (v)	(a)	— (c)
Mnut — Minuette	1	— 5	A	— C
SM — Sub-Minuette	i	— v	a	— c
Mc — Micro	.1	— .5	.A	— .C
SMc — Sub-Micro	.i	— .v	.a	— .c

Review

For two (or more) patterns to be combinable, they **must** be of the same Degree. Anytime an Elliott pattern is assembled, it is automatically implied each segment is of the same Degree (but not necessarily of the same Complexity). For two waves to be of the same Degree, it is imperative they possess some similarities of <u>Price</u> and/or <u>Time</u> (see "Rule of Similarity and Balance"; Chapter 4, page 3). Ideally, a mixture of time and price *similarities* is present in waves of the same Degree. Realistically, when a *same degree* wave is not *similar* in price, the market will compensate by closely matching (or exceeding) the time consumed by the previous move. If time is not within the relational range indicated in the "Rule of Similarity..." section, then price will adjust for the deficiency by closely matching (or exceeding) that of the previous wave. If neither the time nor the price consumed by a wave falls within the "relational range" of an **adjacent** wave, the two waves are almost surely <u>not</u> of the same Degree.

Another method of detecting "same Degree" patterns involves **Complexity Level.** For patterns to be considered the same Degree, they must be of the <u>same</u> Complexity or of an <u>adjacent</u> complexity. *[NOTE: an extremely rare exception to this statement may take place between the center portion of a Triple Combination Correction and one or both of the x-waves which surround it.]* This concept becomes *very* helpful as the patterns being studied unfold beyond the 2nd or 3rd complexity level. It will keep your patterns in proper relationship to each other and help avoid incorrect integration of patterns.

To begin your application of Degree labels, it is suggested the *Sub-Minuette* Title be used on the monowaves of your smallest chart (this should have already been done in the Chapter 5). After moving to the **Compaction** section, the **Degree** of the compacted pattern will be one level above the Degree of the highest Degree labels used <u>before</u> Compaction. This applies every time you compact a pattern. For example, the first wave group you worked with (or are working with) should have been composed of three or five *Sub-Minuette* monowaves. After a group is compacted into a single Structure label, the pattern, as a whole, would be elevated to *Minuette* Degree.

What is a Wave? - revisited

At the very beginning of the book, the term "wave" was described strictly from a **Monowave** perspective. Not until the end of the **Conclusion** chapter could a more *general* description of the term "wave" be given which encompasses patterns of every Degree (mono-, poly-, multi- and macrowave):

> *A <u>Wave</u> is the distance between two adjacent Progress Labels "of the same degree."*

For example, the action between Progress Label's (1) and (2) **or** A and B, is a **Wave**. This differs from the definition given earlier of the more specific *monowave* which was strictly limited to <u>the motion between one change in price direction and the next</u>. Now that you know the general definition of "wave," it should be clear why it was impossible to provide you with a general description earlier.

Flow Chart of Entire Neely Method
of Elliott Wave Analysis

The diagram on the next page is a complete "flow chart" of the **Neely Method** of Elliott Wave analysis. After reading "Mastering Elliott Wave" through the end of Chapter 7, you have been exposed to all of the major steps in this analysis process (each step is clearly deliniated on the flow chart). Chapters 8-12 are designed to provide you with more specific information for the proper designing and testing of each Elliott pattern but are not "named" steps in the general process of analysis. The flow chart should significantly increase your general understanding of how (and in which order) Elliott and Neely concepts should be applied to price action. After you become comfortable with Chapters 1-7, simply following the flow chart of the **Neely Method** may be enough to assist you in the accurate compilation of any Elliott Wave pattern.

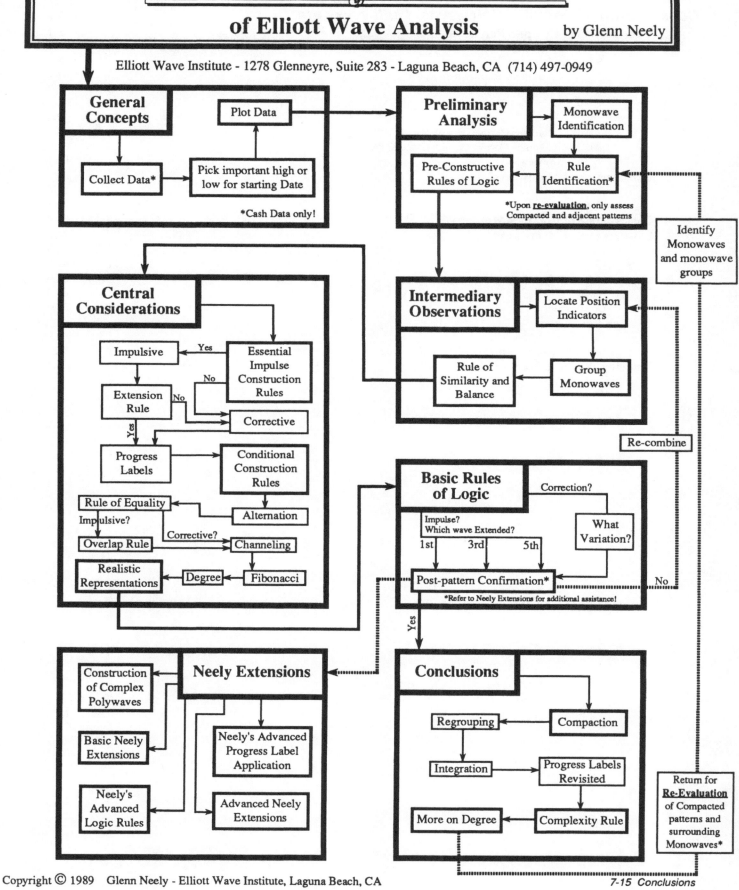

The Neely Method

of Elliott Wave Analysis   by Glenn Neely

Elliott Wave Institute - 1278 Glenneyre, Suite 283 - Laguna Beach, CA (714) 497-0949

General Concepts

Collect Data* → Pick important high or low for starting Date → Plot Data

*Cash Data only!

Preliminary Analysis

Monowave Identification → Rule Identification* ← Pre-Constructive Rules of Logic

*Upon re-evaluation, only assess Compacted and adjacent patterns

Identify Monowaves and monowave groups

Central Considerations

Impulsive —Yes→ Essential Impulse Construction Rules —No→ Corrective

Extension Rule —No→

Yes → Progress Labels → Conditional Construction Rules

Rule of Equality ← Alternation

Impulsive? Overlap Rule —Corrective?→ Channeling

Realistic Representations ← Degree ← Fibonacci

Intermediary Observations

Locate Position Indicators → Group Monowaves → Rule of Similarity and Balance

Re-combine

Basic Rules of Logic

Correction?

Impulse? Which wave Extended?
1st   3rd   5th

What Variation?

Post-pattern Confirmation*

*Refer to Neely Extensions for additional assistance!

No

Yes

Neely Extensions

Construction of Complex Polywaves ←

Basic Neely Extensions → Neely's Advanced Progress Label Application

Neely's Advanced Logic Rules ← Advanced Neely Extensions

Conclusions

Regrouping ← Compaction

Integration → Progress Labels Revisited

More on Degree ← Complexity Rule

Return for Re-Evaluation of Compacted patterns and surrounding Monowaves*

Copyright © 1989   Glenn Neely - Elliott Wave Institute, Laguna Beach, CA

7-15 Conclusions

After identifying one or more Polywave patterns on a real-time price chart, to expand your perspective of the market's future possibilities, you will need to combine a *group of polywaves* (to create a *Complex* Polywave) **or** combine a *combination of mono- and polywaves* to create a Multiwave pattern.

Construction of Complex Polywaves

Complex Polywaves fall into two categories. The first category is the **Standard** type which is an Impulsive or Corrective pattern where the most subdivided wave is a Corrective polywave only. Impulsive polywaves are excluded from the previous sentence because a pattern containing an Impulsive polywave is a Multiwave pattern (or higher).

The other category of Complex Polywave is the **Non-Standard** type. Special rules must be obeyed and certain circumstances must exist to create a Non-Standard pattern. Non-Standard patterns are only possible when combining several polywaves (or higher). They cannot simply occur from a combination of uncompacted monowaves.

Standard Type

The formation of a Standard Impulsive or Corrective **Polywave** is not normally composed strictly of three or five adjacent monowaves. Usually, one of the **Corrective** phases (but not any of the Impulsive phases) of the polywave will itself be a polywave (see Figure 8-1 on page 8-4). This allows for better Alternation between the two *detached* corrective phases of an Impulse pattern (i.e., waves 2 and 4) or the two *adjacent* segments of a Corrective pattern (i.e., waves A and B).

When a Polywave contains one or more Corrective Polywaves, it is considered a **Complex** Polywave. For a pattern to remain a Complex Polywave, no segment labeled with a Structure of ":5" can subdivide. If one or more of the ":5's" in a pattern does subdivide into a Polywave, the pattern should be considered a Multiwave (refer to "Construction of Multiwaves," page 3-16).

Whether a Polywave is strictly composed of monowaves **or** one or two Corrective polywaves, all the same Construction Rules covered in the **Central Considerations** chapter apply.

Non-Standard Type

The presence of a Non-Standard formation is only possible if there are at least two **Corrective** polywave patterns (compacted to their *base* Structure of ":3") which are separated by a mono- or polywave corrective phase. These *base* Structure labels serve the same combining function they did for monowaves in Chapter 4, to assist you in the proper grouping of adjacent wave patterns.

Due to the more elaborate design of polywaves (relative to monowaves), they can interrelate with other polywaves in ways not possible at the monowave level. For example, to detect the internal *Structure* of a monowave requires observing how much surrounding market action retraces it. On the other hand, when working with polywaves, the Structure is already known; there is no need to wait for a reaction. This fact brings up additional interaction possibilities which are covered below.

Additional Retracement Rules

If you are following along with your own chart and see a confirmed, Compacted Corrective polywave which is retraced *less* than 61.8%, or *more* than 161.8% by the next monowave or Corrective polywave, and then another Corrective polywave occurs immediately thereafter, refer to the "***Specifications***" section below for an explanation of what such an arrangement could mean and how to solidify such a grouping into a legitimate Elliott formation.

If neither of the above relational circumstances is present in your polywave grouping, the grouping should be considered of **Standard** design. If that is the case for your wave group, go to the "Standard" heading on page 8-1, then move on to the **Intermediary Observations** chapter. Work with your polywave group as if it were a monowave group; all the same Rules and procedures apply. The size of, or time consumed by, a wave group has virtually no effect on how it is analyzed. Wave patterns lasting several years will still subdivide into Flats, Triangles, Impulsions, etc. In working with such large patterns, *Structure Labels* are the key to maintaining analytical perspective and formational integrity.

Specifications

All **Non-Standard** market action involves **X-waves**. An X-wave is a Corrective pattern (always) which separates two **Standard** Elliott Corrections. Locating behavior indicative of X-waves is the key to finding Non-Standard wave patterns.

How do you recognize X-wave behavior? There are *two* important conditions to watch for.

Condition 1: The strongest signal of an X-wave developing in the market occurs when two *compacted* Corrections (polywave or higher) are separated by an intervening Corrective wave (monowave or higher and of Standard or Non-Standard nature) which retraced the first Corrective phase less than 61.8%. That intervening wave (the x-wave) will usually be of one lower **Complexity** Level than the two Corrections which it separates.

Condition 2: If three Compacted polywave Corrections occur one after the next, with the second Correction 161.8% (or more) of the first, the probabilities are very high the second Correction is an X-wave. In this situation, all Corrections will usually be of the same **Complexity**. If any one of the patterns is of a higher Complexity, it will usually be the last Correction of the entire formation.

If you witness either one of the above **Conditions**, the market is probably creating a Non-Standard formation. At that time it would be necessary to *revert* your Compacted poly- multi-, or macrowave pattern back into its pre-*compacted* Structure Series [i.e., if the first Correction of the group was a Zigzag, and you Compacted it to a "3," reverse the process and return it to its original ":5-:3-:5" condition.]

To continue with the analysis, note which of the two above conditions was met by your wave group. If **Condition-1** was met, move on to the heading *Complex Correction with Small x-Wave* ("a." below). If **Condition-2** was met, go to the *Complex Correction with Large X-Wave* heading ("b." on page 8-11).

a. *Complex Correction with Small x-Wave(s)*

When a **Non-Standard** wave pattern unfolds, the chances are much greater the x-wave will be smaller in price than the previous Corrective phase (less than 61.8%). This Non-Standard variation will often produce the look of an Impulse pattern, but close attention to detail should rule out that possibility (for details, see **Emulation** and **Missing Waves** in Chapter 12).

Each Non-Standard wave pattern, depending on its Structure Series, is given a different name. Listed below are the *Non-Standard Structure Series Combinations* which contain small X-waves. Each Combination listed has the pre-*compacted* Structure of each Corrective phase on the left. The entire formation's *compacted* value is on the right (c.t. = "Contracting Triangle only"). On the far right are the *Figures* which correlate to the formations on the left. In each Figure, a correct and incorrect interpretation is provided to help you avoid misinterpretations when deciphering market action.

Table A

1. (5-3-5) + (x-wave) + (5-3-5) = Double Zigzag = ":3"		Fig. 8-2a
2. (5-3-5) + (x-wave) + (3-3-3-3-3, c.t.) = Double Combination = ":3"		Fig. 8-2b
3. (5-3-5) + (x-wave) + (3-3-5) = Double Combination = ":3"		Fig. 8-3
4. (3-3-5) + (x-wave) + (3-3-5) = Double Flat = ":3"		Fig. 8-4
5. (3-3-5) + (x-wave) + (3-3-3-3-3, c.t.) = Double Combination = ":3"		Fig. 8-5
6. (5-3-5) + (x-wave) + (5-3-5) + (x-wave) + (5-3-5) = Triple Zigzag = ":3"		Fig. 8-6
7. (5-3-5) + (x-wave) + (5-3-5) + (x-wave) + (3-3-3-3-3, c.t.) = Triple Combination = ":3"		Fig. 8-7
8. (5-3-5) + (x-wave) + (3-3-5) + (x-wave) + (3-3-3-3-3, c.t.) = Triple Combination = ":3"		Fig. 8-8

In each sequence above, the Structure of the x-wave is not revealed. This is done because X-waves can be virtually any Corrective pattern without effecting the name or overall look of the larger formation. An X-wave can even be a Non-Standard pattern when the time period covered by the two surrounding patterns is long enough to allow for it. When reviewing the listing in Table B (page 8-11), keep in mind that an X-wave will usually alternate its formation with the Corrective pattern which immediately preceded it. For example, if the first Correction is a Zigzag, the x-wave will probably be a monowave, Flat or Triangle. If the first Correction is a Flat, the X-wave will probably be a monowave, Zigzag or Non-Standard formation (probably not a Triangle). Exceptions do occur, but they are relatively rare.

[Continued on page 8-11]

Figure 8-1

As the time covered by a polywave begins to increase, one of the corrective sections of the polywave will begin to subdivide (either wave-2 or wave-4). Never will one of the odd numbered segments of an Impulse pattern subdivide before one of the even numbered segments (unless the Impulse pattern is Terminal).

In this diagram, wave-a and wave-c are both monowaves. The b-wave was the first corrective phase to subdivide into a Polywave. <u>Note</u>: The b-wave is usually the first section of a correction to subdivide. As a result, the b-wave is usually more complex and time consuming than wave-a.

Wave-4 was the first corrective phase to break down into its own <u>Polywave</u>

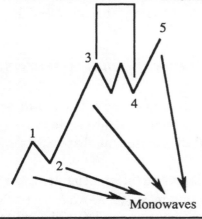

Monowaves

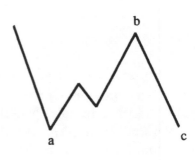

Figure 8-2a

Double Zigzag

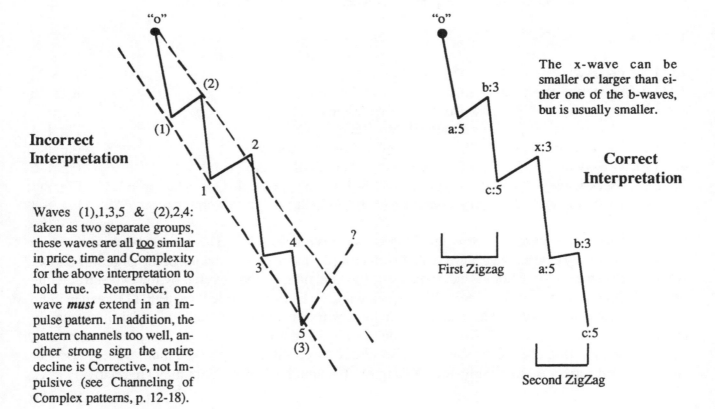

Incorrect Interpretation

Waves (1),1,3,5 & (2),2,4: taken as two separate groups, these waves are all <u>too</u> similar in price, time and Complexity for the above interpretation to hold true. Remember, one wave *must* extend in an Impulse pattern. In addition, the pattern channels too well, another strong sign the entire decline is Corrective, not Impulsive (see Channeling of Complex patterns, p. 12-18).

The x-wave can be smaller or larger than either one of the b-waves, but is usually smaller.

Correct Interpretation

First Zigzag

Second ZigZag

Figure 8-2b

Double Combination
(ending with a Triangle)

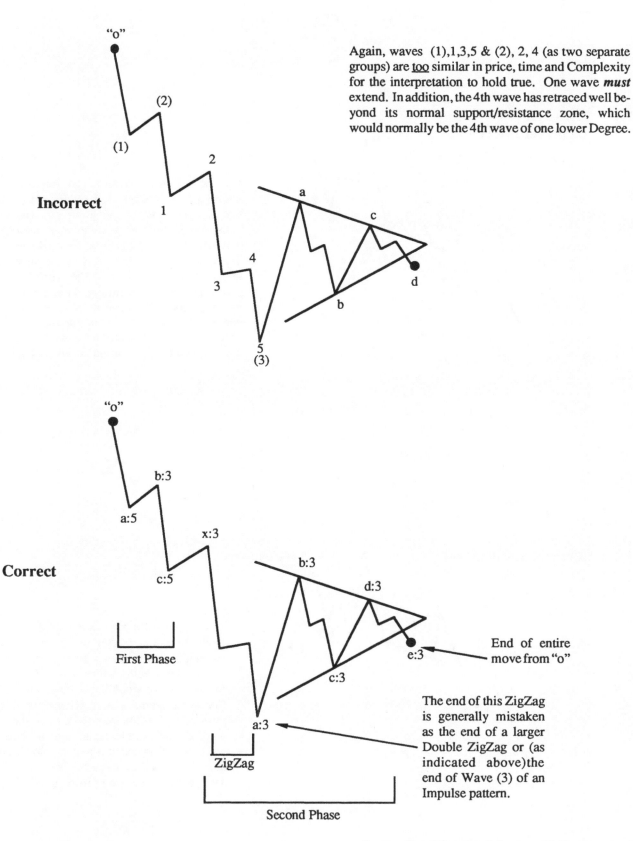

Again, waves (1),1,3,5 & (2), 2, 4 (as two separate groups) are <u>too</u> similar in price, time and Complexity for the interpretation to hold true. One wave *must* extend. In addition, the 4th wave has retraced well beyond its normal support/resistance zone, which would normally be the 4th wave of one lower Degree.

Incorrect

"o"

(1)

(2)

2

1

4

3

5
(3)

a

c

b

d

Correct

"o"

a:5

b:3

c:5

x:3

First Phase

b:3

d:3

c:3

e:3

a:3

ZigZag

Second Phase

End of entire move from "o"

The end of this ZigZag is generally mistaken as the end of a larger Double ZigZag or (as indicated above) the end of Wave (3) of an Impulse pattern.

Figure 8-3

Double Combination
(ending with a Flat)

Incorrect Interpetation

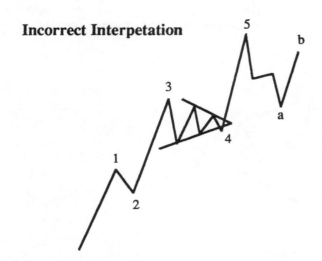

For the following reasons, an Impulse interpretation is eliminated. Wave-(3) is the shortest wave of the group (impossible). Even if wave-(3) were longer than wave-(1), the break-out of the fourth wave Triangle is too large. Any Triangular break-out of more than 200% virtually guarantees a Non-Limiting Triangle is forming, not a 4th-wave or b-wave Triangle. The length and time durations of each advance are too similar for them to be labeled with the same Impulsive degree.

Correct Interpetation

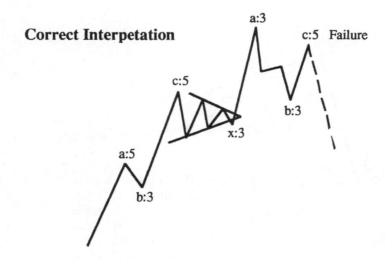

A Triangle (or a monowave) is the most likely formation to occur in the x-wave position if the last pattern is going to be a Flat with a c-wave Failure. The next best choice for the x-wave would be a monowave (as long as the overall formation is not too complicated). The Complexity Level of the x-wave should be at least the same as the c-wave which preceded it and should be no more complex than the most complex Standard pattern in the entire formation.

Figure 8-4

Double Flat

**Incorrect
Interpetation**

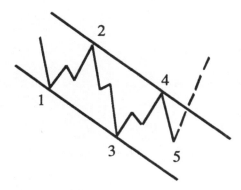

If you were to count this as an Impulsive pattern, you
have learned nothing from this book. This count
breaks all the rules. Wave-2 retraces too much of
wave-1, wave-3 looks corrective instead of Impulsive
and there is no Alternation between waves-2 & 4

**Correct
Interpetation**

This is the only way to
count this formation!

Figure 8-5

Double Combination

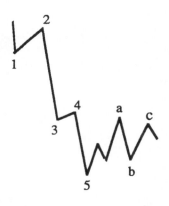

Incorrect Interpretation
Wave-2 retraced too much of wave-1 and the
Rule of Equality is not being adhered to by
waves 1 & 5. The c-wave is too simple in
relation to wave-a, unless a Triangle is forming. Even though the 3rd wave is the longest,
it barely meets the requirements of a 3rd
wave extension. Normally, the extended
wave (wave-3 in this case) will be 161.8% or
more of the next longest Impulse segment.
Here, wave-3 is less than 161.8% of wave-5.

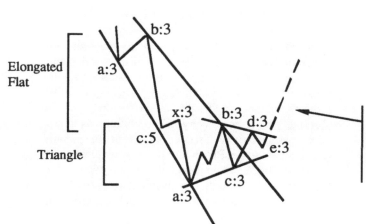

Correct Interpretation
The post-pattern market action would
be the most substantial evidence that a
Complex Correction formed instead
of and Impulse pattern.

Figure 8-6

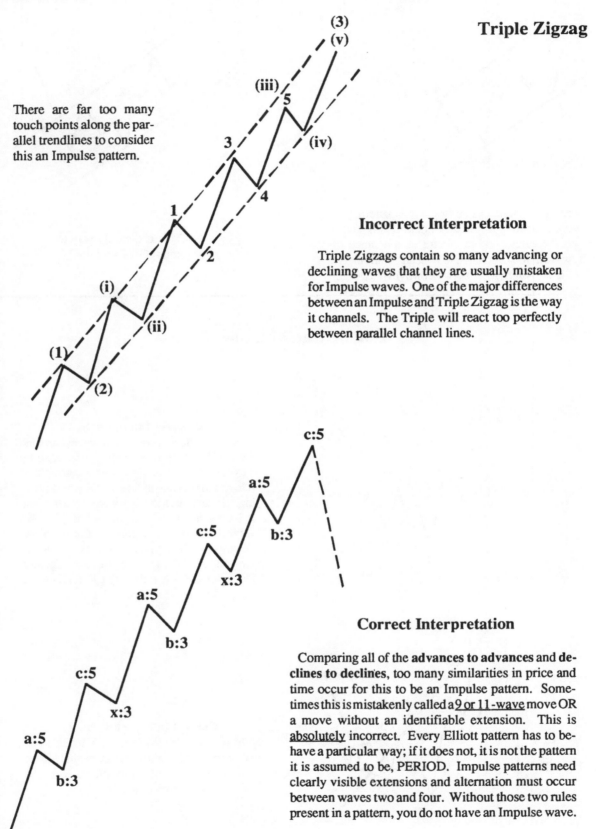

Triple Zigzag

There are far too many touch points along the parallel trendlines to consider this an Impulse pattern.

Incorrect Interpretation

Triple Zigzags contain so many advancing or declining waves that they are usually mistaken for Impulse waves. One of the major differences between an Impulse and Triple Zigzag is the way it channels. The Triple will react too perfectly between parallel channel lines.

Correct Interpretation

Comparing all of the **advances to advances** and **declines to declines**, too many similarities in price and time occur for this to be an Impulse pattern. Sometimes this is mistakenly called a 9 or 11-wave move OR a move without an identifiable extension. This is absolutely incorrect. Every Elliott pattern has to behave a particular way; if it does not, it is not the pattern it is assumed to be, PERIOD. Impulse patterns need clearly visible extensions and alternation must occur between waves two and four. Without those two rules present in a pattern, you do not have an Impulse wave.

Figure 8-7

Triple Combination

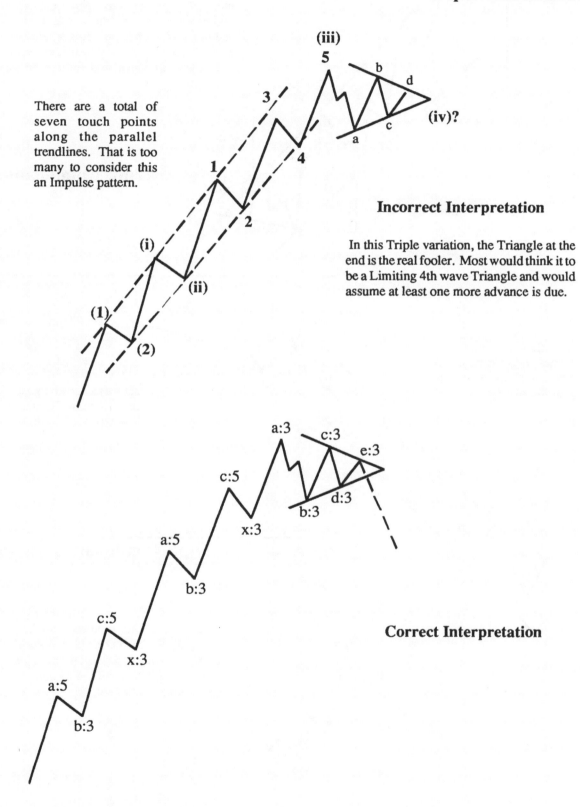

There are a total of seven touch points along the parallel trendlines. That is too many to consider this an Impulse pattern.

Incorrect Interpretation

In this Triple variation, the Triangle at the end is the real fooler. Most would think it to be a Limiting 4th wave Triangle and would assume at least one more advance is due.

Correct Interpretation

Figure 8-8

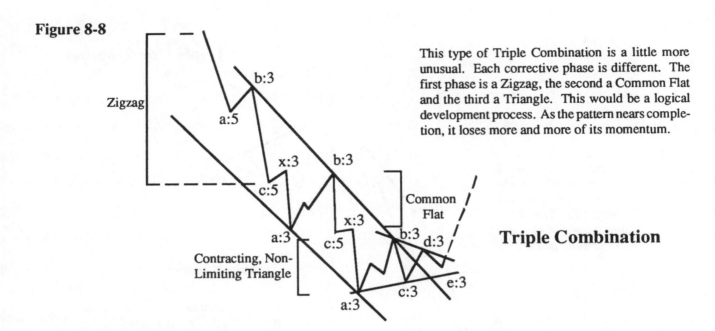

This type of Triple Combination is a little more unusual. Each corrective phase is different. The first phase is a Zigzag, the second a Common Flat and the third a Triangle. This would be a logical development process. As the pattern nears completion, it loses more and more of its momentum.

Triple Combination

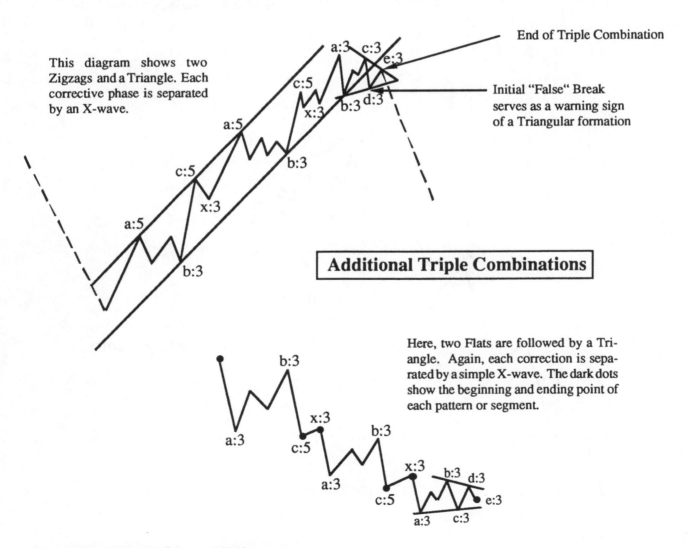

This diagram shows two Zigzags and a Triangle. Each corrective phase is separated by an X-wave.

End of Triple Combination

Initial "False" Break serves as a warning sign of a Triangular formation

Additional Triple Combinations

Here, two Flats are followed by a Triangle. Again, each correction is separated by a simple X-wave. The dark dots show the beginning and ending point of each pattern or segment.

The possible Corrections which can occur in the X-wave position are listed below by their Structure Series:

Table B

1. 5-3-5   Zigzag
2. 3-3-5   Flats (all variations, except Elongated)
3. 3-3-3-3-3   Triangles (Contracting Non-Limiting, only)
4.   3?   This series (mostly included to get your attention) represents a Corrective monowave which *can* be an X-wave in "simple" Double and Triple patterns. Keep in mind, x-waves are almost always the smallest Corrective patterns (time-wise) of a complex formation whether or not they are larger **or** smaller in price coverage than the previous Correction.
5.   When the X-wave is shorter than the previous corrective phase and is not one of the above patterns, its Construction is limited to Non-Standard Category 2 patterns (see Chapter 10, the **Power Ranking** diagram. [As a reminder, to properly associate a complex X-wave with the previous Correction, make sure the LOGIC Rules are followed for proper integration of market action.]

b. Complex Correction With Large X-Wave(s)

When the X-wave of a Complex formation is <u>larger</u> than the previous Correction (price-wise), the entire formation will be categorized as a Double or Triple *Three* pattern. Listed below are the Non-Standard wave patterns, along with their names and *compacted* wave Structures, in which the X-wave is larger than the previous Corrective phase (in the listing, c.t. = "Contracting Triangle only!").

1. (3-3-5) + (x-wave) + (3-3-3-3-3, c.t.) = Double Three Combination= 3   Fig. 8-9
2. (3-3-5) + (x-wave) + (3-3-5) = Double Three = 3   Fig. 8-10
3. (3-3-5) + (x-wave) + (3-3-5) + (x-wave) + (3-3-3-3-3, c.t.) = Triple Three Com.= 3   Fig. 8-11
4. (3-3-5) + (x-wave) + (3-3-5) + (x-wave) + (3-3-5) = Triple Three = 3   Fig. 8-12

The Non-Standard wave patterns listed above are in order of *most* to *least* likely. Triple Threes, as mentioned earlier, should be considered rare. When you see one of these patterns, it probably will be constructed like one of the two variations listed on page 8-15.

Review

Once you have decided what <u>Non-Standard</u> wave pattern variation the market has formed, detour to the **Logic Rules** chapter and check the list of attributes associated with that particular formation. Examine the "realistic" diagrams in the Non-Standard section to see if the pattern you are working with resembles the one in the book. Remember, exact correlations are not only <u>unnecessary</u>, but very improbable. Afterward, move on to the **Conclusion** chapter to finalize your assessment.

Figure 8-9

Double Three Combination
(in this form, it is rare)

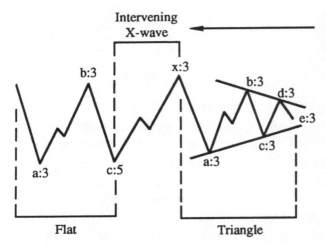

Intervening
X-wave

b:3

x:3

b:3
d:3
e:3

a:3   c:3

a:3   c:5

Flat   Triangle

Note: the x-wave exceeded the end of wave-b slightly. The stronger the move to follow this entire formation, the higher the x-wave will go.

This pattern would be difficult to count incorrectly, so a faulty interpretation is not included.

Figure 8-9
Continued

Double Three Combination
(Running variation)

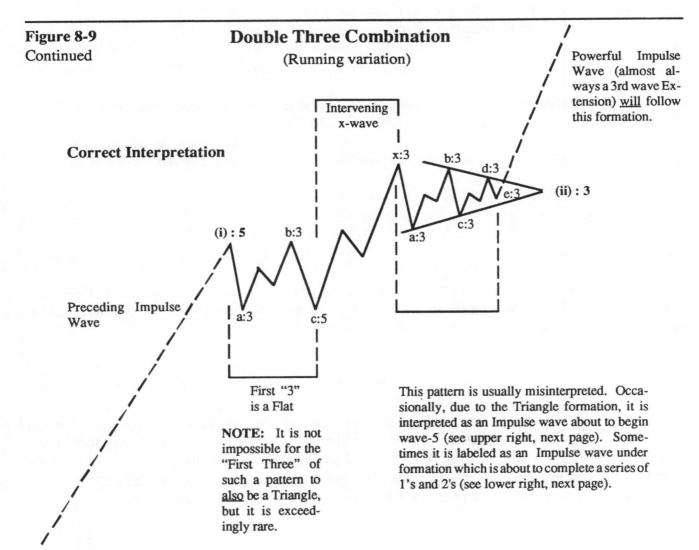

Correct Interpretation

Intervening
x-wave

Powerful Impulse Wave (almost always a 3rd wave Extension) <u>will</u> follow this formation.

x:3   b:3
d:3
e:3

(ii) : 3

(i) : 5   b:3

a:3   c:3

a:3

a:3   c:5

Preceding Impulse Wave

First "3" is a Flat

NOTE: It is not impossible for the "First Three" of such a pattern to <u>also</u> be a Triangle, but it is exceedingly rare.

This pattern is usually misinterpreted. Occasionally, due to the Triangle formation, it is interpreted as an Impulse wave about to begin wave-5 (see upper right, next page). Sometimes it is labeled as an Impulse wave under formation which is about to complete a series of 1's and 2's (see lower right, next page).

Figure 8-9
Continued

Double Three Combination
(continued from previous page)

This incorrect interpretation of a Double Three Running Correction is very common due to the association many make of Triangles with 4th waves. Complicating the situation, a Triangle in one of these patterns frequently does not "overlap" the first a-b-c Correction, thus creating the illusion of an Impulse wave with a first wave extension. The signal that this interpretation is incorrect is the Structure of wave-3. In this diagram it is Corrective, eliminating the possibility that this interpretation is correct, **unless** the market is forming a Terminal Impulse. The thrust out of the Triangle will answer that question. If the thrust is bigger than wave-3, the pattern is a Double Three Running Correction; if less, a Terminal Impulse.

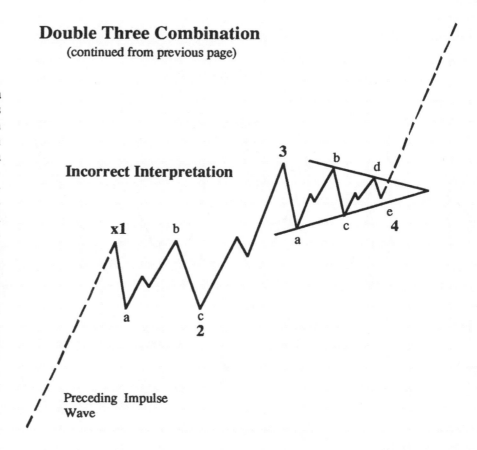

Incorrect Interpretation

Preceding Impulse Wave

Figure 8-9
Continued

Incorrect Interpretation

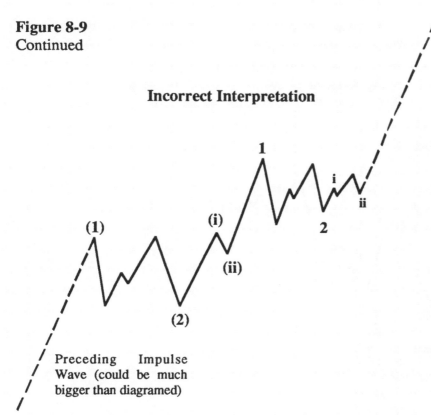

Preceding Impulse Wave (could be much bigger than diagramed)

This interpretation is so full of errors it is hard to know where to begin. To have a reliable series of 1's and 2's (of decreasing Degree), before a powerful "3rd wave of a 3rd wave" move begins, it is mandatory the market develop in a parabolic fashion. Each smaller Degree 2nd wave should take less time and price, be of a more Powerful Construction and retrace less (percentage-wise) than the previous larger Degree 2nd wave. In a like manner, each smaller Degree 1st wave should take less time and price and advance at a greater slope than the previous, next higher Degree 1st wave.

As is obvious, most of the above parameters do not hold true in the figure on the left. **NOTE:** A true Series of 1's and 2's (simultaneously visible on the same chart) beyond the <u>second set</u> is very rare.

Figure 8-10

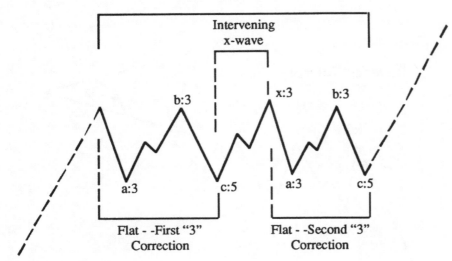

Double Three
(in this form it is rare)

Intervening
x-wave

b:3

x:3   b:3

a:3   c:5   a:3   c:5

Flat - -First "3"   Flat - -Second "3"
Correction   Correction

This pattern is unusual in any position except wave-b of a Zigzag, or as a smaller x-wave. Even then the pattern will have a strong tendency to drift with the trend *of one larger Degree.*

Figure 8-10
Continued

Double Three
(this version is the most likely)

This pattern is very similar to the Double Three Running Correction illustrated at the bottom of page 8-12

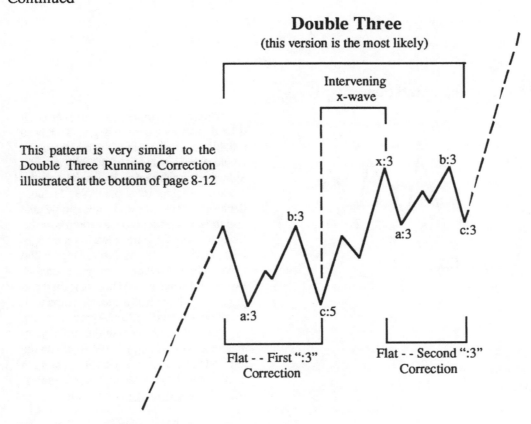

Intervening
x-wave

x:3   b:3

b:3   a:3   c:3

a:3   c:5

Flat - - First ":3"   Flat - - Second ":3"
Correction   Correction

Figure 8-11

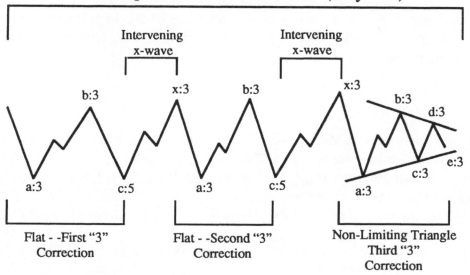

Triple Three Combination (very rare)

Intervening x-wave

Intervening x-wave

b:3   x:3   b:3   x:3

b:3   d:3

a:3   c:5   a:3   c:5

a:3   c:3   e:3

Flat - -First "3" Correction

Flat - -Second "3" Correction

Non-Limiting Triangle Third "3" Correction

Not only is a Triple Three Combination extremely rare, but for it to occur over an almost perfectly horizontal plane should be considered basically impossible. If you ever see one of these formations, it should definitely drift with the trend *of one larger Degree.*

Figure 8-12

Triple Three (very rare)

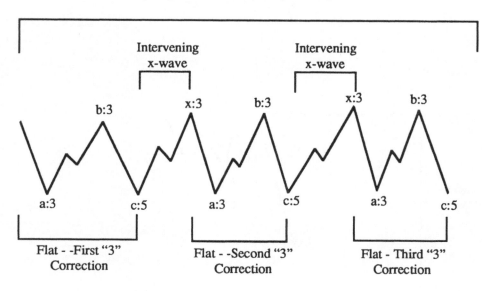

Intervening x-wave

Intervening x-wave

b:3   x:3   b:3   x:3   b:3

a:3   c:5   a:3   c:5   a:3   c:5

Flat - -First "3" Correction

Flat - -Second "3" Correction

Flat - Third "3" Correction

All of the above script for the Triple Three Combination also applies to the Triple Three.

Construction of Multiwaves

Slightly different from Complex polywaves, a Multiwave requires one of the Impulse segments subdivides into a polywave. Depending on which **Class** the pattern you are analyzing falls in, divert to the Impulsive or Corrective heading below.

Impulsive

In a polywave, all ":5's" are monowaves. In a Multiwave, one (or more) of the ":5's" is a polywave. To construct a Multiwave, several conditions are required. They are as follows:

1. Of the three thrust waves in an Impulse pattern (1, 3, or 5), one - <u>and only one</u> - **must** be a polywave. The other two thrust waves <u>should</u> be monowaves.
2. At least **one** of the Corrective phases (either wave 2 or 4) **must** be a polywave; the other may be a mono- or polywave.
3. The Correction (2 or 4) which takes the longest period of time should occur <u>immediately</u> before or after the Extended wave. If the 1st wave extended, wave-2 should take the most time. If the 5th wave extended, wave-4 should take more time than wave-2. If the 3rd wave extends, it makes no difference which wave (wave-2 **or** wave-4) takes more time; simply make sure Alternation is present between the two waves.

Figure 8-13a displays a *very* common Multiwave pattern. The 3rd wave is the Extended and Subdivided wave *(see Extensions vs. Subdivisions, page 8-21, for detailed coverage of the Subdivision concept and the Rules which govern its operation)* creating the <u>only Trending polywave</u> of the pattern. Waves-1 and 5 are the two Trending monowaves (see Rule 1 above). Wave-4, the largest corrective pattern, occurred immediately after the Extended 3rd wave (see Rule 3). The diagrams in Figure 8-13b also show the same Rules being followed by Multiwaves with 1st and 5th wave Extensions.

Figure 8-13a

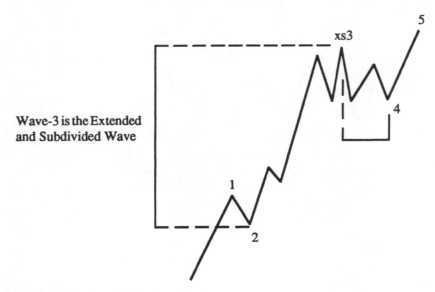

Wave-3 is the Extended and Subdivided Wave

Figure 8-13b

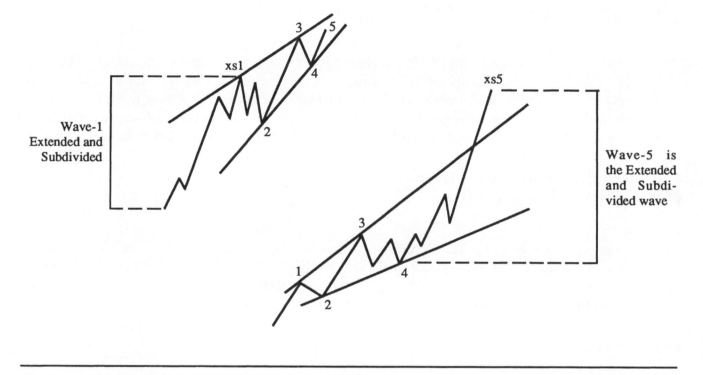

Corrective

A Corrective Multiwave has requirements imposed on it similar to the Impulsive variety. They are:

1. One or two of the ":5's" in the larger pattern **must** visibly subdivide into a polywave (see Figure 8-14). If there is only one subdivided ":5," it must be wave-c of a Flat or Zigzag (all varieties).
2. The probabilities <u>strongly</u> favor the b-wave of a Multiwave will be a <u>Corrective</u> polywave.

   Again, as indicated in the Impulsive section, all other Rules listed from the beginning of the book which refer to Corrections should apply to Multiwaves of all varieties.

Figure 8-14

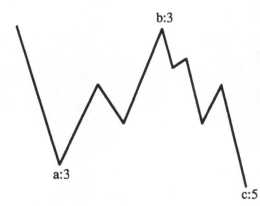

In the Figure on the left, the c-wave (a ":5") subdivides. That is the minimum requirement for a Multiwave to be created. The Figure on the right has two Subdivided ":5's," but the pattern is still a Multiwave. Obeying Rule 2 (above), the b-wave in both patterns is a polywave.

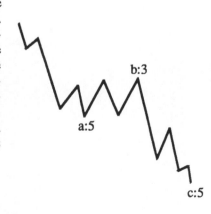

Construction of Complex Multiwaves

The discussion on **Complex Multiwaves** is basically the same as Complex Polywaves. The only difference is Complex Multiwaves are made from a collection of Multiwaves, not a collection of Polywaves. If you witness a Corrective Multiwave which is retraced 161.8% or more by the next correction, OR is retraced less than 61.8% by the next corrective phase, then a Complex pattern is forming. Apply the same Rules and principles as those discussed in the Complex Polywave section.

Construction of Macrowaves

As a market continues to unfold, a group of Multiwaves will eventually need to be formed into a Macrowave. Fortunately, the processes explained in the Multiwave section under the Impulsive and Corrective headings are the same as those that need to be used to form Macrowaves. The only differences are those listed below under the Impulsive and Corrective headings.

Impulsive

The minimum construction necessary to form a Macrowave is at least one of the ":5's" (wave-1, 3, or 5) must be a Multiwave and one of the other ":5's" must be a Polywave. Usually there will be two polywaves, but on rare occasions, the smallest wave might be a monowave. This creates a condition of "Triplexity," a word coined by the author which means "three Levels of Complexity contained within the same pattern." From observation, it appears the only time Triplexity is possible is when the 5th wave Extends and Subdivides in an Impulse wave or when wave-c is the most complex pattern of a Flat or Zigzag. For an illustration of Triplexity in action, refer to the bottom of page 7-10.

Corrective

To form a Macrowave Correction, at least one wave **must** be a Multiwave and one wave a Polywave. If only one Multiwave occurs in a Correction, it should be wave-c of a Zigzag or Flat. All other rules applying to Multi- and Polywaves would also apply for Macrowaves.

More on Alternation

In the **Central Considerations** chapter, the concept of Alternation was presented. The simple aspects of the **Rule** (Time, Price, Severity) should be readily understood by now. The more difficult aspects of Alternation (Intricacy and Construction, first mentioned on page 5-5) will be covered here to make sure there is no misunderstanding on the subject. Starting with **Intricacy**, this deals with the number of subdivisions present in one pattern compared to those of an adjacent pattern. This concept is difficult to apply to highly developed patterns, but is a very useful and critical test at the polywave and multiwave level. It is always preferable that one wave subdivides and one does not. Below are examples of how this happens in Corrective and Impulsive formations (see Figure 8-15).

Figure 8-15

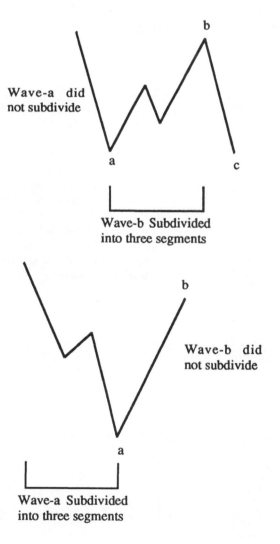

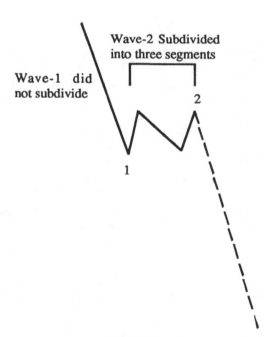

Unlike a corrective phase, wave-1 cannot subdivide without wave-2 subdividing (unless the market is forming a Terminal pattern). Waves-3 & 4 can Alternate on an Intricacy basis in the same way waves-1 & 2 did above.

Construction

If adjacent waves in an Impulsive or Corrective pattern both subdivide, other forms of Alternation need to be considered to keep the count accurate. One of the forms in which Alternation can occur is Construction. If one pattern is a Zigzag, expect the next pattern to be anything but a Zigzag (see Figure 8-16a). If the market forms an Impulse pattern, expect the next move, *of the same Degree*, always to be a Corrective pattern (see Figure 8-16b below).

Figure 8-16a

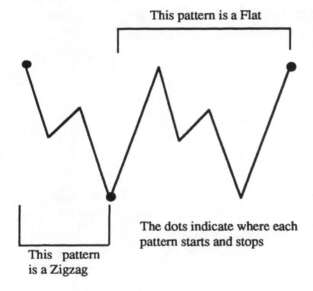

This pattern is a Flat

The dots indicate where each
pattern starts and stops

This pattern
is a Zigzag

Figure 8-16a

Impulsive (:5)

Corrective (:3)

Both of theses patterns exhibit Alternation of Construction. In the upper diagram, the Alternation is between two different corrective patterns. In the lower diagram, the Alternation is between an Impulsive pattern and a Corrective pattern. Obviously, the lower diagram presents a form of Alternation which is much easier to detect.

More on Extensions

Extensions vs. Subdivisions

Most Elliotticians think of the term *Extension* as describing two inextricably linked factors; that of length **and** the number of subdivisions present in a pattern. Through many years of study, I have discovered the two factors, Extensions and Subdivisions, to be separate phenomena. The term Extension should only be thought of as describing the *longest **trending** wave* (wave-1, 3 or 5) of an Impulse sequence. It should not automatically be assumed that the longest wave in a pattern will be the most subdivided. On rare occasions, the longest wave will be one of the simpler patterns (based on Complexity ratings) and the subdivided pattern will be the second longest segment. This indicates the Extension Rule should be applied separately from the Subdivision Rule (see Independent Rule, page 9-7). Granted, the two rules (Extension & Subdivision) will usually be simultaneously present in the same wave (90% of the time), but not always. Figure 8-17 shows how a pattern looks when both Rules are affecting the same wave. Figure 8-18 shows how the two Rules might act <u>independently</u> of each other. Figure 8-19 shows how the Rules *should not* be applied and where they *cannot* act independently of each other.

Figure 8-17

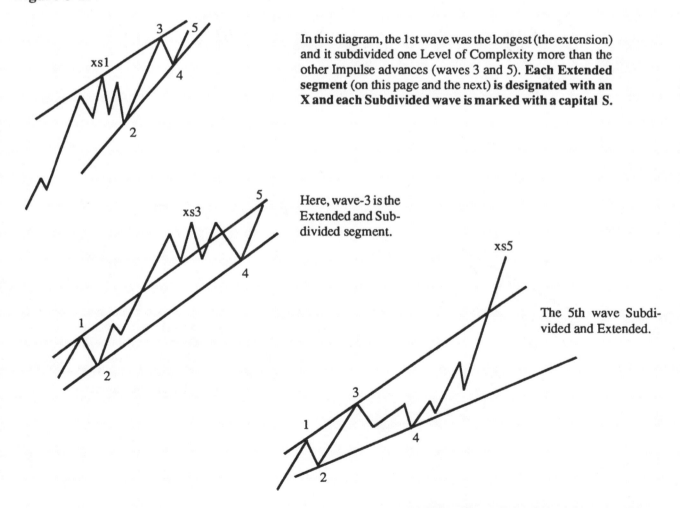

In this diagram, the 1st wave was the longest (the extension) and it subdivided one Level of Complexity more than the other Impulse advances (waves 3 and 5). **Each Extended segment** (on this page and the next) **is designated with an X and each Subdivided wave is marked with a capital S.**

Here, wave-3 is the Extended and Sub-divided segment.

The 5th wave Subdivided and Extended.

Figure 8-18

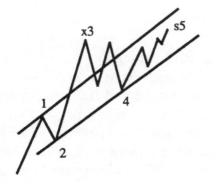

Most common manifestation of separate Extended and Subdivided waves. Wave-3 is Extended, Wave-5 is Subdivided.

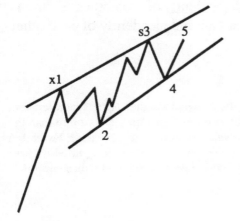

Next most likely occurrence:
Wave-1 is Extended
Wave-3 is Subdivided

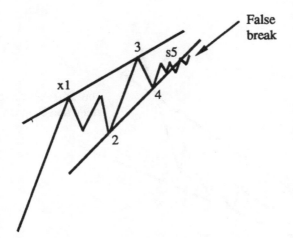

Least likely occurrence. If you were to witness this setup, the final thrust (wave-5) would probably be a Terminal Impulse. As a rule, when the final thrust is a Terminal pattern, the base line (across the two corrective phases, wave-2 and wave-4) will be broken several times before the pattern terminates.

Figure 8-19

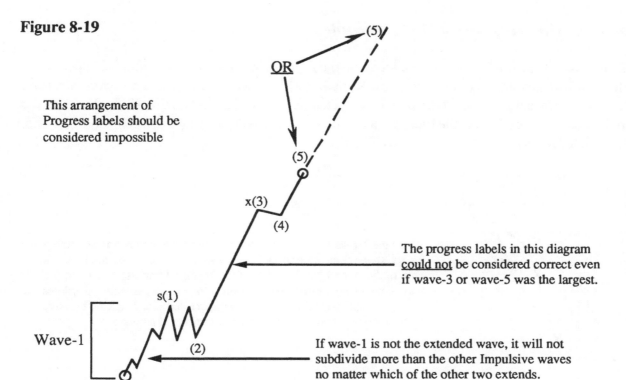

This arrangement of Progress labels should be considered impossible

The progress labels in this diagram could not be considered correct even if wave-3 or wave-5 was the largest.

If wave-1 is not the extended wave, it will not subdivide more than the other Impulsive waves no matter which of the other two extends.

It would not be impossible for market action like that above to unfold, but the Progress labels would have to be changed. It is impossible for the high (circled) to be the end of the Impulse wave. The net diagram below illustrates the correct labeling for such a situation. The Degree of the labels may differ in real-time, but the relationship between Degrees will not.

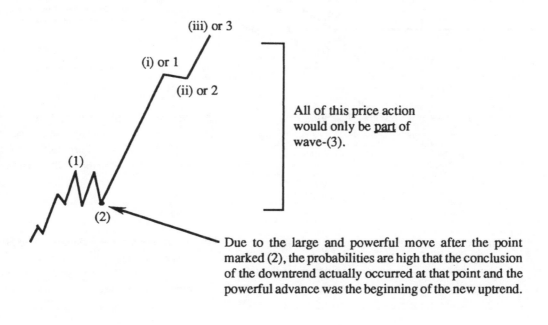

All of this price action would only be part of wave-(3).

Due to the large and powerful move after the point marked (2), the probabilities are high that the conclusion of the downtrend actually occurred at that point and the powerful advance was the beginning of the new uptrend.

Importance of Knowing which Wave Extended

The Extended wave in an Impulse sequence is the <u>most significant</u> factor in deciding the appearance, relationships and behavior of an Impulse pattern. Knowing which wave Extended instantly provides you with a wealth of information on how the pattern will channel and which Correction (2 & 4) will be the most complex. Let's take a look at the four major Impulse variations depending on which wave Extends, beginning with a 1st wave extension in Figure 8-20.

Figure 8-20

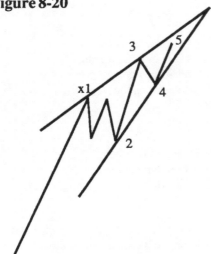

When the 1st wave is the longest, the pattern will assume an ascending wedge shape. The upper trendline is usually drawn across the terminal ends of the 1st and 3rd waves. Unlike a Terminal Impulse, where the 5th wave usually breaks the upper trendline, this pattern generally completes its 5th wave below and away from the upper trendline. When the 1st wave is the longest, wave-2 should be more complex than wave-4.

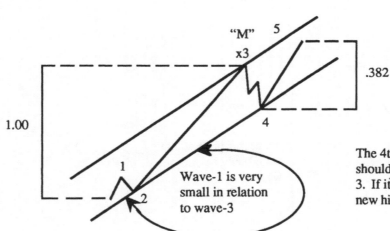

> There are three significantly different formational variations possible when the 3rd wave extends. Below is the first variation, the last two are on the next page.

The 4th wave (in this case) at its lowest point should not retrace more than 38.2% of wave-3. If it does, wave-5 will probably not make a new high, thereby creating a 5th wave Failure.

This setup occurs when wave-1 is "microscopic" in relation to the Extended 3rd wave. The 5th wave will usually relate to the whole move (beginning of wave-1 to end of the end of wave-3) by 38.2%. If it is more than 38.2%,(such as 61.8%) the move is just a Zig-Zag and what you thought was wave-1 & 2 are probably part of a previous pattern, or less likely, they were incidentally visible parts of the first advance to point "M" (see "Missing Waves," page 12-34).

Figure 8-20
continued

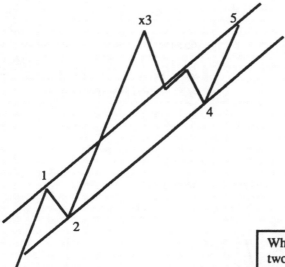

When a pattern channels like one of the two diagrams on this page, and it completes a larger pattern, it is very possible Wave-5 may not exceed the end of Wave-3. If it does not, the 5th-Wave would be labeled a *5th Wave Failure*.

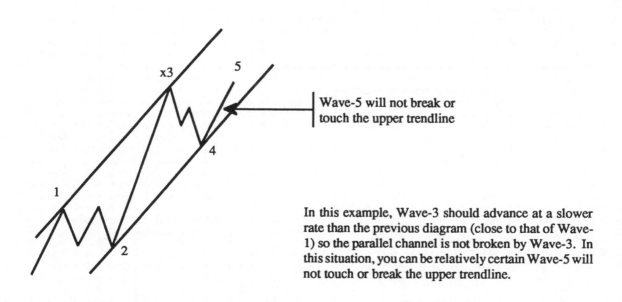

Wave-5 will not break or touch the upper trendline

In this example, Wave-3 should advance at a slower rate than the previous diagram (close to that of Wave-1) so the parallel channel is not broken by Wave-3. In this situation, you can be relatively certain Wave-5 will not touch or break the upper trendline.

Figure 8-20
Continued

The 5th wave (when extended) will quite often break the upper trendline ("*false break*") only to quickly retrace 61.8%-95% of the entire 5th wave.

When the 5th Wave is the longest, waves-1 & 3 will usually be equal or related in time by 61.8%. The 3rd wave must be slightly longer than the 1st wave, but it should not complete beyond **161.8% of wave-1 added to the top of wave-1.** Usually wave-1 & 3 relate internally* by about 161.8% (price-wise).

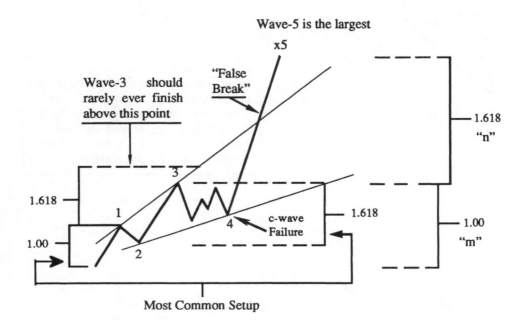

Wave-5 is the largest

x5

"False Break"

Wave-3 should rarely ever finish above this point

3

1.618

1

1.618

"n"

4   c-wave Failure

1.618

1.00

"m"

1.00

1

2

Most Common Setup

A 5th wave Extension will usually be 161.8% of the entire move measured from the very beginning of the pattern to the top of wave-3, illustrated by "m" & "n's" **External** relationship*. Occasionally, the 5th wave will relate **Internally** to "m" by adding the equal length (or 161.8% of the length) to the end of wave-4. Less likely relationships include the 5th wave being 100% of "m" or 261.8% of "m."

Wave-4 should be more complex and time consuming than wave-2 when the 5th wave extends. One odd thing about wave-4 (when wave-5 extends) is it generally retraces a significant amount (40-61.8 % of wave-3. For wave-4 to counteract a significant retracement, it will usually end in a c-wave failure (see diagram above) or complete above its lowest point by way of a Complex correction concluding with a Non-limiting Triangle).

* For a complete discussion of External Relationships, refer to pages 12-22 through 12-34

Knowing Where to Start a Count

As the **Complexity** of a pattern increases beyond the 1st-Level, Trending waves <u>within</u> Impulse patterns begin to develop into their own Impulsive polywaves. Due to the difficulties which arise when trying to decipher moves beyond Level-2 Complexity, knowing when an Impulse sequence is complete requires strict adherence to numerous rules, channeling techniques, and Fibonacci relationships (all to be outlined in this book). Proper analysis requires knowing how Impulse patterns begin and end **and** how they combine with other patterns. To do this you must first correctly decipher small Impulse patterns.

When first trying to analyze a market's price action, You may be tempted to start with long-term price charts. In doing this, you could easily misinterpret the position of the market due to a common mistake; that of starting the count from a major high or low (see Figure 8-21). To start an interpretation from the

Figure 8-21a

An inexperienced analyst would be tempted to begin the analysis from the circled low and possibly label the move upward as indicated in Figure 8-21a. For several reasons this would be incorrect.

1. Both waves-2 & 4 indicate strength. That conflicts with the Rule of Alternation.

2. Wave-2 retraces more than 61.8% of Wave-1 contradicting the signs of strength indicated by an Irregular correction.

3. Waves-3 & 5 are too close in price coverage. This conflicts with the *"Rule of Extension"*

4. Finally, the low from which the analysis began <u>was not</u> the termination of the down move.

Figure 8-21b

If you are analyzing a market's price action for the first time, you could easily misinterpret the action due to a common mistake; that of starting the count from a major high <u>or</u> low. Believe it or not, most large Elliott patterns **<u>do not</u>** finish at the highest or lowest price obtained by the market.

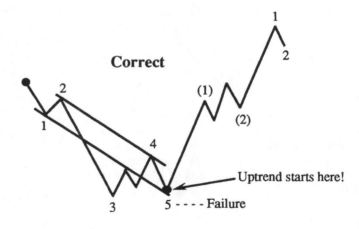

Figure 8-21b indicates the correct interpretation of the price action. Due to the 5th wave Failure, the upside count starts <u>above</u> the lowest price level.

Figure 8-23

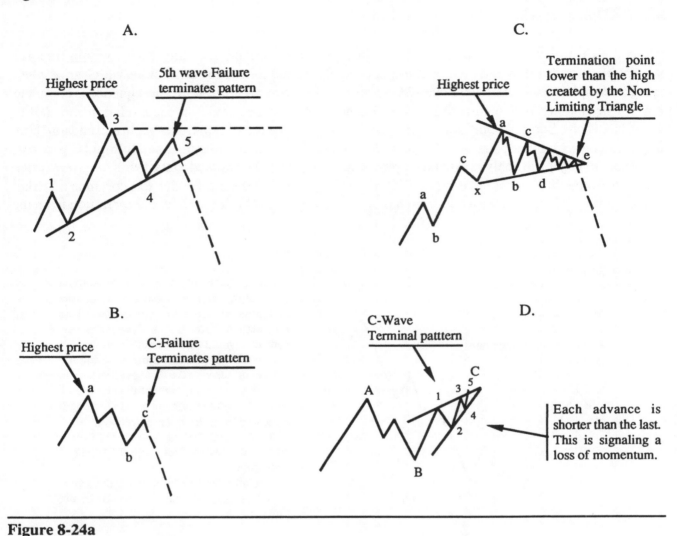

A.

Highest price

5th wave Failure terminates pattern

C.

Highest price

Termination point lower than the high created by the Non-Limiting Triangle

B.

Highest price

C-Failure Terminates pattern

D.

C-Wave Terminal pattttern

Each advance is shorter than the last. This is signaling a loss of momentum.

Figure 8-24a

Figure 8-24b

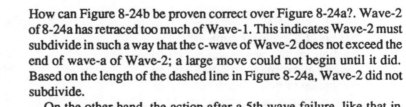

An incorrect analysis of this move could easily make you think 2nd waves frequently retrace almost all (i.e. more than 61.8%) of Wave-1. Actually, a 2nd wave retracement of wave-1 exceeding more than 61.8% (even if the termination of wave-2 settles at a lesser level) is extremely rare.

How can Figure 8-24b be proven correct over Figure 8-24a?. Wave-2 of 8-24a has retraced too much of Wave-1. This indicates Wave-2 must subdivide in such a way that the c-wave of Wave-2 does not exceed the end of wave-a of Wave-2; a large move could not begin until it did. Based on the length of the dashed line in Figure 8-24a, Wave-2 did not subdivide.

On the other hand, the action after a 5th wave failure, like that in Figure 8-24b, would produce a violent reversal of trend due to the weakness an upside failure implies. Since a severe drop (depicted by the dashed line) did occur and the Impulse pattern was retraced rapidly, it is confirmed the 5th-Failure scenario is correct (Figure 8-24b).

improper point makes it very difficult to correctly predict what will take place in the future. Believe it or not, most large Elliott patterns do not finish at the highest or lowest price on a chart. This is an inherent part of the Wave Theory which is frequently misunderstood and is essentially unique to Elliott. I say it is unique since most techniques look at the highest high or lowest low as the important price or time measuring point for further calculations. Often, with Elliott, points <u>below the high</u> *or* <u>above the low</u> are more important for specific calculations than the actual high or low. The concept of a pattern finishing **after** a high or low is one of the significant reasons many Elliotticians experience difficulty applying Elliott to longer time periods.

If you start a count from a point on a chart which is not the termination of an Elliott pattern, your analysis could run astray for long periods of time. Eventually you should be able to recognize the mistake and make the appropriate adjustments. Unfortunately, by the time a mistake is detected, the market move could be nearing completion or you may have taken several losses due to trades initiated based on inaccurate assumptions. All this is, of course, due to starting a count from a "visually" important top or bottom which is not the termination of an Elliott pattern.

The more Elliott patterns you combine into larger patterns, the more likely the larger pattern will finish below (above) the highest (lowest) price level obtained (see Figure 8-21b at the bottom of page 8-27). Why does this happen? At the end of a major trend, the market begins to lose momentum. It is this loss of momentum which sometimes prevents the market from making a new high or low right at the end of a pattern. Loss of momentum generally manifests itself in one of four forms:

   a. An Impulse pattern will contain a **5th-Wave Failure**
   b. A Flat pattern will finish with a **C-Wave Failure**
   c. A complex or rare formation will end with a **Contracting, Non-Limiting Triangle**
   d. An Impulsion will conclude with a **Terminal pattern**
       (Each incident listed above is diagramed on the previous page (Figure 8-23) using the same alphabetic lettering.)

Three of the four situations above (a, b & c) always create a situation where the highest or lowest price on the chart **is not** the end of the Elliott pattern. The Terminal pattern, if its 5th wave subdivides, can also conclude *below* the highest high or *above* the lowest low if wave-5 exhibits a loss of momentum in one of the three circumstances listed above (scenario **a,b,** or **c**).

If a pattern is completely retraced and the highest price achieved by the market is not the end of the Elliott pattern, the termination point will occur <u>after</u> the high, not before. Always be on the lookout for a **secondary** spike or trough following an important high or low. It may be warning that an Elliott pattern terminated after the top or bottom. Also, watch for a significant amount of consolidation near, <u>but after</u>, an important high or low. The consolidation may be a Non-Limiting Triangle which terminates the trend after the high or low.

Starting counts from improper points can often lead to incorrect conclusions about the correct application of Elliott Rules. Figure 8-24 demonstrates how that may take place. Another incorrect conclusion most analysts make is in reference to the series of 1's and 2's of decreasing degree which occasionally takes place as the market approaches the center of a 3rd wave Extension, the most frequent wave to extend (see Figure 8-25, next page).

Figure 8-25

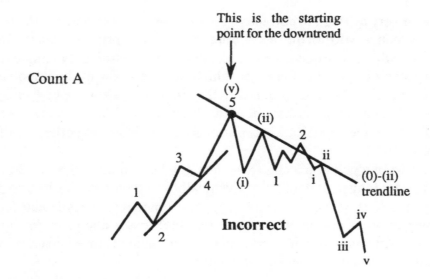

Count A

This is the starting point for the downtrend

(0)-(ii) trendline

Incorrect

There are a number of significant **Logic** and **Rule** errors involved in count "A"; they are as follows:

1. The length of waves 1, 3 & 5 are all too close to the same price distance.
2. Waves 2 & 4 do not alternate well.
3. During the decline, wave-2 takes more time and percentage retracement than wave (ii). This indicates the market is not getting weaker (which is mandatory as you approach the middle of a declining 3rd wave) but is getting stronger.
4. The 0-(ii) trendline was broken by the 2nd-wave indicating wave-(ii) is not finished or it is incorrectly interpreted.

Count "B" presents the proper way to label the above move. Notice the market did not begin to decline quickly until time "ran out" at the apex point of the triangle. This is typical of this type of **Non-Limiting Triangle.**

Figure 8-25
(continued)

Count B

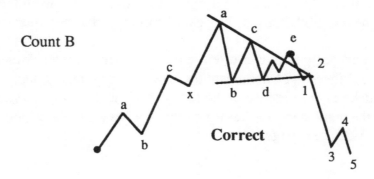

Correct

Basic Neely Extensions

Mentioned in the beginning of the book, many of the **Neely Extensions** have simply been integrated into the text where appropriate. Virtually all of Chapter 3 discussed techniques which I developed over years of teaching the Telephone Course. The step by step, objective process of analysis - which is the essence of "Mastering Elliott Wave" - was also created during the teaching of the same course. A multitude of new, more descriptive words were coined or redefined (Monowaves, Polywaves, Complexity, Compaction etc.) to assist in more accurate communication on the subject. The Progress Label and Logic Rule sections, the concept of a NEW type of Triangle with its own, unique formational rules and the significant advances in the quantification of Triangular activity are all important additions to the Elliott Wave Theory and are what comprise the **Neely Method**. Furthermore, all the rules in Chapter 9 also belong in the above list.

While the Rules which follow are not quite as important as the ones discussed up to this point, they are very helpful in increasing your pattern interpretation confidence level. Quite often, the utilization of these more subtle Rules will help you decide which of a list of possibilities is the most probable.

Trendline Touchpoints

This rule will help you quickly differentiate between <u>Impulsive</u> and <u>Complex Corrective</u> activity. The Rule states that, **"in a pattern consisting of five segments** (which has six possible 'same degree' touch points), **only a total of four can simultaneously touch the two opposing trendlines** (see Figure 9-1). This rule is applicable to Impulsive (Trending and Terminal) and Triangular patterns since both are composed of five segments. NOTE: This Rule was partially covered under the Triangle heading (Central Considerations, page 5-25).

Figure 9-1

Ssome leeway to the **Touch Points** Rule may occasionally be required during the formation of non-limiting Triangles.

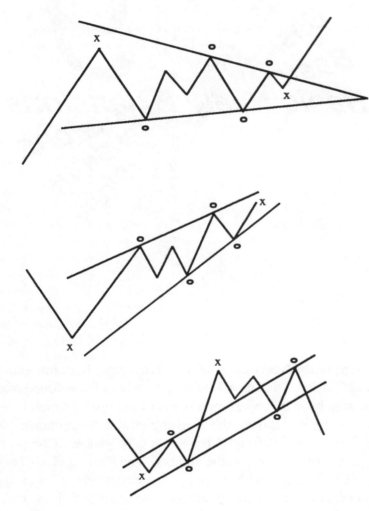

The dots are the four points which touch opposing trendlines, the x's do not.

In this diagram, of the six points which make-up an Impulse wave, only a _total_ of four (the dots) touch the upper and lower trendlines. The x's indicate the two points which do not touch a trendline.

Again, only four (a different four) points of the same degree touch the parallel trendlines (marked with dots). The two that do not touch either one of the trendlines are marked with an "x".

How can this Rule be used to best advantage? Over long periods of time, when a pattern is developing on a Complex Poly-, Multi- or Macrowave level, the existence of <u>more</u> than four touchpoints on the parallel trendlines is a strong indication of Corrective, <u>not</u> Impulsive, activity. Usually, the Corrective pattern would be a Double or Triple Zigzag or Combination. Even though it is not typical, more than four touch points <u>could</u> occur along the two trendlines (which may be converging, diverging, or parallel) in an Impulse pattern , but not all of those touch points will be of the same Degree. Remember, the Rule only applies to waves of the *same degree*.

The rule also applies to other **Standard** corrective patterns. All Standard Corrections have four possible touch points. Only three of the four points should touch parallel trendlines. If four points of the "same degree" touch parallel trendlines in any **Standard** correction (except Triangles), it is a reliable indication the Correction will be part of a **_Complex_, Non-Standard** formation. This rule is most useful during Zigzags (see Figure 9-2).

Figure 9-2

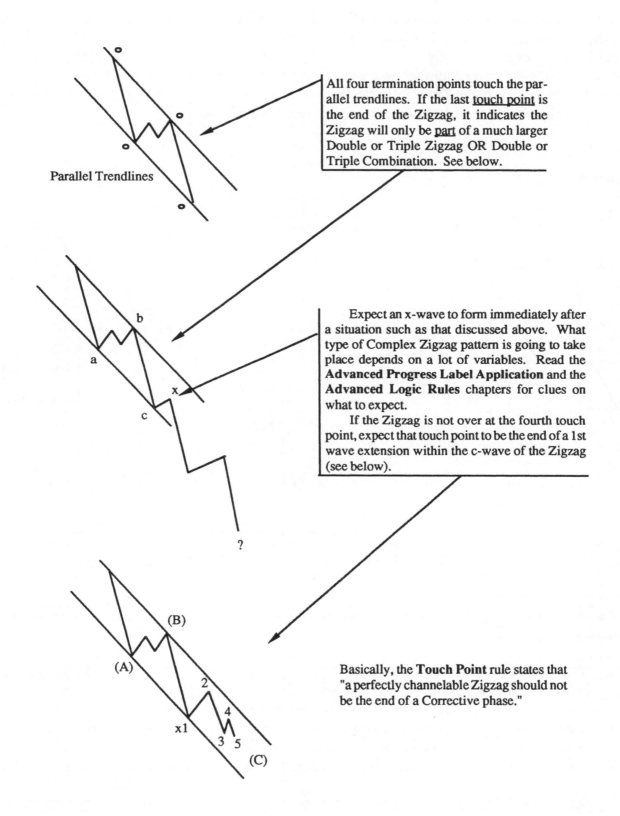

All four termination points touch the parallel trendlines. If the last <u>touch point</u> is the end of the Zigzag, it indicates the Zigzag will only be <u>part</u> of a much larger Double or Triple Zigzag OR Double or Triple Combination. See below.

Parallel Trendlines

Expect an x-wave to form immediately after a situation such as that discussed above. What type of Complex Zigzag pattern is going to take place depends on a lot of variables. Read the **Advanced Progress Label Application** and the **Advanced Logic Rules** chapters for clues on what to expect.

If the Zigzag is not over at the fourth touch point, expect that touch point to be the end of a 1st wave extension within the c-wave of the Zigzag (see below).

Basically, the **Touch Point** rule states that "a perfectly channelable Zigzag should not be the end of a Corrective phase."

Time Rule

Time plays an important function in producing the correct look for a wave pattern. Elliott discovered that the two unextended waves in an Impulse pattern often conform to the same time duration. Also, he noticed that waves a & c in a Zigzag tended toward equality in time. After years of observation, I have discovered there are many additional ways time can be used to enhance the analytical process.

In its simplest form, the *Time Rule* can be stated as such: <u>No three adjacent waves (of the same Degree) can be simultaneously equivalent in time</u> (see Figure 9-3).

Figure 9-3

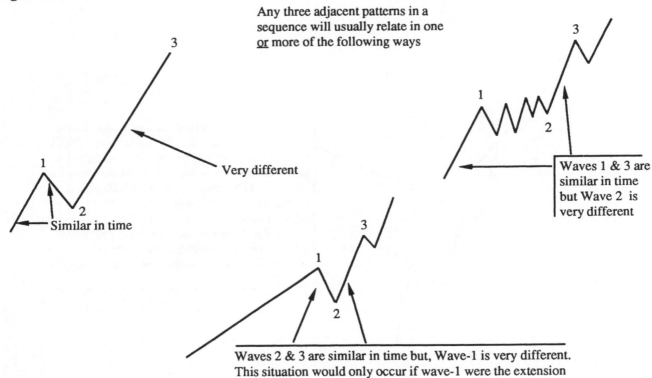

Any three adjacent patterns in a sequence will usually relate in one or more of the following ways

Very different

Similar in time

Waves 1 & 3 are similar in time but Wave 2 is very different

Waves 2 & 3 are similar in time but, Wave-1 is very different. This situation would only occur if wave-1 were the extension

Observation: <u>Any three</u> adjacent patterns in a sequence will usually relate by time in one of the following ways:

A. If the first two segments of a pattern are equal in time, the third will consume much less or much more time than either one of the other two taken separately. Often, the third segment will consume the accumulated time of the two smaller patterns.
B. If the second segment is much longer in time than the first, the third segment will equal the first or relate to it by 61.8% or 161.8%.
C. If none of the waves are of the same time duration, they may all relate by a Fibonacci ratio.

See Figure 9-4 for an example of how the above Rules might effect the development of a Flat.

Figure 9-4

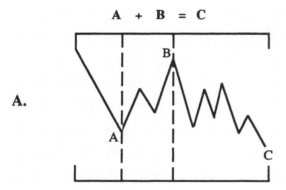

A.

A + B = C

If the last wave of a Correction (wave-c) elongates <u>or</u> is a Terminal pattern, frequently wave-a & b will be similar in time and the last section (wave-c) will be much longer. In this diagram, the combined time of the smaller waves equals the larger wave.

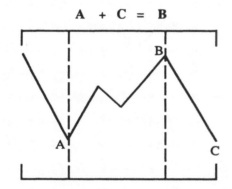

B.

A + C = B

The most typical set-up for a Flat. Here waves a & c are equal in time while wave-b is much longer.

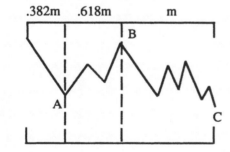

C.

.382m   .618m   m

Frequently, if no two waves are equal in time, all waves will relate by 61.8% or 38.2%.

Application of the Time Rule

If two adjacent patterns take equal amounts of time followed by a third segment which also takes the same amount of time, you can confidently assume (on any scale above simple polywave development) that the third segment is not complete or that the three segments being viewed are not all of the same Degree (see Figure 9-5, next page).

Figure 9-5

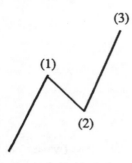

(3)
(1)
(2)

Incorrect

All these moves take the same amount of time. This virtually guarantees the third advance is only <u>part</u> of a much bigger advance.

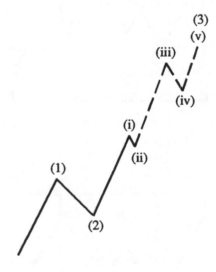

(3)
(v)
(iii)
(i)
(ii)
(iv)
(1)
(2)

Good Possibility

This interpretation eliminates the problem of three same degree waves occurring one after the next with all of them taking the same amount of time. The third segment (wave-1) is a lower degree pattern, so now only two adjacent waves are equal in time, which is how it should be.

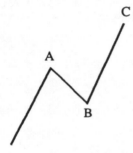

C
A
B

Also Incorrect

For the same reasons as those mentioned in the top diagram, this action cannot be the the a-b-c of a Zigzag since all waves (of the same degree) are of the same time duration.

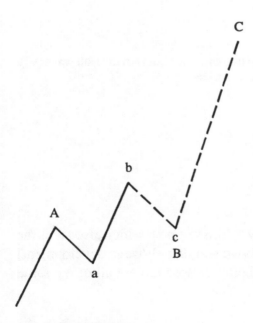

C
b
A
c
B
a

C
5
3
1
4
A
2
B

Also Possible

Correctly interpreted, the market could be forming a Running Correction (left) or may be subdividing for wave-c (right).

Independent Rule

The Independent Rule applies to **all** <u>other</u> Rules of the Theory that already have been and will be presented in this book. All Elliott Rules and guidelines (and the Neely Extensions) should be applied as separate entities to *every* wave pattern deciphered (as long as there is enough detailed price action to apply them). In other words, do not get in the habit of thinking certain characteristics always go hand-in-hand. Example: Most Elliotticians believe Extensions and Subdivisions are synonymous. Usually, the two concepts are present in the same pattern at the same time, but not always. Each Rule, that of Extensions and Subdivisions, should be considered *independently*. If they are simultaneously present in the same move, fine. If they are not, the validity of the proposed count is not in jeopardy.

Simultaneous Occurrence

This Rule was needed to solidify all the other rules presented in this book. To define *Simultaneous Occurrence*, all Elliott rules applicable to a particular situation have to be "simultaneously" present in order to create a reliable interpretation. To apply this rule, begin with the rules and techniques described in **Preliminary Analysis**, continue to through the **Intermediary Observations, Central Considerations**, etc. If any Rule normally associated with the pattern on your chart does not manifest, it is likely some other formation is developing. The count which meets *all* or most of the applicable parameters should be considered the best interpretation. Do not forget, even post-pattern behavior is part of the investigative process needed to find the foremost choice.

Exception Rule

ASPECT 1

Occasionally, at important market turning points or under "unusual" conditions, one of the Rules which is usually critical to proper pattern construction may not apply or be followed. An important Rule would be any found from Chapter 3 up to (but not including) the *"Breaking Point"* section under Impulsions or Corrections in Chapter 5. Situations which substantiate "unusual" conditions are:

A. The end of a Multiwave **or** larger pattern.
B. The 5th wave or C-wave of a Terminal (diagonal triangle) pattern.
C. A move which <u>is</u> or <u>ends</u> with a Contracting or Expanding Triangle.

Not mentioned until now, Triangles and Terminal patterns are the exception to virtually all Rules, situations and conditions (i.e. time, price, channeling, Progress Labels, Fibonacci relationships, etc.). Whenever an important Rule gets broken, it is likely that a Triangle or a Terminal pattern had something to do with it.

If a major Rule is not represented in the market action being observed, the interpretation might still be acceptable as long as there is substantiation for the failure of the Rule. Substantiation might come from one of the conditions above (A, B, or C). The *extreme* importance of the Rules discussed up to the "Breaking Point" headings in Chapter 5 precludes the possibility that two important Rules be broken simultaneously. If more than one *important* Rule is "broken" the count should be eliminated.

Another aspect of the Exception Rule states:

"The failure of a Rule occurs for a specific reason; this reason usually creates another Rule."

Below are two examples of how this happens:

1. The 2-4 trendline of an *Impulse* wave should **never** be broken by any part of wave-3 and hardly ever by any part of wave-5. If the 2-4 trendline **is** broken by part of wave-5, this activates a Rule which says "the market must be forming a Terminal Impulse pattern." A Terminal pattern is a legitimate reason for the break of a true 2-4 trendline.

2. When the "thrust" out of a Triangle does not stop advancing or declining when the time zone occupied by the apex point of the Triangle is reached, then the market is forming a Terminal pattern or the previous Triangle was of the Non-Limiting type.

Necessity of Maintaining "Structure" Integrity

Structure is the critical consideration in putting together a "Progress Label" scenario. Only through strict attention to detail, and adherence to past wave Structure, can you expect to accurately interpret a market with any consistency. With larger patterns, analysis can become more difficult, unless proper care has been taken to dissect each subsegment and classify it as Corrective or Impulsive (:3 or :5). Even if it is currently not clear which pattern a market is forming, adherence to past, provable price "structure" is mandatory. Eventually, the Elliott pattern which has been under formation all along will become obvious. This usually occurs as the pattern approaches a conclusion.

Locking in Structure

After having identified a reliable Elliott pattern, and then proceeding through the process of Compaction, it is very important the simplified Structure of the move (being just a :3 or :5) is not changed at some later date. This would be a drastic mistake. Generally, you should never change the Structure of a properly Compacted pattern. Often, you might be tempted to do so in order to create an interpretation you would like to believe is unfolding, but as a Rule, never do it.

Contrary to what appears a common belief among Elliotticians and especially the general public, the Structure of a formation cannot be legitimately changed at the whim of the analyst to suit current opinion, market fundamentals, technical indicators, etc. Wave patterns develop based on the net result of all the buy and sell orders reaching the floor of an exchange. Those buy and sell orders are sent to the floor by traders around the country who are analyzing Fundamentals, Technicals, Astrological aspects, Volume, Open Interest, Bullish Sentiment, etc. Since a market cannot move unless there is someone bidding to buy or someone asking to sell (all other events notwithstanding), when the two come together it creates an "official" supply/demand price level. As a result, there is no disputing that **price action** is the net result of all external market influences and therefore is the best indicator of future market performance.

The level at which a market trades does not occur by chance. The trade takes place because it satisfies both sides of the equation, the buyer and the seller. Waves are the by-product of these transactions connected end-to-end showing a continuous chain of events. Transforming the structure of a pattern to fit what you feel is going on **or** because you are confused about the market's position is a very poor reason to alter the previously established "Structure" of a count. Changing previous wave "structure" to make the count finish the way you "believe" it should **or** to get the pattern to complete sooner will almost always result in an incorrect interpretation.

Under the Wave Principle, extreme confidence in a change of price direction is only possible as the market terminates the last wave segment. The only way to stay on track between one important termination point and the next is to do the following:

a. Mark the termination point of **_every single_** monowave on your chart with a dot (this step can be eliminated as you become more experienced).

b. Study the **Retracement Rules** and then using the **Pre-Constructive Rules of Logic,** carefully place the appropriate Structure Label(s) at each turning point.

c. Look for group Structure combinations among adjacent monowaves which adhere to **acceptable** (Standard or Non-Standard) Elliott Series'.

d. Once a Series is identified, move to the appropriate section (Impulsions or Corrections) of Chapter 5.

e. Proceed through all the Essential Rules which apply to that general formation category.

f. Place **_Progress Labels_** on the wave pattern and continue through the "Conditional Rules" section.

g. Check all the applicable Rules and guidelines (under Advanced Progress Label Application & the Advanced Rules of Logic) to confirm all minor details and subtleties of wave behavior are present (Alternation, Fibonacci relationships, etc.). If it passes those tests, go to the next step.

h. Compact the pattern to its base structure, ":3" or ":5" (see Compaction, Chapter 7).

i. If you are working strictly with monowaves, raise the Complexity Level of the Compacted pattern to Level-1 (underline the :5 or :3). If you have developed beyond the monowave stage, raise the Complexity of the Structure Label to the appropriate level based on the techniques discussed in the "Complexity" section of Chapter 7.

j. After Compaction and Complexity designation, the process begins over again. Return to step-b to confirm if the newly compacted pattern effects or changes any potential Structure Labels for the monowaves (or higher) surrounding the compacted pattern.* Then move on to step-c, etc. Continue returning to step-b until you have numerous compacted patterns each marked with a single Structure Label. These single Structure Labels will eventually form identical Series just as the monowave groups once did. The only difference, each Structure Series will be representing a polywave (or higher) instead of a monowave. All the Rules discussed up until now, and throughout the book, will work for these more complex patterns in the same fashion as they did for the simple formations (except where noted). Work with the compacted Series' just as if they were monowave Structure Labels. Look for Series combinations, apply the "Rule of Similarity...", etc. The only additional step necessary in working with the more complex patterns is to check for and follow the Rules under the "**Construction of Complex Polywaves, Multiwaves etc.**"

The compacted Structure Labels should be placed on your charts which cover a period of time longer than the original monowave chart. On your longer-term chart, only mark the single Structure of each compacted polywave, do not bother with each individual monowave's Structure.

The Wave Theory is a relative phenomenon. It does not matter how complex or simple a pattern is; what matters is whether the pattern is Corrective or Impulsive in nature. That will tell you how to handle, combine, interpret and trade market action.

*This is describing the process of Reassessment which is covered in greater detail at the top of page 7-3 in Chapter 7.

Until the market reaches the end of a Series, guessing on the proper Progress Labels which apply is just that, a guess. You do not have to know the Progress Labels of a wave pattern to trade it profitably! Trade with the trend which, based on the Wave Theory, is in the direction of the ":'5's" (Impulsions). Until you have at least three or five Structure Labels to work with, it is impossible to complete an Elliott pattern and thus impossible to be completely certain of the name of the pattern actually unfolding.

In conclusion, you should not have any preconceived notions of a markets' future behavior. The only objective way to analyze a market is to stay strictly with previously determined wave Structure and only combine the patterns when an identifiable Series is complete and when the pattern meets all important criteria. Do not waste time trying to continually guess, and trade on, what you think is the current position of the market.

Flexibility of "Progress" Labels (expansion of a pattern)

It should now be evident why the earlier sections of the book dealt with Structure, instead of Progress labels. It may not be clear, though, why Structure should still be the main focus of your attention.

As mentioned in the last section, Progress Labels are the final touch on a wave pattern, not the first step. What was not mentioned; "after Progress labels have served their purpose in solidifying a pattern, they are no longer helpful in clarifying the larger picture."

Once a pattern is complete, **Compaction** (see Chapter 7) is necessary to reduce the larger formation to its base Structure. This base Structure is needed to integrate the pattern with other Compacted segments into a larger formation. Knowing the Progress labels of each individual section of the previously Compacted pattern has no further value. You must use the base Structure of the compacted pattern on a longer-term chart to decide how the Compacted segment becomes part of a larger Series.

Advanced Logic Rules

One of the last, critical considerations of an interpretation's validity involves the incorporation of *Logic* into the market action. This is an area of the Wave Theory which is completely overlooked by the majority. The *Logic* rules are the product of exhaustive market research and observation by the author, from an Elliott Wave standpoint, for over a half-decade. The essence of the **Logic Rules:** **Post-pattern market action *must* adhere to specific types of behavior which is predicated upon the pattern just completed.**

For example: after completing a Terminal Impulse pattern, the reaction should be quite violent and should retrace *all* of the Terminal pattern within less than 50% of the time consumed by the Terminal Impulse. Lack of obedience to this Rule indicates a significant flaw with the current interpretation. If you thought you had uncovered a Terminal pattern in a market, but post-Terminal behavior did not react as described, a Terminal Impulse pattern <u>never</u> existed in the configuration predicted.

From a slightly broader perspective, the **Logic** Rules pertain to the integration of <u>separate</u> Elliott patterns in a reasonable, coherent, and consistent manner. The majority of rules which make up the "logic complex" revolve around the inherent strength and weakness implications of each Corrective pattern and it's variations. The table below covers many situations in which pattern implications are crucial to the correct interpretation of a move.

Pattern Implications

All Elliott patterns imply and transfer specific amounts of **"Power"** to future market action. In addition, many require particular post-pattern behavior including the minimum price length the next pattern should achieve and the minimum time it should consume.

The table below lists each Corrective pattern, along with its **Power** rating. Power ratings indicate whether strength or weakness is inferred by the pattern in question. Using a numerical scale from -3 to +3, the lowest rating belongs to market action which is *very* detrimental to upside price potential. The

+3 ranking is for patterns which are *very* beneficial to the continuation of a market advance. To decide on the subtle "Power" differences between two corrections given the same rating, here is what you do. When the rating is positive, the Correction listed farthest away from neutral is the stronger of the two. If the rating is negative, the pattern farthest away from neutral will be the weaker correction.

Retracement Based on Power Ratings

Having an idea of the "power" of a correction is *extremely* important in the **Logical** integration process. It provides you with an accurate idea of whether a Correction will, should, might, should not or will not get completely retraced by the next move. Some corrections are so weak (or strong) they **cannot** terminate a move. Any formation which concludes upward and has a +1 to +3 rating should not be fully retraced by the next pattern (of the same Degree). **Caveat:** *It has been the author's experience that the legs of Triangles and Terminals do not transfer implications to the next leg (there is always one exception to every formational Rule). This indicates if a Double Zigzag is an entire segment of a Triangle or Terminal it might be retraced completely by the next leg "of the same Degree." In addition, the next move may be simpler than the Double Zigzag. As a result of this phenomenon, an important Rule can be derived. If a pattern with a (+1, +2, +3) or (-1, -2, -3) power rating is completely retraced by a pattern with a Power Rating closer to zero but which is still of the same Degree, the market is signaling a Triangle or Terminal pattern if forming (all variations; Expanding or Contracting, 1st, 3rd or 5th Extended).*

POWER RATINGS —	When the pattern completes upward:		When the pattern completes downward:	
1. Triple Zigzag	+3		-3	
2. Triple Combination	+3		-3	**Non-Standard**
3. Triple Flat	+3		-3	**Formations**
4. Double Zigzag	+2		-2	(Category 1)
5. Double Combination	+2		-2	
6. Double Flat	+2		-2	
7. Elongated Zigzag	+1	(in a Triangle = 0)	-1	
8. Elongated Flat	+1	(in a Triangle = 0)	-1	
9. Zigzag	0		0	
10. B-Failure	0		0	**Standard**
11. Common	0		0	**Formations**
12. C-Failure	-1	(in a Triangle = 0)	+1	
13. Irregular	-1	(in a Triangle = 0)	+1	
14. Irregular Failure	-2	(in a Triangle = 0)	+2	
15. Double Three	-2		+2	
16. Triple Three	-2		+2	
17. Running Correction	-3		+3	**Non-Standard**
18. Double Three Running Correction	-3		+3	**Formations** (Category 2)
19. Triple Three Running Correction	-3		+3	

Realize, a pattern's Power rating can only be properly applied if you fully *compact* all patterns under formation. By compacting a move, you arrive at the largest, allowable formation possible. It is the compacted formation which the Power ratings will reliably affect. For example, let us assume a Double Zigzag recently completed. If the Double Zigzag occurs anywhere from the first to the second to last formation of a larger pattern, the Power rating should be reliable. At any one of those points, the Compaction process would only reveal a completed Double Zigzag as a separate pattern. But, if the Double Zigzag occurred as the last part of a more Complex formation (such as a Terminal Impulse), the Power rating of the Double Zigzag would not be reliable since the largest pattern completed is a Terminal Impulse, not a Double Zigzag. In other words, when a pattern is the last part of a larger formation, the smaller pattern is unimportant. Only take into account the Power rating of the largest pattern.

The **higher** the absolute value of the Power Rating, the **less** likely the move will be completely retraced. Below are the anticipated retracement ratios implied by each possible **Power** rating:

1. **A ("0") rating does not imply any specific retracement value, virtually any retracement level is possible from just above 0% to beyond 100% of the previous pattern.**
2. **A (+1,-1) rating would allow for no more than about 90% retracement by the next** <u>completed</u> **pattern** *of the same Degree.*
3. **A (+2,-2) would indicate no more than about 80% retracement.**
4. **A (+3,-3) rated pattern should be retraced the least, around 60-70%.**

Listed below are <u>all</u> Elliott Corrective patterns (except Triangles, which will be covered shortly). When the last segment of one of the Corrections listed on the previous page *terminates* upward on a chart, the pattern designated #19 would be considered the weakest, while pattern #1 would be the strongest. If the last segment of one of the Corrections *terminated* downward, the reverse is true.

Under the next heading, "special situation" notes are listed for each Elliott pattern. These guidelines will direct you in "solidifying" price action into *absolutely* reliable Elliott Wave patterns.

All Corrections (except Triangles)

The largest price moves in a market happen <u>after</u> Corrective action. This makes understanding the *implications* of a Correction far more important than those of an Impulse pattern.

Triple Zigzag

This is the most powerful corrective pattern that can occur. If its movement is downward, it implies the market is currently very weak. If its movement is upward, the market is currently strong. A Triple Zigzag will hardly ever be seen but, when it does occur, it is usually the longest segment of a Terminal or Triangle. When part of a Terminal pattern, it should definitely be the Extended segment. Based on market position, if a Terminal is not possible, the only choice is that the Triple Zigzag is the largest segment of a Triangle. If part of a Flat or Contracting Triangle, a Triple Zigzag can **never** be completely retraced by the pattern which immediately follows it *of the same Degree.*

Triple Combination

This pattern can be a combination of Zigzags, Flats (frequently elongated) and Triangles. Virtually always, a pattern which falls into this category terminates with a Triangle. Either one of the two x-waves can also be Triangles, but are not required to be. The first two Corrections (the one which starts the move and the one right after the first x-wave) should not be Triangles. When a Triple Combination occurs *in* a Triangle, it will be the largest wave of the Triangle in price and probably time also. It can occur in only one other pattern, a Terminal Impulse. When a Triple Combination occurs as the extended 5th wave of a Terminal, it is the only time the market *should* completely retrace a Triple Combination, even though the retracement will be of a *larger Degree*.

Triple Flat

The pattern which follows a **Triple Flat** should not completely retrace it <u>unless</u> the Triple Flat is the last leg of a 5th Extension Terminal (i.e. wave-5). Even then, the retracement pattern would be *of one larger Degree*. They should not be completely retraced by the next move *of the same Degree*. These patterns are <u>very rare</u> and it is unlikely you will ever see one. It is only included here for thoroughness.

Double Zigzag

This pattern should not be completely retraced by the move to follow unless, once again, it is the last segment of a 5th Extension Terminal.

Double Combination

Just like a Triple Combination, a Double Combination will almost always end with a Triangle, OR during simpler (level-1 Complexity pattern) patterns, a c-wave failure. This pattern *can* be completely retraced if it <u>completes</u> a larger formation (such as a Terminal Impulse - *1st, 3rd or 5th wave Extended*), but usually will not be. If the Corrective phase, after wave-x, finishes with a "severe" C-Failure or a Running Triangle, then it is possible the entire correction *will* be completely retraced.

Double Flat

These are not very common patterns, but they do occur. The pattern to follow will probably not completely retrace the Double Flat unless the Double Flat finishes with a "severe" C-wave Failure *or* concludes a larger formation (such as a Terminal Impulse, any variety). The move following one of these patterns should be a little faster than the move following a Double Zigzag.

Elongated Zigzag

Named by the author, an Elongated Zigzag pattern occurs *almost* exclusively in Triangles and Terminal Impulse patterns. It will usually be the entire leg of such a pattern. It could occur as any wave in a Contracting Triangle (except wave-e) or any wave in an Expanding Triangle (except wave-a). It should **never** be completely retraced by the wave *of the same degree* that immediately follows it.

Elongated Flat

This is one of the most interesting and distorted Elliott patterns. The fact that this pattern occurs almost *exclusively* in Triangles is a discovery of the author's. It occasionally can be found in Terminal Impulse patterns. When in Triangles, it is *almost* always <u>an entire leg</u> of the Triangle. If not the entire leg, then it will occur as a "one lower degree" segment of a complex correction which <u>is</u> the entire leg of a Triangle.

Zigzag

This ranks among the top three, most common patterns. To be considered simply a Zigzag, the c-wave should not be less than 61.8% of wave-a *nor* more than 161.8%.

Wave-c longest:

Of the three Zigzags listed here, this has the weakest implications if it is moving down and the strongest if it is moving up. It is not likely to be completely retraced *unless* the next wave (of the same degree) is more complex and time-consuming *or* the Zigzag <u>completes</u> a Corrective phase.

Wave-c equal:

This pattern may or may not be completely retraced. It is one of the most frequently occurring corrective patterns. Virtually anything could happen after this pattern.

Wave-c shorter:

This pattern is almost sure to be completely retraced unless it is followed by an x-wave. It does not really give any clues of what to expect from the next pattern.

B-Failure

A B-Failure is the most *neutral* pattern existing under the Wave Theory. Nothing in particular can be expected following this pattern since virtually anything is possible.

Common

Again, this is a pretty neutral pattern. It can occur virtually anywhere a Correction is allowed. It does not imply anything about future market action, but is <u>definitely</u> more powerful than a Zigzag.

Irregular

Despite what many readers may believe, this pattern is actually quite <u>abnormal</u> and infrequent. It creates a state of "self contradiction." Why? When the b-wave of a pattern exceeds the beginning of wave-a, it demonstrates the power of the trend of one higher degree. When the c-wave turns around and exceeds the end of wave-b, it nullifies the power exhibited by the b-wave creating an illogical condition. This behavior is acceptable in Terminals and Triangles, not in Flats (which is what an Irregular is).

C-Failure

This pattern **must** be completely retraced by the move to follow of the same degree. If the C-Failure completes the Corrective phase, the Impulse wave to follow should be larger than the previous Impulse wave which is traveling in the same direction and is of the same degree. If the C-Failure is not the end of the larger Correction, the rally to follow should be the x-wave of a Double Three Running Correction, *or less likely*, the b-wave of a Running Correction.

Irregular Failure

This pattern **must** be completely retraced. From this point onward, the patterns listed begin to indicate "Power" in the opposite direction that they conclude. Usually this pattern will occur as wave-2 (of a Trending pattern) which will be followed by an Extended 3rd wave. The Impulse wave after an Irregular Failure should be bigger than the previous Impulse wave usually by a factor of 1.618.

Double Three

As a general rule, the longer this Correction takes, the more powerful the move to follow should be. Double Threes are more common as b-waves than 2nd waves. WHY? If a second wave is a Complex (Non-Standard) Correction, the 3rd wave is most likely going to be the Extended wave. The inherent **Power** of that Extension will affect the formation of wave-2. Generally, this power "stretches" the Double Three in the direction of the 3rd-wave extension to follow, making it a Double Three Running Correction, not just a Double Three.

The price action for a c-wave would not normally be as large as a 3rd wave Extension. The c-wave's inherent lack of power usually avoids the stretching effect of the Double Three and therefore prevents the Double Three from turning into a Running pattern.

Double Three Running Corrections imply a great deal of strength for the trend of one larger degree; therefore, you are only likely to see one in the 4th wave position **if** it is followed by a 5th wave extension. The Impulse move after a Double Three should be at least 161.8% of the previous Impulse wave (i.e., if an Impulse wave occurred immediately before the Double Three).

Triple Three

Triple Threes' are a virtually non-existent phenomenon. If you witness one, the Impulse move to follow should be *at least* 261.8% of the previous Impulse wave (if there was one). Almost without exception, Double and Triple Threes drift slightly in the opposite direction of their last wave.

Running Correction

The movement to follow a Running Correction *must* be the extended wave of an Impulse pattern *or* an Elongated c-wave of a Flat or Zigzag. After a Running Correction is complete, the market should commence with the next Impulse wave; it should not continue to form a more complicated Double *or* Triple Three. The Impulse wave after a Running Correction should be *more* than 161.8% of the previous Impulse wave; frequently it is 261.8% or more .

Double Three Running Correction

This is a pattern which seems to cause more confusion and controversy than any other. The funniest part of this story is that a Double Three Running is virtually never interpreted correctly. During their formation, some Elliotticians count this pattern as a series of 1's and 2's. Some will interpret it as a Terminal Impulse pattern. Some will call it a complex Correction such as a Double Zigzag. The most important aspect of correctly deciding between all of these unrelated formations is wave Structure (:3 or :5). Strict adherence to Structure is an area many analysts tend to overlook. They casually put together an interpretation based on general shape and appearances, not on concrete structural evidence. To foil this less diligent group, there are patterns which can **Emulate** the appearance of other patterns causing drastic errors in judgment. Fortunately, close attention to detail will usually reveal the correct interpretation. Channeling is another key to avoiding the misinterpretation of this pattern (see page 12-2).

The move after a Double Three Running correction *must* be Impulsive. This Impulsion should exceed 161.8% of the previous Impulse wave. More than 261.8% is common. Basically, the only place this pattern will occur is as wave-2; furthermore, the formation almost always concludes with a Triangle.

Triple Three Running Correction

This formation is incredibly rare and for good reason. The power built up by a Double Three Running Correction is so great it is unrealistic to expect the market to continue to postpone the day of reckoning. Remember, a market never develops beyond the Triple Three Corrective phase; as a result, the market **cannot** "continue creating x-waves indefinitely," as one of my clients once asked. If you were to see a Triple Three Running Correction, expect the Impulse move following to be a minimum length of 261.8% of the previous Impulse wave.

Triangles

The unusual nature of Triangles necessitates they fall into a category all their own. [Non-Limiting Triangles are a category developed by the author.]

Contracting

A *"thrust"* (powerful, violent price movement) always follows a Contracting Triangle which, depending on the variation, may be large or small, long or short-lived. The thrust out of a **Contracting** Triangle should always exceed the highest or lowest price obtained during the formation of the Triangle unless it drifts against the direction of wave-b. Depending on subtle formational differences, the **Contracting** Triangle may be a 4th or b-wave (see Limiting) **or** an x-wave **or** the last phase of a Complex Correction (see Non-Limiting, page 10-8).

Triangles create (or identify), important support or resistance zones in the market. Once a Triangle has finished, the price level at which the apex point of the Triangle occurred will provide significant *resistance* to any advance or *support* to any decline. When the market breaks through one of these "barriers" it will generally be violent. The important support or resistance created (or recognized) by the Triangle is almost always a reliable point to look for Fibonacci relationships, especially if the Triangle is the b-wave of a Zigzag.

The influence of the Triangle's apex point (horizontally, _price-wise_), will usually hold for only two or three violations. If the market has moved up or down through the price level occupied by the apex point more than three times, the market is indicating that price level has lost its significance.

I. *Limiting*

Unlike most Corrections, whose Implications depend a great deal on what the previous Impulse or Corrective pattern did, **Limiting** Triangles are "self-descriptive." Post-triangular action in a **Limiting** Triangle is dictated by the width of the widest leg of the Triangle and the placement of the apex point created by the converging trendlines which form the Triangle.* These two factors provide a reliable price/time window which readily verify (or nullify) a **Limiting** Triangle's validity. If the Triangle being analyzed is truly **Limiting**, the market action _after_ the thrust should return to the breakout point and will usually exceed it.

Depending on the slope of the trendlines which encase the Triangle, the thrust could be as much as 261.8% of the widest leg of the Triangle (added to the end of wave-e) or as little as 75%. The thrust should terminate almost exactly into the _time period_ (not price level) where the converging trendlines crossed, creating the apex point. If all indications are that the Triangle is a b-wave or a 4th wave, but the post-thrust action has not returned to the breakout point or beyond and the market action has exceeded the original thrust's highest or lowest price level, there must be a Terminal pattern forming (applies to both situations) **or** an x-wave unfolding (only applies if the Triangle is a b-wave).

If working with a **Limiting** B-wave Triangle and the "post-thrust" action does not exceed the breakout point shortly after the apex time zone is reached, an x-wave is probably forming after the thrust.

a. *Horizontal*

The thrust out of the Triangle should be approximately _equal_ to the widest wave of the Triangle, plus or minus 25%. This Triangle variation _implies_ normal market conditions.

b. *Irregular*

Since the b-wave in one of these patterns exceeds the end of wave-a, it implies the pattern is more powerful than a Horizontal (in _either_ price direction). The thrust out of this variation can be as much as 161.8% of the widest leg of the Triangle.

c. *Running*

This is the most powerful Triangular pattern. The thrust can be as much as 261.8% of the widest leg. Market conditions are unusually strong (probably climactic) _if_ the Triangle is drifting upward or weak _if_ the Triangle is drifting downward.

II. *Non-Limiting*

A Non-Limiting Triangle imposes _no specific_ post-triangular price or time limitations on market action. The only way to decide what the pattern "implies" is to study the pre-Corrective action. If a Correction _completes_ with a Triangle it means the Correction must be complex in construction (any

Double or Triple pattern). The Impulse segment *of the same Degree* after such a pattern (if one occurs), should be at least the same length as the <u>previous</u> Impulse wave.

The "thrust" out of a Non-Limiting Triangle can, and usually will, be *much* larger than the width of the Triangle. Despite this fact, during the early stages of the thrust, the market will usually try to emulate "normal" post-triangular behavior by initially breaking out approximately the width of the Triangle and then reacting. Where the two variations differ is in what happens next. As mentioned in the **Limiting** section, the apex point of the Triangle (price-wise) should be touched or broken after the thrust. In a Non-Limiting Triangle, the reaction will not return to the apex point (price-wise) and will eventually exceed the end of the initial thrust. Another major difference between Limiting and Non-Limiting patterns is the Non-Limiting "thrust" **will not** terminate during the time period occupied by the apex point.

Expanding Triangles

The interesting characteristic of **Expanding** patterns is that they basically *imply* the opposite attributes of a **Contracting** formation. The "thrust" (actually a misnomer), should be <u>less</u> than the widest leg of the Triangle. If the pattern takes place in the b-wave position, a C-wave Failure is inevitable. Even if an Expanding Triangle completes a larger pattern, it is unlikely the next move will completely retrace the e-wave of the Triangle. If it does, the retreacement should take more time than the e-wave.

I. Limiting

There is very little difference between the varieties of **Expanding** Triangles. The **Limiting** and **Non-Limiting** categories were kept to identify between the 4th-wave/b-wave variety and all other variations. The only *implication* that is reliable about this pattern is it should not be completely retraced.

II. Non-Limiting

This variation would define those **Expanding** Triangles that occur as x-waves, OR the first or last phase of a complex Corrective pattern. Its most common position is the first phase of a Complex Correction. In that position it will not be completely retraced by the next move *of the same Degree*.. If it is the last phase of a Complex Correction (which is not likely, then it has to be completely retraced.

Impulsions

The *implied* future behavior of a market is not quite as useful when working with **Impulsions**. About the only predictable action after an Impulse pattern regards how much the next move, *of the same Degree*, should retrace the Impulsion. It would be very difficult to predict what type of Correction is going to take place after an Impulsion. The listing to follow covers *post* Impulsive behavior almost strictly from the vantage of retracement expectations.

* A concept first presented in the book *Elliott Wave Principle, Key to Stock Market Profits*, by Frost & Prechter: New Classics Library, Gainesville, GA

Trending

After a Trending Impulse wave concludes, it should never be completely retraced unless it is wave-5 or wave-c of a larger pattern. If the Trending Impulse is wave-a, wave-1 or wave-3 of a larger pattern, the action to follow should not retrace more than 61.8% of the Impulse wave.

If retracement of the previous Impulse pattern exceeds 61.8% and the previous Impulse pattern is wave-1 of the next larger Degree, expect a complex (relative to wave-1) time consuming 2nd-wave correction to develop which contains a c-wave Failure. If the Impulse pattern finalized was wave-3 of the next larger pattern and wave-3 was retraced close to or slightly more than 61.8% by wave-4, expect a 5th wave Failure to occur. If the 4th-wave recovers from the severe retracement of wave-3 **and** it concludes at a 61.8% retracement of wave-3 (or less), a 5th wave Extension is a possibility if wave-4 is more complex and time consuming than wave-2 <u>and</u> wave-3 is not more than 261.8% of wave-1.

1st Wave Extension

In an Impulse pattern with a 1st Wave Extension, the action to follow should definitely retrace to the termination of wave-4 (within the recently completed Impulsion). If the Impulsion completed wave-(1) or wave-(5) of a larger Impulse wave, the market should fall into (or beyond) the price zone for wave-2 of the previous Impulsion.

3rd Wave Extension

If the 3rd wave extends within an Impulse wave, the reaction should at least return to the 4th wave zone (anywhere from the highest to lowest price of that zone) of the recent Impulse wave. If the entire Impulse pattern (with the 3rd wave Extension) completes wave-(5) of a larger pattern, the whole Impulse wave should be retraced more than 61.8%. If the entire Impulse completes wave-(1) or wave-(3) of a larger Impulse, the pattern (with the 3rd wave Extension) should be retraced <u>less</u> than 61.8%.

5th Wave Extension

The 5th Wave Extension is the only extended wave which *should* be consistently retraced more than 61.8% by the next wave *of the same Degree* no matter what part of a larger scale the 5th wave is. A 5th wave Extension should not be completely retraced by the next move of the same Degree unless the 5th wave concluded wave-c of a Correction. A 5th Extension is one of the few patterns which is not likely to be completely retraced by even the next move of *one <u>higher</u> Degree*.

Terminal

The market action after a Terminal Impulse must retrace the entire pattern in 50% or less of the time consumed by the Terminal pattern. Usually, all that is required is 25% of the time (give or take a few percentage points). A Terminal pattern always **completes** a larger formation and the high or low it creates should hold for approximately twice the time period (or more) covered by the Terminal. If the Terminal is the 5th wave of an Impulse pattern, usually the larger Impulse pattern will also be completely retraced.

One of the major reasons the Elliott Wave Theory allows you to predict future price action with such a high degree of accuracy (at times) is due to its strict limitation on where certain types of market behavior can and cannot occur. The information in this section should help to further clarify a market's current position and assist you in predicting its future activity. Specific pre- and post-market action requirements are listed in this chapter to aid you in correctly linking one pattern with the next. This area of the Theory should be dealt with only after *all* the elementary aspects of analysis are considered (i.e., Retracement, Structure Series, Channeling, Alternation, Fibonacci relationships, etc.).

The proper application of *Progress Labels* requires a vast understanding of market behavior. The minute you use a *Progress Label* on market action you instantly imply certain types of time, price, structure, complexity, volume, velocity and momentum characteristics which must be inherent in the market action labeled. These are not characteristics the beginning student could be expected to instinctively understand. Obviously, the use of Progress Labels is much more involved than just looking to see how much a move gets retraced.

To confidently place *Progress Labels* on price action you must understand the subtle implications and slight differentiations of each Elliott pattern and its variations. As the market approaches the middle of a Trending or Corrective Elliott pattern, it is especially important to deal with the basics of market "Structure." Why? Because it is at these times, more often than not, you will not be able to limit *Progress Label* possibilities enough to arrive at only one scenario (or possibly, any scenario at all).

Only after you understand the rudiments of deciphering monowave market action as either Trending or Corrective in nature, after you learn to Serialize the Structure of the monowaves and then learn how to roughly correlate the action to Standardized Elliott patterns, are you ready for this section on *Advanced Progress Label Application*. Progress labels are the *final* check on a proposed wave count. To properly apply Progress Labels, it is required you check through a sometimes lengthy list of essential and subtle criteria. This list will solidify the proposed combination of Trending and Corrective segments into larger, legitimate Elliott patterns.

Each *Progress Label's* **innate** characteristics allow the knowledgeable Elliottician to finalize current market structure and forecast future activity. They invoke form to a market's action and, due to their ordered occurrence (i.e., 1,2,3,4,5,a,b,c), provide a road map (albeit, sometimes rough) of future market behavior. *[If this subject is new to you, it should be obvious why this area of the Theory could not be discussed until now, after an understanding of each label's implications was presented to you.]*

The next section is intended to help you accomplish the task of transforming your chart's wave "Structure" into "Progress" labels as quickly and reliably as possible. A check list of attributes, which must be considered every time you place a particular *Progress Label* on real-time market action, is included in this next section. Each listing contains an array of characteristics which describe the implications of each Progress label on past and future market action. Try to understand the concepts behind these listings; they are all based on logical induction and deduction. Once the concepts are understood, it will be as if you had memorized this entire section because you will be able to logically explain the interactions of Elliott Wave price patterns.

Attempting what has never been attempted before, each *Progress Label* has been listed under each type of Elliott pattern. It is through this list that I hope to eliminate the belief in the investment world that Elliott Wave Theory allows *too* many possibilities and has too many exceptions to be viable. Actually, the reverse is true, but until now no source provided you with this *"process of elimination"* information.

Impulse Patterns

An Impulsion can transpire only in specific Progress Label positions. Never can waves-2, 4, b, d, e, or x be an Impulse pattern.

Trending

Wave-1 Extended

The move that follows Extended wave-1 (wave-2) *cannot* retrace much more than 38.2% of wave-1. Wave-2 will not likely be a Zigzag pattern. If you do see a Zigzag form after the 1st wave Extension, most likely it will be the completion of only wave-a of a larger Flat correction for wave-2. Wave-2 *cannot* be a Running correction. Wave-5 **must** be the shortest of the three thrust waves (1,3,5). Probabilities greatly favor that wave-2 will be more complex and time-consuming than wave-4 *and* it is likely that it will be the most complex and time consuming pattern in the entire series (1-5). If the pattern is above polywave development, the probabilities greatly favor that wave-1 will be the subdivided wave (the most complex of the three thrust segments). If wave-1 is not the subdivided pattern, then it **must** be wave-3.

Wave-1 Non-extended

Wave-2 can retrace as much as 99% of wave-1. If it does, and wave-1 is a polywave or higher, the 2-wave *will* subdivide into an a-b-c affair in which the c-wave will fail (the 2-wave may subdivide whether the 1st wave does or not).

If you are witnessing the *first* wave-1 develop after a prolonged advance *or* decline, there are no specific price level requirements for the termination of wave-1. If it is wave-1 of a larger 3rd or 5th wave, then wave-1 should approach (and preferably exceed) the termination point of the last Impulse wave of one larger degree. The 3rd wave *must* be longer than wave-1. If the pattern is above Complexity level 1, the third *or* fifth wave will be the subdivided segment, *not* wave-1. This is not to imply wave-1 will not subdivide, it just means it will not be the subdivided wave of the group. In other words, wave-3 or 5 will be *more* subdivided.

Wave-2

If wave-1 turns out to be (or is believed to be) the longest wave in the sequence, the second wave cannot retrace much more than 38.2% of the first wave. If the first wave is not the longest wave, the lowest point of wave-2 can retrace as much as 99% of wave-1. If wave-1 is a polywave or higher, the 2-wave **must** subdivide into a polywave or higher pattern. If the 2-wave subdivides and wave-a (in wave-2) retraces more than 61.8% of wave-1, the entire correction will inevitably turn out to be a *Double Failure* or a *C-Failure,* with the C-failure occurring at a point 61.8% or less of wave-1.

Wave-3 Extended

This is the most likely wave in a sequence to extend. If wave-(3) is also the subdivided wave, the strongest tendency is for the 3rd wave of wave-(3) to also extend. To get a 3rd wave extension immediately after wave-2 completes, without forming a smaller 2nd wave on the same chart, the 2nd wave should be of **Power** level 1 (+ or -) or higher. If the 3rd wave is a polywave, consecutively smaller 2nd-waves (before the explosive "3 of 3" move) will be similar in construction. Under those specific conditions, it is mandatory that the smaller 2-waves take less time, less price and retrace less (as a percentage of wave-1) than the larger 2-waves. Also, the larger 0-2 trendline should not be broken by the smaller 2-wave correction. Always be aware that wave-5 *can* Fail **if** the 3rd wave extends.

Wave-3 Non-extended

When wave-3 is **not** the longest of the three thrust waves, either wave-1 *or* wave-5 *will be* shorter in price than wave-3. If wave-3 is shorter than wave-1, then wave-1 will be the extended wave and wave-5 will be shorter than wave-3. If wave-5 extended, wave-1 would be shorter than wave-3. When the first wave extends, wave-3 should complete no further away than 61.8% above (or below if the market is trending down) from the end of wave-1.

Wave-4

If the 5th wave extends, wave-4 should be more complex, time-consuming, and possibly of one greater complexity level than wave-2. If wave-1 extends, wave-4 should be simpler in price, time and structure than wave-2. If the fifth wave extends, wave-4 should retrace a greater percentage of wave-3 than wave-2 does of wave-1. Frequently wave-4 will retrace as much as 50-61.8% of wave-3 when the 5th wave is the extension. If the fifth wave fails in the Impulse sequence, the fourth wave should be the most complex pattern and should retrace more than 38.2% of wave-3 (as much as 61.8% is allowable).

Wave-5 Extended

Wave-5 should be at least equal to the price distance from the beginning of wave-1 (point "0") through the end of wave-3 added to the end of wave-4. The **maximum** length of the 5th extension should not exceed 261.8% of the length from "0" to the end of wave-3 added to the end of wave-3. When considering waves-2 and 4, wave-4 should take the greatest amount of price/time and have the most complex structure. Wave-1 should progress at the sharpest angle with wave-3 following closely behind and wave-5 possessing the slowest rate of acceleration. A 5th wave extension cannot be completely retraced *unless* it is the c-wave of a correction or is the end of a larger 5th wave Extension.

Wave-5 Non-Extended

The 5th wave should be retraced close to 100% or more by the next corrective phase. If wave-1 extended in the sequence and the sequence concludes wave-(1) or wave-(a) *of a larger Degree*, the correction after wave-5 should drop into the 2nd wave price zone. If the sequence completes wave-(3) *of a larger Degree*, the correction after wave-5 will probably stop in the fourth wave zone.

5th Wave Failure

As a general rule, 5th wave Failures are only possible when the 3rd wave is the Extended move. The 4th wave should be the most complex of the two corrective phases. Wave-4 should retrace more of wave-3 than wave-2 does of wave-1. Almost always, wave-1 & 5 will be practically identical in price and time. On fewer occasions, waves 1 & 5 will relate in price and/or time by 61.8%. A 5th wave can Fail in an Impulse pattern only under one of the following circumstances:

A. If the Impulse pattern (which contains the 5th wave Failure) is itself the 5th wave of a larger Impulse pattern (see Figure 11-1).

B. If the Impulse pattern (containing the Failure) is the c-wave of one (and only one) larger Degree (see Figure 11-2).

C. Under extremely rare circumstances you might see a larger 3rd wave experience a smaller 5th wave Failure. For this to take place, the market would need to be forming a top or bottom that is very significant. "Significant" is, of course, a relative term. In this situation it would mean a Multiwave pattern, or preferably higher. If you witness an event such as this, the larger 5th Wave will also have to Fail, and it must Fail even further away from a high or low than the wave-5 Failure within the 3rd wave. To have wave-5 Fail at the end of a larger 3rd wave would indicate exceptional counter trend power. After such an event, a long-term top or bottom should be in place. Note: *Even though it cannot be absolutely ruled out, the author has never seen a 5th wave Failure at the end of a larger 3rd wave. Through logical deduction and a long-term familiarity with wave behavior, it is reasonable to assume that the conclusions above are valid for such a situation.*

Figure 11-1

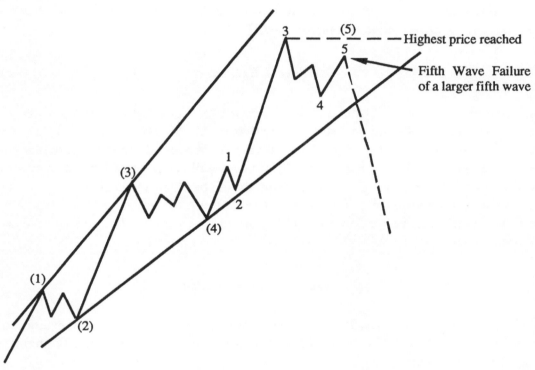

Figure 11-2

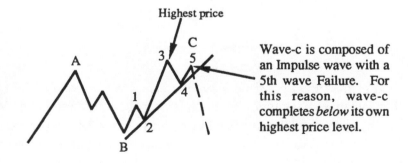

Terminal

This type of pattern was referred to by Elliott as a "Diagonal Triangle." To avoid any incorrect association of "Diagonal Triangles" with "Horizontal Triangles," I decided it was necessary to retitle the pattern. The new phrase, "Terminal Impulse," is a more accurate description of the pattern and its implications. I apologize to the reader who is accustomed to the old term, but I had to approach this subject with the idea that the reader is completely inexperienced. This new term should provide you with a better understanding of the phenomenon and its place in the larger scheme. Since Terminal and Triangle patterns consist of the same Structure Series, they tend to exhibit similar behavior characteristics. The deciding factor is that Terminal patterns adhere to all of the Essential Impulse Construction Rules; it is impossible for a Horizontal Triangle to follow all the same Rules. *[For diagrams of Terminal pattern variations, refer to Chapter 5, pages 14-15]*

Wave-1 Extended

A 1st wave extension in a Terminal pattern is *by far* the most common setup. Measured from its termination point, wave-2 should not retrace more than 61.8% of wave-1. Wave-3 should not be much more than 61.8% of wave-1, but at least 38.2% of wave-1. Wave-5 **cannot** be more than 99% of wave-3, but will more likely be only 61.8% to 38.2%. Wave-4 will usually be 61.8% of wave-2 in price *and* equal or relate by 61.8% in time. The 2-4 trendline should be clean and easy to identify. The only way the 2-4 trendline can experience a 'false break' (under these circumstances) is if the 5th wave is an Elongated Flat, a correction experiencing a C-Failure or a Horizontal Triangle or a c-wave Terminal within the 5th wave.

Wave-1 Non-extended

Wave-2 can retrace as much as 99% of wave-1. When the 1st wave does not extend, the **Terminal** pattern will most likely be the c-wave of a correction (excluding the c-wave of a Triangle), **not** the fifth wave of an Impulse pattern.

Wave-2

If the first wave extends, wave-2 should not terminate at a point lower than 61.8% of wave-1. If the first wave does not extend, wave-2 can retrace up to 99% of wave-1. If the first wave extends, wave-2 should consume more time and price than wave-4 and be the most complicated of the two corrections.

Wave-3 Extended

This constitutes one of the rarest possible wave patterns. When wave-3 "extends" in a Terminal pattern (unlike *all* other instances), it cannot be much longer than the first wave. Wave-2 must retrace *more* than 61.8% of the first wave and the fourth wave should retrace 38.2% (preferably less) of the third wave. The 2-4 trendline should perform as usual. A 1-5 trendline *should* be drawn which will be broken by the third wave. The fifth wave should not be more than 61.8% of wave-3. You will probably only see this pattern as the c-wave of a sequence, **not** the 5th wave of an Impulse pattern.

Wave-3 Non-extended

If wave-3 does not extend, the probabilities are considerably in favor of the first wave being the extension. If the first wave extended, wave-3 should not be much more than 61.8% of the first wave and wave-5 should not be much more than 61.8% of wave-3. If wave-1 is smaller than wave-3 (but not less than 61.8%) the 5th wave will probably extend. This would entail wave-4 overlapping wave-1, wave-4 being larger in price and/or time than wave-2. The 5th wave should be at least equal to 1-3 in price added to the low of wave-4. The 5th wave should not be more than 161.8% of waves 1-3 added to the top of wave-3.

Wave-4

Wave-4 cannot retrace more than 61.8% of wave-3 *unless* the fifth wave is going to extend and even then, it is *extremely rare*. If the fifth wave extends, wave-4 will probably take more time and price than wave-2 and consist of more subdivisions than wave-2. Wave-2 and 4 will likely relate by 61.8% in price and/or time.

Wave-5 Extended

This setup should only occur if the **Terminal** comprises the 5th wave of a larger Impulse pattern in which that 5th wave is also the extended wave **or** when the Terminal is the c-wave of any correction *except* a **Horizontal Triangle**. To segregate this pattern from an Expanding Running Triangle, some important observations are necessary. Unlike an Expanding Running Triangle, a **Terminal** *must* drift slightly upward if the trend of one larger degree is up, making higher highs for waves 1-3 & 5 and higher lows for waves 2 & 4 (vice versa if the trend is down). In an Expanding Running Triangle, the first leg must be shorter than the second. The 5th wave of the 5th Extension Terminal pattern should be at least 100% of waves-1 & 3, while wave-3 should not be more than 161.8% of wave-1. Wave-4 should retrace at least half of wave-3 and *can* retrace up to 99% (even though that is highly unlikely).

Wave-5 Non-Extended

If the 5th wave is not extended, it should not be more than 61.8% of wave-3. It should not be of the highest complexity of the three thrust waves. The 5th wave should break beyond the trendline drawn across waves 1 & 3. The 2-4 trendline should be "clean" and will not be broken until after wave-5 is complete (unless wave-5 is a Triangle). Wave-4 should consume less time and price than wave-2.

Corrective Patterns

To avoid repeating the same statements under each section, it should be known that no Corrective pattern can occur in any **Trending Impulse** Progress Label position (i.e., wave-1, 3, 5, wave-a in a Zigzag, or wave-c in a Zigzag or Flat [all variations]). The corrective patterns begin on the next page with **Flats.**

Flats

Flat patterns have many ways of manifesting themselves depending on the subtle differences of strength between one pattern and the next. All the patterns that fall under the broad heading of **Flats** are composed of the same **Structure Series** [3,3,5] (i.e. two adjacent corrective segments followed by a third Impulsive segment). Previously, we dealt with minimum requirements and maximum limits of Flat patterns. In this section, a little more detail will be covered along with the specific implications of each situation.

Variations of the Flat theme appear when the b-wave begins to retrace more than 100% or less than 100% of wave-a. The larger wave-b is in relation to wave-a, the less wave-c will retrace of wave-b <u>and</u> the more similar wave-c and wave-a will be to each other. The less wave-b retraces of wave-a, the larger wave-c will become. In Figure 11-3, the three diagrams show how wave-c can act based on various b-wave retracements of wave-a <u>from 100% to 61.8%</u>. (Keep in mind, these are minimum requirements for the length of the <u>c-wave in relation to wave-b</u> depending on how much wave-b retraces of wave-a.)

B-Failure

This term describes a pattern where the a-wave is Corrective, but the b-wave retraces <u>only</u> 61.8 - 81% of wave-a (see Figure 11-4).

The inability of wave-b to retrace more than 81% of wave-a indicates temporary weakness in the market. When wave-b is that weak, you can expect wave-c (as long as it is Impulsive) to retrace at least 61.8% of wave-b.

A B-Failure almost always takes place when the a-wave is a Double Zigzag or Double Combination (completing with a Triangle). It is characterized by the b-wave retracing at least 61.8%, but less than 81% of wave-a. The c-wave in a B-Failure pattern *must* retrace all of wave-b or else the pattern falls into the Double Failure category. No specific Fibonacci price relationships are required between waves-a,b or c, but wave-a & c may relate by 61.8%. Waves-a & b should alternate in time, construction and complexity to the extent possible based on the time consumed by each. This pattern can occur as:

    Waves: Any 2,4,a,b

    Waves: c,d,e of a Horizontal Triangle

    Part of a rare Running Double Flat

C-Failure

A **C-wave Failure** occurs whenever wave-c fails to retrace all of wave-b. This usually, but not always, occurs when wave-b retraces all, or almost all, of wave-a. A **C-wave Failure** is definitely a sign of counter trend strength (refer to Figure 11-5).

When a C-wave Failure occurs, the c-wave should definitely not be the shortest "time pattern" in the correction and very frequently it will consume the longest period of time of the three Corrective segments. Additionally, it can be equal to wave-a or wave-b (see Figure 11-6, bottom of next page).

When a C-wave Failure occurs, it quite frequently will be a Terminal Impulse. A Terminal Impulse is the perfect way to consume time without damaging price levels (see Figure 11-7, to of page 11-11).

Figure 11-3

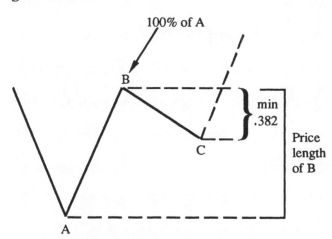

100% of A

min
.382

Price
length
of B

Regarding all the diagrams on this page;
as the B-wave retraces less and less of
wave-A, the C-wave will tend to increase in length and take more time.

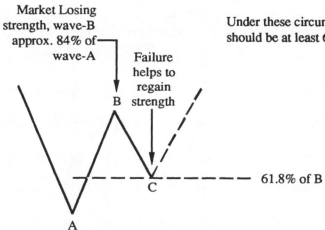

Market Losing
strength, wave-B
approx. 84% of
wave-A

Failure
helps to
regain
strength

61.8% of B

Under these circumstances the C-wave
should be at least 61.8% of wave-B

The slope of wave-C becomes steeper as it gains
strength from the weakening B-wave. Remember, there is no reason why the C-Wave could
not be longer than illustrated here, these are designed to show <u>minimum allowable</u> limits.

When wave-B retraces 61.8% (or less) of wave-A, it
becomes almost mandatory wave-C retrace all of wave-B.
That does not mean it has to get to the lowest point of wave-
A. Remember, wave-a can complete above <u>its own</u> low.

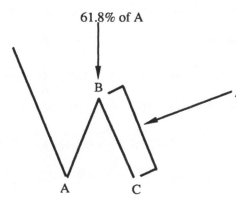

61.8% of A

At this point Wave-C is 100% of Wave-B

Figure 11-4

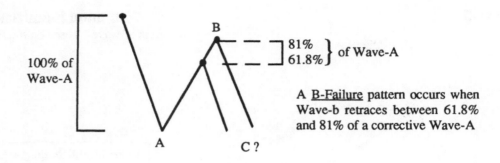

100% of
Wave-A

B

81%
61.8% } of Wave-A

A B-Failure pattern occurs when
Wave-b retraces between 61.8%
and 81% of a corrective Wave-A

A   C ?

Figure 11-5

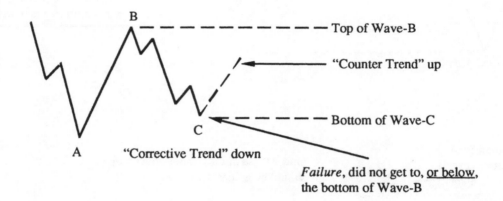

B — — — — — — — — Top of Wave-B

— "Counter Trend" up

— Bottom of Wave-C

C

A   "Corrective Trend" down

Failure, did not get to, or below,
the bottom of Wave-B

Figure 11-6

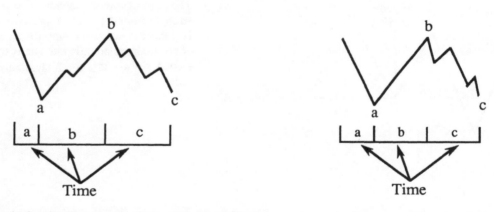

b

a   c

| a | b | c |

Time

The a-Wave is the shortest wave in the pattern (time-wise).

b

a   c

| a | b | c |

Time

Waves-b & c took the same amount of time

Figure 11-7

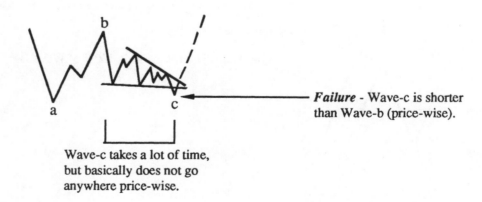

b

a   c

Failure - Wave-c is shorter
than Wave-b (price-wise).

Wave-c takes a lot of time,
but basically does not go
anywhere price-wise.

C-wave Failures which are less than 61.8% of wave-b should be considered *extremely* rare. When it does happen, the b-wave should take the most time, with wave-a & c equal in time (Figure 11-8).

In a C-Failure pattern, the b-wave should subdivide more completely than wave-a. Usually it will be a Double Zigzag. The c-wave should be either 61.8% of wave-a **or** complete in a zone 61.8% of wave-a taken from the beginning of wave-a. The b-wave <u>must not</u> exceed the highest point of wave-a. If it does, refer to **Irregular Failures**. The c-wave must, of course, be Impulsive (5). This pattern can occur as:

Wave: Any 2, 4, a, b
Wave: 5 (of a Terminal Impulse)

Figure 11-8

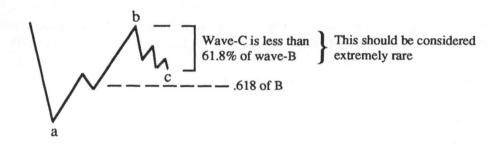

Wave-C is less than 61.8% of wave-B } This should be considered extremely rare

Common

Figure 11-9 illustrates the typical shape of a Flat pattern. In a **Common** pattern, all of the waves will be approximately equal in price (see Figure 11-9).

The **Common** can occur at virtually anytime in a sequence where the pattern needs to be corrective. Wave-b <u>should not</u> be more than 100% of wave-a, but *must* retrace at least 81% of wave-a and wave-c must retrace all of wave-b. Wave-c should move slightly beyond the end of wave-a, but not by more than 10 or 20%. It is not a *terrifically* strong pattern, but it does imply more strength than a Zigzag.

Alternation of time, construction, and complexity are the most important considerations. Wave-b will, the majority of the time, be the most time-consuming of the three segments and will be more subdivided (complex) than wave-a. Wave-c *must* be Impulsive.

This pattern can occur in the following places:

Waves: Any 2, 4, a, b
Waves: c,d,e of a Horizontal Triangle **or**
Part of a rare Running Double Flat.

Figure 11-9

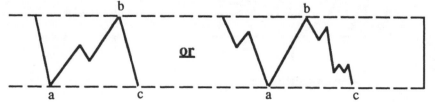

All waves are approximately equal in vertical price points

Double Failure

This describes the infrequent situation where the b-wave fails to retrace more than 81% of wave-a <u>and</u> the c-wave fails to retrace all of wave-b (see Figure 11-10). It can generally only take place if the a-wave is a Double Combination, followed by a strong Zigzag for wave-b. Wave-c would fail to retrace all of wave-b. Another possibility is that wave-a will be a Triple Combination and the b-wave will be a Double Zigzag or Combination. This would prevent the c-wave from retracing all of wave-b (reasons for this were discussed in Chapter 10, Advanced Logic Rules); refer to Figure 11-11 for details.

Frequently, the a-wave of a Flat will help to create the *appearance* of a Double Failure (see **Emulation**). This usually occurs when the a-wave is a Double Combination pattern completing with a Horizontal Triangle (study Figure 11-12).

In a general sense, this pattern will resemble a Horizontal Triangle because of the continual exhibition of contraction. Close attention to price detail is essential in deciding whether you are in a B-Failure or Horizontal Triangle. In a B-Failure, the c-wave is an Impulse pattern. In a Horizontal Triangle, the c-wave is a Corrective pattern. Virtually without exception, the a-wave in one of these patterns will be a Double Zigzag *or* Double Combination (a pattern concluding with a Non-Limiting Triangle). The c-wave will generally be 61.8% of wave-a *or* terminate in a price zone 61.8% of wave-a subtracted from the beginning of wave-a. Places of acceptable formation:

Wave: Any 2,4
Wave: a (in an Irregular or Triangle pattern)
Wave: b (of a Zigzag, Common, or Elongated pattern)
The first corrective phase of a rare Running Double Flat

Figure 11-10

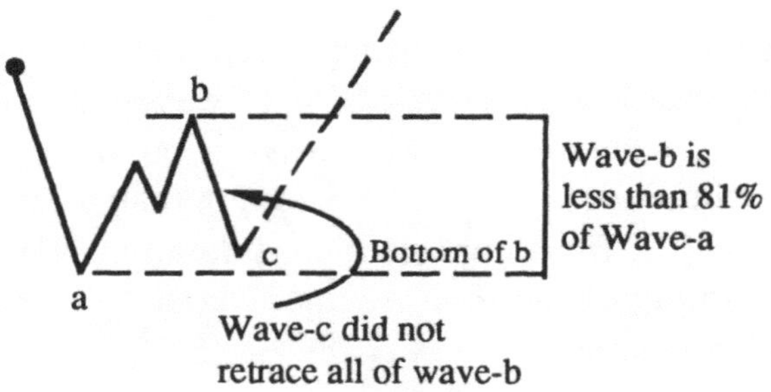

Wave-b is less than 81% of Wave-a

Bottom of b

Wave-c did not retrace all of wave-b

Figure 11-11

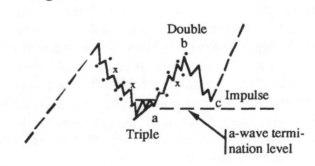

Double

Impulse

Triple

a-wave termination level

Figure 11-12

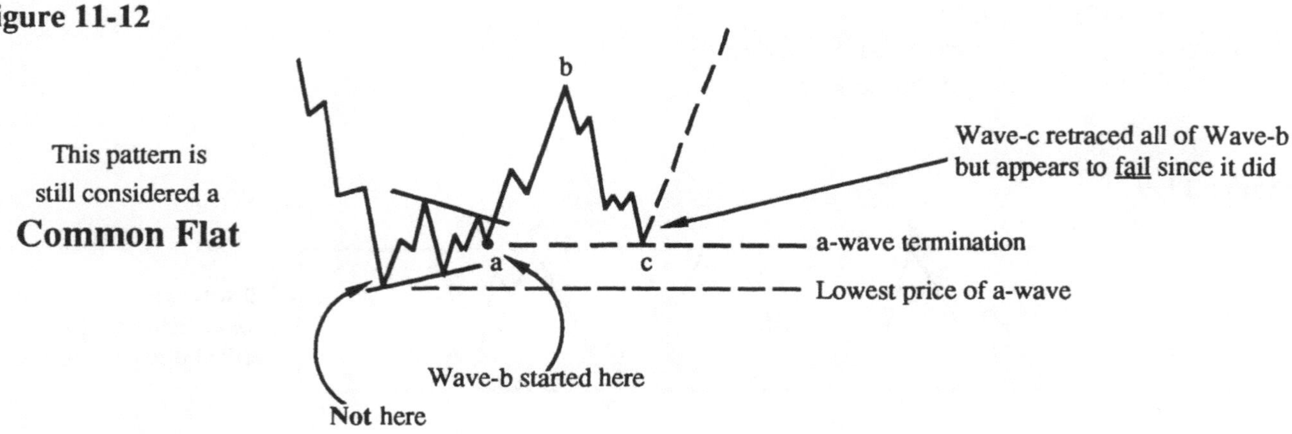

This pattern is still considered a
Common Flat

Wave-c retraced all of Wave-b but appears to <u>fail</u> since it did

a-wave termination

Lowest price of a-wave

Wave-b started here

<u>Not</u> here

Elongated

An **Elongated** pattern adheres to all of the general specifications of a Flat pattern with the following additional criteria:

1. Wave-c *must* be <u>more</u> than 138.2% of wave-b (preferably, more than 161.8%); see Figure 11-13.
2. Wave-a and wave-b should be similar in price <u>and/or</u> time (one of the few times that **Alternation** <u>may not</u> have a major impact on these two adjacent waves), while wave-c consumes a significant amount more of both (refer to Figure 11-14).

Figure 11-13

Figure 11-13 (more complicated version)

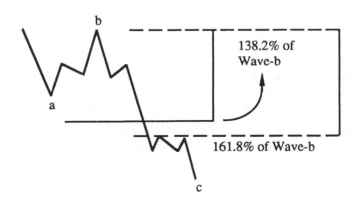

Figure 11-14

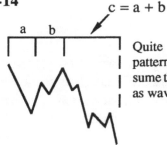

$c = a + b$

Quite often in an elongated pattern, the c-wave will consume the same amount of time as waves a & b combined.

It has been discovered by the author that Elongated patterns take place only under special circumstances. These patterns (with virtually no exceptions) occur as the entire leg of a Triangle *or* as a segment *within a leg* of a Triangle (refer to Figure 11-15, top of next page)

The reason these patterns appear <u>almost</u> exclusively in Triangles is due to the way Triangles behave. During the early stages of a Triangle, market action tends to be very volatile. The market has the propensity to hug one trend line for a while, then in a jolt of buying or selling, jump to the opposite side of the formation (at the top of page 11-15, Figure 11-16 illustrates this concept).

Figure 11-15

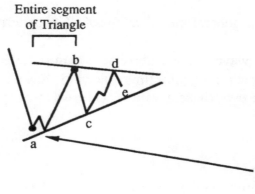

Entire segment
of Triangle

The b-Wave is one entire segment of a five segment Horizontal Triangle.

The b-wave is the motion between the two circled points. It is an *Elongated Flat.*

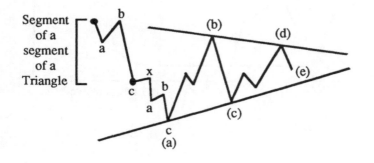

Segment of a segment of a Triangle

The action between the circled points constitutes an elongated pattern <u>within</u> a single segment of a Triangle. In other words, the elongation is one <u>degree</u> lower than the degree of the Triangle.

Elongated Flats are very important, early warning signs of triangular formation. Once you witness one of these patterns, it is generally wise not to trade until you are sure the Triangle is complete. To form this pattern, wave-a & b *must* be similar in price (wave-b *must* retrace at least 61.8% of wave-a) and will usually be close in time (or related by 61.8%) with the c-wave much larger in price. Points of occurrence are as follows:

Wave: 1,3 or 5 of a Terminal Impulse
Wave: a,b,c or d of a Horizontal Triangle
Wave: "e" of an Expanding Horizontal Triangle; it is possible (but not likely) that wave-e
        could be Wave-5 of an Expanding Terminal Impulse
Wave: It could be the first corrective phase of a Double Flat which will probably be
the entire leg of a Horizontal Triangle *or* Terminal Impulse

Figure 11-16

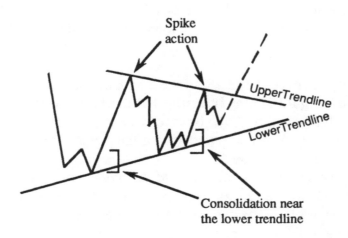

Spike action

UpperTrendline

LowerTrendline

Consolidation near
the lower trendline

Irregular

This is one of the easiest patterns to define and recognize. Some readers may be astounded to read this, but on a real-time, properly plotted Cash chart, this pattern is <u>not</u> very common. Following are the parameters which outline this type of pattern.

Minimum Requirements:

1. Wave-b must be at least 101% of wave-a. (see Figure 11-17)
2. Wave-c must be at least 101% of wave-b. (see Figure 11-18)

Figure 11-17

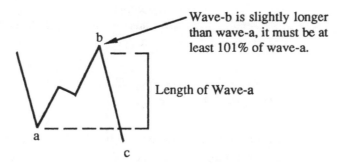

Wave-b is slightly longer
than wave-a, it must be at
least 101% of wave-a.

Length of Wave-a

Figure 11-18

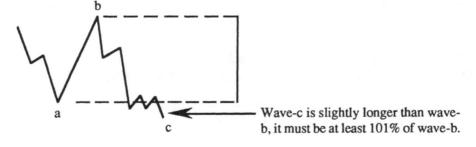

Wave-c is slightly longer than wave-b, it must be at least 101% of wave-b.

As the b-wave gets longer (in relation to wave-a), the probabilities dropped dramatically that wave-c will be longer than wave-b. An Irregular pattern must adhere to the following Rules. **Maximum** Limits:

1. The b-wave should not exceed 138.2% of the length of wave-a. If it does, the pattern will <u>probably</u> not be an **Irregular**, but an **Irregular Failure** (see next heading). In other words, wave-c will not retrace all of wave-b. If the b-wave is *more* than 138.2% of wave-a, the c-wave is not likely to retrace all of wave-b; in that case the pattern will have to be called an **Irregular Failure** (refer to Figure 11-19)
2. The b-wave in an **Irregular** Correction *should* subdivide more than wave-a (see Figure 11-19).

The b-wave will almost always be a Zigzag (or less likely, a combination of corrections), while wave-a will usually be some type of Flat. The c-wave must be Impulsive. If wave-c is corrective, you may be experiencing an Expanding Triangle **or** a Contracting Triangle with a large c-wave. This pattern can be found as:

Wave: Any 2,4,a,b
Wave: c,d,e (in a Horizontal Triangle)

Figure 11-19

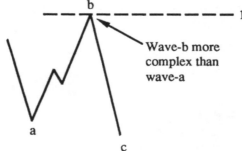

138.2% of Wave-a

Wave-b more complex than wave-a

As long as wave-b does not exceed 123.6% of wave-a, it is still possible for wave-c to make a move beyond the end of wave-a. If it is any longer, the c-wave will have a tough time being bigger than wave-b (thus moving beyond the end of wave-a).

Irregular Failure

This pattern starts to become <u>very</u> likely when the b-wave exceeds 138.2% of wave-a (see Figure 11-20). As the b-wave becomes longer, wave-a and wave-c become more similar (see Figure 11-21). This pattern signals *post* market-action strength. Wave-b *must* be larger than wave-a in price. Wave-c *must not* retrace all of wave-b. Wave-c will generally relate to the a-wave (price and time) by equality *or* the Fibonacci ratio. This pattern should occur as one of the Progress Labels below:

Wave: 2,4 (before an extended wave)
Wave: b (before an Elongated c-wave of a Flat)
Wave: b (of a Zigzag in which wave-c is *at least* 161.8% of wave-a)

Figure 11-20

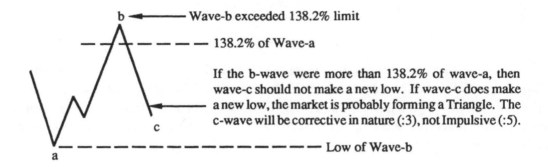

b ◄————— Wave-b exceeded 138.2% limit

— — — — — — 138.2% of Wave-a

If the b-wave were more than 138.2% of wave-a, then wave-c should not make a new low. If wave-c does make a new low, the market is probably forming a Triangle. The c-wave will be corrective in nature (:3), not Impulsive (:5).

— — — — — — — — — — — — — — Low of Wave-b

Figure 11-21

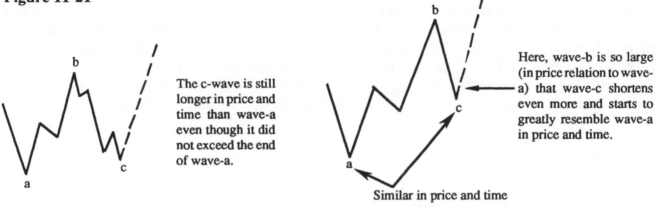

The c-wave is still longer in price and time than wave-a even though it did not exceed the end of wave-a.

Here, wave-b is so large (in price relation to wave-a) that wave-c shortens even more and starts to greatly resemble wave-a in price and time.

Similar in price and time

Running

The most powerful *Standard* Correction, the **Running,** imposes very specific requirements on post market action. It is essential that the movement after a Running correction advances (or declines) quicker than the pattern immediately before the Running correction (of the same degree). The movement immediately after a Running Correction should always be the longest segment of the completed Elliott pattern. Places where Running Corrections *can* occur:

Wave: 2 (right before an Extended 3rd wave)

Wave: 4 (right before an Extended 5th wave) - consider this rare

Wave: b (right before an Elongated c-wave *and* in which the whole a-b-c is probably *the entire leg of a Triangle* or **one** of the completed phases of a Complex Correction which is *the entire leg of a Triangle*)

Wave: b (in a Zigzag that is part of a Triangle [of only one or two Degrees higher])

Wave: x (to be considered *very* unlikely, but could occur right before the longest Corrective group of a Complex corrective series)

In a Running Correction, wave-b must be the largest wave by far. Wave-a and wave-c will tend toward equality in price and time. The b-wave does not have to relate to wave-a but, if it does it will probably be 261.8% of wave-a (for more detail on Running Correction, refer back to Chapter 5).

ZigZags

Unlike Flats, **Zigzags** do not come in as many variations. Their construction and requirements are few, but demanding. The two areas of significant consideration for **Zigzag** patterns are: how long is the c-wave is in relation to the a-wave; and how much does wave-c subdivide in relation to wave-a? The following diagrams have been drawn to show how these and other rules (when considered simultaneously) go to form realistic Zigzag patterns (see Figure 11-22).

Wave-a

Wave-a *must* be an Impulse structure and should not be retraced by wave-b more than 61.8%. If you are certain that the a-wave is **Impulsive**, but the b-wave retraces more than 61.8%, then that retracement will only be the a-wave (of one smaller degree) of wave-B in which the c-wave (of wave-B) will fail. If part of the B-wave retraces 81% or more of wave-a, then you should recheck your analysis of wave-a. It will most likely fall under the **Missing Wave Rule** in which wave-a (that at a glance only looked **Impulsive**) is better interpreted as a Corrective Double Zigzag or Double Combination.

Figure 11-22

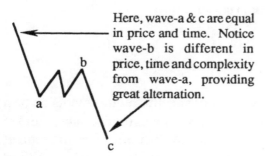

Here, wave-a & c are equal in price and time. Notice wave-b is different in price, time and complexity from wave-a, providing great alternation.

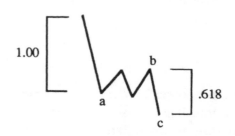

Here wave-c is 61.8% of wave-a.

1.00   .618

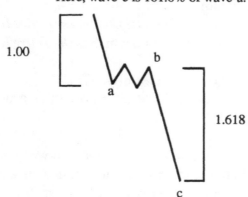

Here, wave-c is 161.8% of wave-a.

1.00   1.618

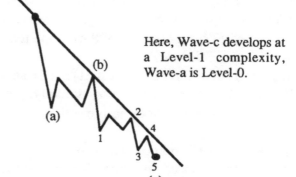

Here, Wave-c develops at a Level-1 complexity, Wave-a is Level-0.

Wave-b

Wave-b absolutely *must* be corrective (3) and should not retrace more than 61.8% of wave-a when measured from its terminus. It should not be a **Running** Correction unless the Zigzag, of which the b-wave is a part, is in a Triangle. If what you think is wave-b turns out to be a **Running** Correction, most likely it is wave-2 of an **Impulse** wave. Wave-b can be almost any other Corrective pattern in a Zigzag except a Double or Triple Zigzag or a <u>Running Double or Triple Flat and their Combinations</u>. If you see one of those patterns forming after an Impulsive a-wave (which is in a Zigzag), then the Complex movement will only be *part* of the b-wave correction, not the whole correction.

Wave-c

When a **Zigzag** (of which wave-c is a part) is not part of a Triangle (of only one or two higher degrees), its price length should fall between 61.8% and 161.8% of wave-a. If the whole Zigzag is part of a Triangle (of only one or two higher degrees), the c-wave *can* exceed those limits, but is not required to do so. If the c-wave exceeds those limits, it is one of the better indications of Triangular formation (of one or two higher degrees).

Triangles

After years of experience with real-time market analysis and trading, it was clear that Elliott's rules concerning **Triangles** were not comprehensive enough to describe all market consolidation phases which occurred between converging trendlines. It became evident there needed to be a subdivision of the two general Triangular categories (Contracting and Expanding). The first sub-category I call **Limiting** Triangles. These make up the well known types that Elliott described: b-waves and 4th-waves. As their name implies, these Triangles have a *limiting* effect on post-Triangular market action.

The other Triangle sub-category I designated **Non-Limiting**. These make up a relatively unknown type which occur in unusual places in a wave sequence. Non-limiting Triangles impose only minor post market action limitations. The formation of these Triangles is slightly different from the Limiting type. Close attention to detail is necessary to properly detect which type of Triangle is forming. Both types will be discussed during the next few pages.

All of the rules which follow are in reference to **Triangles**. Learning to decipher polywave activity is at the root of the entire **Theory**, so most of the rules were designed with polywaves in mind, but the rules can be carried into the multi- and macrowave arenas.

Note: Under no circumstances is wave-d the largest wave of <u>any</u> Triangle pattern. Under no circumstances can three legs (of the same degree) within a Triangle be equal in price (give or take 5%). In Expanding Triangles, only four of a possible five retracements (including the pattern of the same degree just before the Triangle started) **needs** to be 100% or more; one of the five may not be completely retraced. *Never* can any leg (other than wave-e) of a Contracting Triangle be, itself, a Contracting Triangle. When mentioned in later text that an a, b, c, or d-wave can be any Corrective pattern, it means any Corrective pattern <u>***except***</u> a Triangle (unless otherwise mentioned).

Contracting Triangles

Contracting Triangles are the most common Triangular formation. They come in two major categories, Limiting and Non-Limiting, with each category breaking down into several varieties.

I. Limiting

It is crucial, if you want to master Elliott, that you understand the difference between the general make-up of a **Limiting** and **Non-Limiting** Triangle. All **Limiting** Triangles have a very similar construction. One of the most important characteristics is that wave-e, the end of the Triangle, concludes well before the apex point of the Triangle is reached (see Figure 11-23).

Another element, which I have discovered is characteristic of Limiting Triangles, deals with time. The **apex** of the Triangle should complete in a time zone equal to 20-40% of the time consumed by the Triangle added to the end of the e-wave (refer to Figure 11-24).

The apex of the Triangle has the greatest tendency to complete around the 38.2% [$.618^2$ (squared)] time zone. As mentioned earlier, after a Triangle is complete, it produces what is called a "_thrust._" The extent of the thrust is the key factor in the proper categorization of Triangles. In **Limiting** Triangles, the _thrust_ is usually limited to the widest leg of the Triangle (+ or - 25%). Only in rare cases, or when the

Figure 11-23

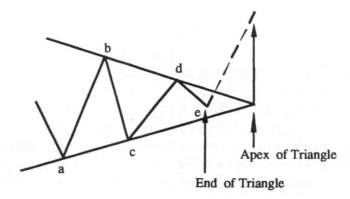

Apex of Triangle

End of Triangle

Figure 11-24

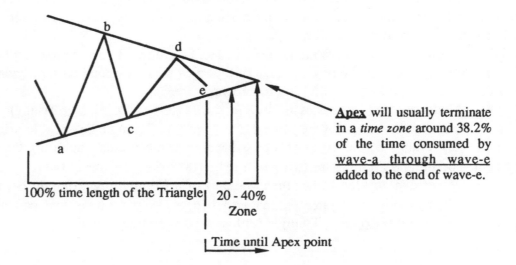

100% time length of the Triangle | 20 - 40% Zone

Time until Apex point

Apex will usually terminate in a _time zone_ around 38.2% of the time consumed by <u>wave-a through wave-e</u> added to the end of wave-e.

Figure 11-25

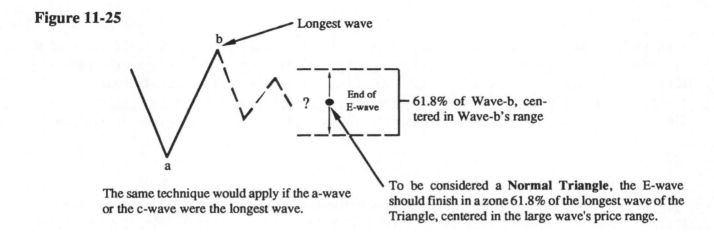

Longest wave

? End of E-wave — 61.8% of Wave-b, centered in Wave-b's range

The same technique would apply if the a-wave or the c-wave were the longest wave.

To be considered a **Normal Triangle**, the E-wave should finish in a zone 61.8% of the longest wave of the Triangle, centered in the large wave's price range.

Triangle falls into the "abnormal" classification, can the "*thrust*" exceed the above limits. A state of abnormality occurs when the Triangle fails to follow the parameters discussed in Figure 11-25. Rare cases occur if the Triangle is the last b-wave or 4th-wave before a final <u>major</u> top or bottom in the market.

Contracting **Limiting** Triangles take place in only two Progress Label positions; as wave-4 **or** wave-b. These are the best known types and are the ones which Elliott discovered. There are other types of Triangles which Elliott did not discover. Immediately following are the different types of Contracting, **Limiting** Triangles. Afterwards, we will proceed into the less known worlds of Expanding and <u>Non-Limiting</u> Triangles.

a. *Horizontal*

Each leg of this type of Triangle should be at least 38.2% of the previous leg (excluding wave-e). Wave-b can be no more than 261.8% of wave-a and wave-c can be no more than 161.8% of wave-b. Wave-d must be shorter than wave-c *and* wave-e must be shorter than wave-d.

Wave-a

Wave-a does *not* have to be the largest segment (price-wise), but it should definitely not be the smallest wave of the pattern. If wave-a is not the largest wave, wave-b almost certainly will be. The a-wave should not be less than 50% of wave-b. If wave-a is smaller than wave-b, wave-a will most likely be a type of Flat and wave-b a Zigzag *or* if wave-a is a simple Zigzag, wave-b will most likely be a complex Corrective pattern (Double Zigzag or Combination, Running Double Flat - a Triple Zigzag or Combination is not very likely to occur in wave-b of this type of Triangle).

Wave-b

If wave-b is smaller in price than wave-a, then all the other patterns *must* be smaller than the previous segment (from left to right). If wave-b is bigger than wave-a, the chances are very small (but do exist) that wave-c may be slightly larger than wave-b and still maintain the **Contracting Triangle** formation. If the c-wave is larger than wave-b, it is *mandatory* that wave-d be smaller than wave-c or else you are entering the Expanding Triangle realm and should move to that section. In other words, in this type of Triangle, once one segment is smaller in price than the last, all the rest must be smaller than the previous. If this is not the case, then the market is not in a Contracting Horizontal Triangle. Maybe it is in some other type of Triangle. **Note**: Wave-b should not be _less_ than 38.2% or _more_ than 261.8% **of wave-a.** There may be exceptions to the foregoing guidelines, but they are rare, so be careful about ignoring them.

Wave-c

Wave-c is the last wave that has a chance (in a Contracting Horizontal Limiting Triangle) to be larger in price than the previous wave (wave-b). It rarely happens, but it is not impossible. If wave-c is the biggest price move, the trendline will be drawn across wave-c & e.

Wave-d

Wave-d _must_ be shorter than wave-c. It should be at least 38.2% of wave-c. It may take more time than wave-c, but should not be of greater complexity level (the same complexity level or one lower level, only, is acceptable).

Wave-e

Wave-e _must_ be shorter in price than wave-d. It can be shorter or longer in time than wave-d, but should not be the longest time Correction in the pattern.

b. *Irregular*

The **Irregular Triangle** is characterized by the b-wave of the pattern, which must be longer (price-wise) than wave-a. After wave-b, all waves are continuously smaller until the Triangle is over.

Wave-a

With few exceptions, the a-wave should be smaller in time than the b-wave. The b-wave should not be more than 161.8% of wave-a (price-wise). Wave-c _must_ move back into the price "zone" of wave-b enough so that the shorter e-wave can also finish in the price zone of wave-b. The a-wave can be any corrective pattern _except_ a Triple Zigzag or an Elongated Flat. To avoid confusion, I will once again mention that wave a,b,c, or d of a Triangle can **never**, itself, be a Triangle. The e-wave is the only segment of a Triangle allowed to form into a smaller Triangle.

Wave-b

The b-wave _must_ be longer than wave-a. It will probably _not_ be much more than 161.8% of wave-a and must not be more than 261.8% of wave-a. If wave-a does _not_ subdivide, the b-wave will most likely be a Zigzag. If wave-a subdivides into a Zigzag, then wave-b will probably be a Double Zigzag. If wave-a was a Flat, wave-b will probably be a single Zigzag. The b-wave should not be more than 61.8% of the previous Impulse wave. The b-wave can be any Flat or Zigzag pattern. It *can* be any Complex Correction such as a Double or Triple or a Combination.

Wave-c

Wave-c *must* be shorter than wave-b, but *must* retrace at least 38.2% of wave-b. The lowest point of wave-c will probably fall back into the price zone of wave-a. If wave-b was a Double Zigzag, wave-c should be a Zigzag *or* a Flat with an Elongated c-wave. If wave-b was a Zigzag, then wave-c will probably be a Flat (of any type) or a monowave.

Wave-d

Wave-d *must* be shorter than wave-c. It may consume more time than wave-c. It *must* retrace at least 38.2% of wave-c. The d-wave can be any Corrective pattern as long as it Alternates with wave-c.

Wave-e

The e-wave *must* be smaller than wave-d and the probabilities greatly favor that the e-wave will be the smallest wave (price wise) in the Triangle. It does not have to retrace the d-wave by any specific amount, but it must move in the opposite direction of the d-wave by at least one tick and *must* terminate in the price zone of wave-d.

Unlike the other four segments in the Triangle (waves a,b,c,d), the e-wave can, itself, be a Triangle, but, e-wave Triangles are far more common at the end of **Non-limiting** Triangles than at the end of **Limiting** Triangles. The terminal point of the e-wave will probably fall at an important Fibonacci point to the pattern of one larger degree. If the pattern of one larger degree is a Zigzag (which means the Triangle is a B-wave), then the e-wave terminus will probably be at 61.8% of the entire move from the start of the Zigzag to its end added to the start of the Zigzag. If the larger pattern is a Flat, the e-wave terminus will probably be at 61.8% of wave-a of the Flat. If the Triangle is the 4th wave of an Impulse pattern, the e-wave terminus will likely occur at 38.2% or 61.8% of the a-wave of the Triangle.

c. *Running*

The **Running Triangle** is detected from the b & d-wave action. Wave-b will be longer than wave-a and wave-d will be longer than wave-c. The thrust out of this type of Triangle will be more than normal. It should be at least 161.8% of the widest segment of the Triangle, but should not exceed 261.8%. When this type of Triangle occurs, frequently it signals that the end of a significant advance *or* decline is about to take place. If the thrust out of the Triangle were more than 200%, then you can be relatively sure that a significant top *or* bottom will take place.

Wave-a

Wave-a should be no less than 38.2% of wave-b. Wave-a will probably be a monowave or a Flat. If it is a Zigzag, then wave-b will need to be a Double (and maybe a Triple) Zigzag. In this type of Triangle, wave-a **cannot** be a Double or Triple Zigzag, a Triangle, or a Flat with an Elongated c-wave. The most common type of Correction for wave-a would be a monowave *or* any Flat (excluding Elongated Flats). Wave-a *must* not retrace more than 61.8% of the previous pattern (if an Impulse wave preceded the Triangle). To do so would indicate weakness and would be a contradiction of the strength i*mplied* by the **Running Triangle**. Normally, the a-wave should not retrace much more than 38.2% of the previous Impulse pattern (if one is present).

Wave-b

If there is an Impulsive a-wave before a Running b-wave Triangle, wave-b should *not* be more than 261.8% of wave-a. Unless all the segments of the Triangle are monowaves, wave-b should be a Zigzag, a Double Zigzag, or on a *rare* occasion, a Triple Zigzag. Wave-b *must* be the largest segment of the Triangle, so obviously wave-c cannot retrace all of wave-b. If wave-a is a Flat, wave-b will probably take less time than wave-a. If wave-a is anything else (including a monowave), wave-b should consume more time.

Wave-c

Wave-c *must* be shorter than wave-b. Wave-c cannot be more complicated than a Double Zigzag or Double Combination. It **cannot** be a Triple Zigzag. The probabilities are very high that the terminus of wave-c will fall back into the price zone covered by wave-a.

Wave-d

Wave-d *must* be larger (in price) than wave-c. This, of course, means that it will exceed the terminus of wave-b. Wave-d should *not* be a Triple Zigzag. It could be virtually any other corrective pattern. Wave-d must not be completely retraced by wave-e. Due to the strong move that should occur immediately after wave-e, the terminus of wave-e should not be at a point which would put it at more than a 61.8% retracement of wave-d. If part of wave-e retraces more than 61.8% of wave-d, the e-wave will develop into a Flat with a C-Failure or on a *rare* occasion, the e-wave could be a Contracting Triangle as long as it did not congest directly into the apex point of the converging lines (see the discussion on the construction of **Limiting Horizontal Triangles**).

Wave-e

Wave-e *must* be smaller than wave-d in price. It could take a longer period of time than wave-d, especially if it were a Triangle, as long as it did not correct into the apex of the larger Triangle (see the construction parameters of a Horizontal Triangle). If the **Running Triangle** created a larger b-wave, it would be very common for the terminus of wave-e to occur at a 61.8% or 38.2% position to the entire larger Corrective advance (Figure 11-26, top of next page).

Figure 11-26

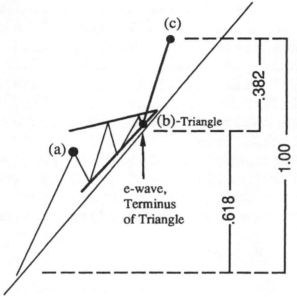

Notice the thrust is larger than any segment of the Triangle. The thrust is approximately 161.8% of the widest wave of the B-wave Triangle. This should be considered normal for a Running Triangle.

II. Non-Limiting Triangles

Some of my most important discoveries concern **Non-Limiting** Triangles (Triangles that fall outside of the 4th & b-wave realm). Understanding how a Non-Limiting forms is crucial to the proper analysis of complex Corrections. In addition, it virtually eliminates being tricked into thinking a Triangle, that is part of a Double or Triple Combination, is a 4th or b-wave. Non-Limiting Triangles occur at the end of complex Corrections (Double and Triple Combinations of all types), as X-waves, Wave-e of a larger Triangle, or as the 5th wave of a Terminal Impulse pattern. The most reliable indication of a **Non-Limiting** Triangle occurs when the market contracts right into the apex of the converging trendlines (see Figure 11-27). This condition is best met when the e-wave, of the larger Non-Limiting Triangle, is also a Non-Limiting Triangle. As a general rule, Non-Limiting Triangles are easier to spot than Limiting Triangles. This is due to their tendency to consolidate longer, letting everyone "know" a Triangle is forming. It is then up to you to decide if the break out will be approximately equal to or greater than the widest wave of the Triangle and whether the thrust will be up or down.

Figure 11-27

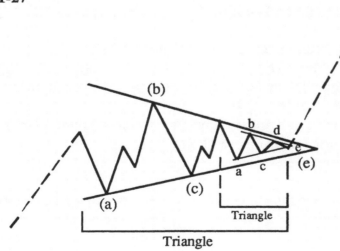

The "thrust" will be bigger than the width of the Triangle. The potential after one of these Triangles is basically unlimited.

Due to Elliott's original guidelines, many students get caught incorrectly positioned in the market when a Non-Limiting Triangle occurs. Elliott never wrote about Triangle's which could occur in any other position except the b-wave and 4th-wave Progress Label position. This leaves many practitioners in a dangerous situation. They assume all Triangles are 4th-waves or b-waves. Remember, b-waves and 4th-waves do not provide thrusts of much greater distance than the widest segment of the triangle. Non-Limiting Triangles are not bound to that Rule. For this reason, a person trying to pick the end of the thrust, out of what is *assumed* to be a b-wave or 4th-wave Triangle, may get caught in a persistent up or down move if the Triangle actually was a **Non-Limiting** Triangle.

All the above rules which apply to Limiting Triangles under each specific category also apply to **Non-Limiting** Triangles of the same type. For example, each leg in an Irregular Limiting would follow the same rules as those outlined for Irregular Non-Limiting Triangles. There are only a few additional behavior characteristics which need to be mentioned; they are listed by **Progress Labels.**

Wave-a

In these patterns, wave-a is almost always the most violent segment of the Triangle (i.e., it covers the most price distance in the shortest period of time). Usually, the a-wave will be the most violent market action to have occurred for quite sometime. The reason for this appears to be that these Triangles retrace Corrective patterns, not Impulsive patterns. A severe and violent retracement of a Correction would be more acceptable and logical than the same event after an Impulse pattern. If wave-a *is* violent, then wave-b should break down into a much more complex and slower developing wave.

Wave-e

The last segment of a Non-Limiting Triangle, wave-e, has a strong tendency to be a Non-Limiting Triangle itself. This creates a Triangle at the end of a larger Triangle. That is the best way for the market to converge right into the apex of the larger Triangle's contracting trendlines.

Of the three types of Non-Limiting Triangles, the **Horizontal** is the one you will most often see in real-time market action. For a more complete explanation of how Triangles should look and behave pre- and post- *thrust*, refer to all sections that mention **Triangles** in Chapters 5 & 10.

NOTE: Any Non-Limiting Triangle (except when it is an x-wave) *must* terminate a larger corrective phase. If you spot one in any position (other than the x-wave position), you can compact the entire larger move into a single '3'.

Some common and uncommon places of **Non-Limiting** Triangular occurrences are:

1. Wave-e of a Horizontal Triangle (return to the previous Figure 11-27).
2. The last corrective phase of a Double or Triple Three (see Figure 11-28).
3. The last corrective phase of a Running Double Three, one of the trickiest and most common points of occurrence (see Figure 11-29).
4. The last corrective phase of a Double or Triple Combination (refer to Figure 11-30).
5. The 5th wave of a Terminal Impulse (study Figure 11-31).
6. The x-wave of a complex Correction (see Figure 11-32).

If you did not notice, the above statements for numbers 1-5 indicate points of Triangular occurrence that terminate a pattern. Number 6 lists the only time a Non-Limiting Triangle can occur in the middle of a move.

Figure 11-28

Triple Three

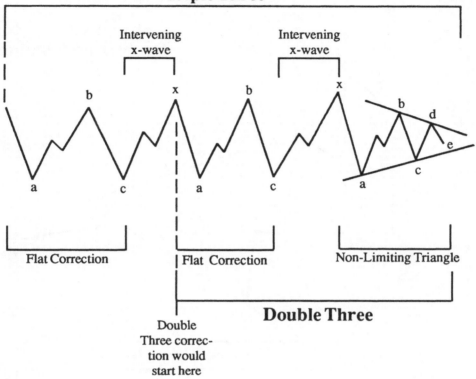

Intervening
x-wave

Intervening
x-wave

b   x   b   x   b
d
e
a   c   a   c   a   c

Flat Correction   Flat Correction   Non-Limiting Triangle

Double Three

Double
Three correc-
tion would
start here

Figure 11-29

Powerful Impulse Wave
<u>will</u> follow this formation

Intervening
x-wave

**Double Three
Running Correction**

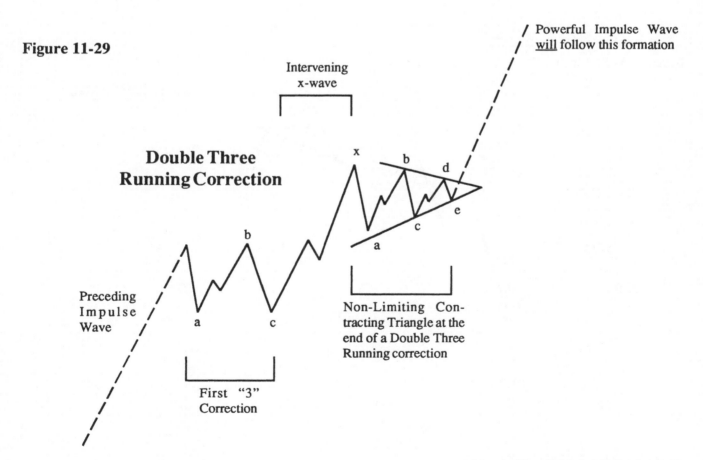

x   b   d
e
a   c

b

Preceding
Impulse
Wave

a   c

Non-Limiting Con-
tracting Triangle at the
end of a Double Three
Running correction

First "3"
Correction

Figure 11-30

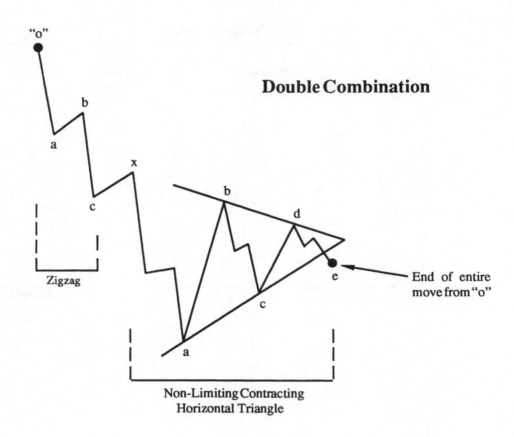

Double Combination

"o"

b

a

c

Zigzag

x

b

d

End of entire
move from "o"

a

c

e

Non-Limiting Contracting
Horizontal Triangle

Figure 11-30 continued

Triple Combination

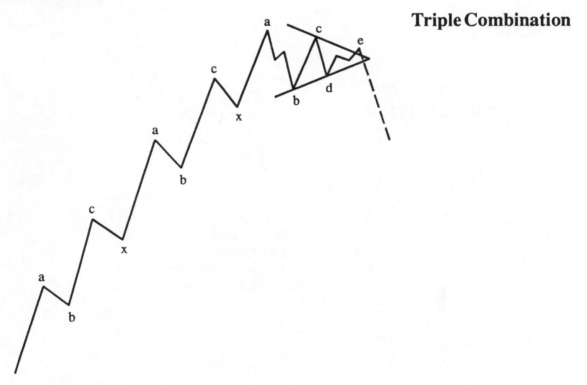

a

c

c

e

b

d

x

a

b

c

x

a

b

Figure 11-31

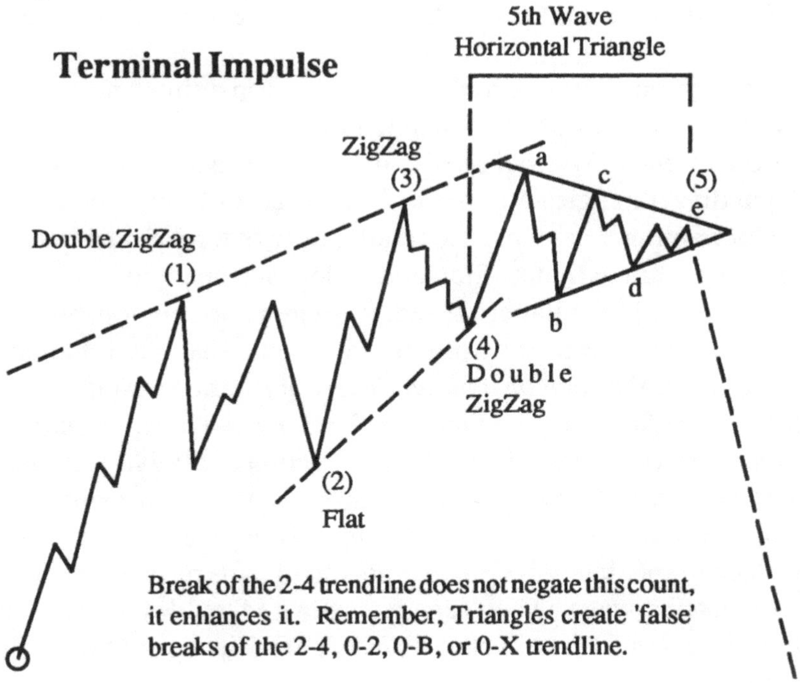

Terminal Impulse

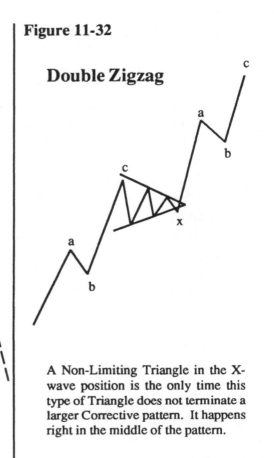

5th Wave
Horizontal Triangle

ZigZag
(3)

Double ZigZag
(1)

(4)
D o u b l e
ZigZag

(2)
Flat

(5)

Break of the 2-4 trendline does not negate this count, it enhances it. Remember, Triangles create 'false' breaks of the 2-4, 0-2, 0-B, or 0-X trendline.

Notice wave-4 drops below the top of wave-1 (overlap). Waves 2 & 4 alternate from multiple standpoints. Waves 1, 3, & 5 also alternate.

Figure 11-32

Double Zigzag

A Non-Limiting Triangle in the X-wave position is the only time this type of Triangle does not terminate a larger Corrective pattern. It happens right in the middle of the pattern.

Expanding Triangles

Of all the wave patterns under Elliott's Theory, the **Expanding** Triangle creates the most treacherous trading environment. Much worse than its non-trending mirror image (the Contracting Triangle), this pattern continually creates the illusion the market is breaking out. This is quickly met with a reversal and then the break of a support or resistance level on the other side of the trading range. The phrase "whip-sawed" was probably created by a trader experiencing this formation.

To improperly diagnose this pattern can cause severe financial "pain." Unfortunately, there is no way to anticipate the formation of one of these patterns. On the other hand, strict and proper application of all the rules and techniques discussed in this book should keep you out of the market since you will not be able to decipher what is taking place. Lack of a clear pattern formation is the ultimate "no action" indicator.

From the very few **Expanding** Triangles I have witnessed over the past eight years, a clear, specific consensus of their nature has been difficult to distill. Instead of listing their behavior on a wave-by-wave basis under each variation, more general descriptions have been settled upon.

Before moving on, the major formational parameters of all Expanding Triangles need to be covered so you can check your wave group against these "standard" guidelines to see if your assumption is on the right track.

1. Either wave-a or wave-b will be the smallest segment of the Triangle.
2. The e-wave has a tendency to "blow off," being much longer in price and time than the other waves.

3. In the same way (and for the same reasons) a Contracting Triangle can occur as the e-wave of a larger Contracting Triangle, an Expanding Triangle can take place as the e-wave of a larger Expanding Triangle.

4. As a general statement, watch for a progression from simple to complex (on a time and subdivision scale) as each larger wave of the Expanding Triangle unfolds.

5. The pattern after the Triangle (wave-c or wave-5) should not completely retrace the e-wave.

6. One of the strangest aspects of Expanding Triangles appears to be their lack of numerous Fibonacci relationships typical of Contracting Triangles. Generally, only one relationship can be found, and you usually have to really search to find it. Included under each heading is the one relationship which seems to normally occur in the pattern listed. If there is no relationship indicated, it is because I have not been able to identify any present in the ones that were studied. *Believe it or not, lack of Fibonacci relationships can be a valuable test of the authenticity of an Expanding Triangle. It is very difficult for five adjacent waves to form in such a way as not to have any (or even only one) Fibonacci relationships. From the information available, it seems correct to say if there is more than one relationship present in a group of Expanding monowaves (or higher), a pattern other than an Expanding Triangle is probably taking place.*

7. Expanding Triangles cannot take place immediately before any powerful pattern such as a 1st, 3rd or 5th wave extension, or an Elongated c-wave. They cannot terminate Double or Triple or Combination Running Three Corrections. It also cannot be the b-wave of a Zigzag, and will not likely occur as the x-wave of any pattern.

8. In order to draw the trendlines to view the pattern correctly, the typical **b-d trendline** should be employed, with the opposing trendline always drawn across wave-a and wave-c.

I. Limiting

The Expanding **Limiting** Triangle can be in the 4th or b-wave position. If it is a b-wave, it can only be part of a Flat. This Triangle as the b-wave of a Zigzag, the x-wave of a Complex Correction, or the b-wave of a larger Triangle appears to be impossible. Of the two Progress Label possibilities, the b-wave Expanding Triangle seems to be far more common than the 4th wave of the same makeup.

The Triangle's apex point, created by the diverging trendlines, occurs backwards in time. Its position, in relation to the whole Triangle, is important in the verification of it's **Limiting** status. This is how the calculations should be done:

1. Measure the entire time duration of the Expanding Triangle from the beginning of wave-a to the end of wave-e.

2. Take 20% of that amount and add it to the end of wave-a going backwards in time. If the apex point occurs *before* that time period, the Triangle should be considered "limiting" in its *implications*. In other words, it will be a b-wave or 4th wave pattern, and the "thrust" out of the Triangle will be *very* limited. It should be less than the length of the e-wave.

The following are general formation rules which should be present in all variations of **Limiting** Triangles:

1. Waves-a & e will usually relate by 161.8% (wave-e, of course, the larger of the two).
2. Either wave-a or wave-b will be the shortest leg of the Triangle.
3. Only wave-b or wave-d can **"Fail"** to exceed the end of the previous wave.

a. *Horizontal*

These are the distinctive features which make up an **Expanding Limiting Horizontal Triangle** pattern:

1. Wave-a is the smallest segment or the Triangle.
2. Each wave, after wave-a, is slightly larger than the previous, so there is no "failure".
3. The e-wave should be the most violent, complex and time-consuming pattern of the group.
4. The only Fibonacci relationship that appears to be reliable in this variation is between wave-a and wave-e. The e-wave tends to be 161.8% of wave-a. If the e-wave is really explosive, it might relate by 261.8% to wave-a.

b. *Irregular*

An Irregular Expanding Triangle is the most common Expanding Triangle variation. It is differentiated by wave-b "failing" to move beyond the end of wave-a. If the b-wave fails, waves-a & e will likely relate by 161.8%.

c. *Running*

This is the second most likely type of Expanding **Limiting** Triangle. It is characterized by wave-d "failing" to exceed the termination of wave-c. If the d-wave "fails," the pattern will have a slightly skewed upward or downward appearance; also, waves-a & e will probably relate by 261.8%.
[For diagrams of all the above Expanding Triangles, see Chapter 5, page 32.]

II. Non-Limiting

Non-Limiting Expanding Triangles should basically follow the same variational parameters of the Limiting Expanding Triangle. There are only a few differences which need to be covered:

1. The chances are good in a Non-Limiting Triangle that *no* Fibonacci relationships will be present between the various segments. If there is a relationship, it will probably be between wave-a & e. Wave-e would be 261.8% of wave-a.
2. The apex point of the Expanding Non-Limiting Triangle (which occurs backwards in time and can only be detected after the Triangle is complete) should be much closer to the start of the Non-Limiting than of the Limiting Triangle. Listed below are the calculations that need to be preformed to test for Non-Limiting status:
   a. Measure the entire time duration of the Expanding Triangle from the beginning of wave-a to the end of wave-e. Calculate 20% of that amount.
   b. Add that time amount to the a-wave (going backward in time). If the apex point falls within that time zone (between 20% and the start of wave-a), the Triangle is Non-Limiting.

Channeling (unique applications)

The importance of **Channeling** in the decision process is grossly misunderstood or ignored by the majority. It seems that most analysts pay little or no attention to Channeling or treat it as if it were a minor tool of the Wave Theory. **Channeling** ranks among the top, essential considerations of pattern formation. Frequently, from a Channeling standpoint alone, it is possible to decide whether a move is Impulsive or Corrective. It is critical in the confirmation of *when* a move is finished *or* is about to finish. It is extremely helpful in deciding what type of pattern the market is forming and which segment of an Impulse pattern is likely to be the Extended wave. It is essential in the detection of the end of wave-2 and wave-4. Proper use of Channeling can virtually guarantee when the market is creating a Terminal Impulse wave, sometimes well in advance. It can provide reliable clues to when the market is experiencing Triangular activity. From the other side, market action can dictate when an assumed 2-4 or 0-B trendline is real, thus helping to substantiate your interpretations. In Chapter 5, "Central Considerations," some Channeling ideas were covered for waves-2 & 4 of an Impulse pattern. When Channeling an Impulse wave there are additional considerations, those will now be discussed.

Wave-2

If you have detected an Impulsive polywave (or higher) and feel it might be wave-1 of a larger Impulse pattern, the following Channeling rules can be employed *(in Figure 12-1, we are assuming that wave-1 is trending upward)*. After the market has formed a corrective polywave in the opposite direction of wave-1 and has turned back up (Figure 12-1, diagram A), draw a line <u>from</u> point "0" <u>across</u> the lowest point of the downmove which is assumed to be wave-2 (diagram B). As long as the "0-2" trendline is not broken, you can assume wave-2 is over and that it completed right at the point the trendline touched the decline.

Figure 12-1

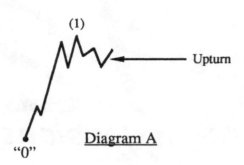

(1)

Upturn

"0"   Diagram A

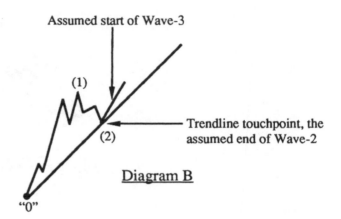

Assumed start of Wave-3

(1)

Trendline touchpoint, the assumed end of Wave-2

(2)

"0"   Diagram B

If the trendline is broken before the supposed 3rd-wave is at least 61.8% of wave-1, or the second decline breaks back below the top of wave-1 at the same time the trendline is broken, then you can be confident wave-2 is still in progress (Diagram C). Why? If an advance, which is moving away from wave-2, does not have the strength to stay above the "original" 0-2 trendline, then it is not Impulsive in nature and **cannot** be part of Impulse wave-3. Once wave-3 is complete, it is allowable for the 4th-wave to break the 0-2 trendline, but not necessary. All previous discussions on proper Impulse wave formation must also be adhered to.

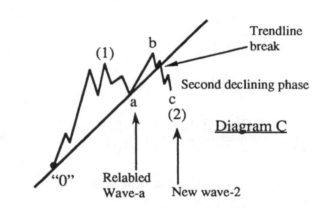

(1)   b   Trendline break

Second declining phase

a   c

(2)   Diagram C

"0"   Relabled Wave-a   New wave-2

Detection of a Running Double Three (a 2nd wave pattern)

After the fact, a **Running Double Three** does not significantly affect the overall interpretation of an Impulse pattern. During their formation, if you do not understand the importance of channeling, you may miss the big move which takes place after a Running Double Three concludes. A Running Double Three is not a difficult pattern to detect. The problem appears to be that most people think Running Corrections do not happen very often. In my experience, Running corrections happen quite often (in Impulsions) and are distinctly different from the (1)-(2), 1-2, i-ii, etc., series for which they are frequently mistaken; channeling is the key to recognizing these Complex formations.

Using the same idea presented in the "wave-2" section, *no part* of wave-3 should break a **true** 0-2 trendline. If the advance, after what you think is wave-2, is followed by a corrective phase which breaks the 0-2 trendline and the advance was not substantial enough to be wave-3, or the advance is actually a corrective rally, the market is experiencing a Running Double Three (see Figure 12-2).

Another *important* reason this could not be interpreted as a 1,2,3,4, with the 5th to follow is the lack of Alternation between the assumed 2nd and 4th-wave.

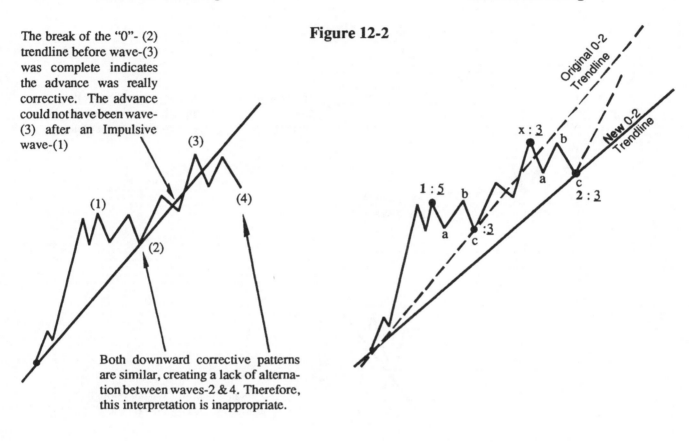

Incorrect Labeling　　　　　　　**Correct Labeling**

Figure 12-2

The break of the "0"- (2) trendline before wave-(3) was complete indicates the advance was really corrective. The advance could not have been wave-(3) after an Impulsive wave-(1)

Both downward corrective patterns are similar, creating a lack of alternation between waves-2 & 4. Therefore, this interpretation is inappropriate.

Wave-4

The illustrations in Figure 12-3 depict the best known type of Elliott trendline, the 2-4 trendline. Similar in concept to the 0-2 trendline, no part of wave-3 or wave-5 should break the 2-4 trendline (see diagram A) *unless* the 5th-wave is a Terminal pattern.

Once the 5th-wave is complete, the market should immediately (within the time-period consumed by wave-5 or less) break the 2-4 trendline and retrace most **or all** of wave-5. Diagram B illustrates what that would look like. If the market does not meet these requirements, the 2-4 trendline is drawn incorrectly (meaning wave-2 and/or wave-4 are not in the position you assumed) **or** the 5th is developing into a Terminal pattern (see Diagram C).

If the assumed 2-4 trendline is broken <u>before</u> the market exceeds the end of wave-3, and a violent reaction does not immediately and significantly move the market away from the end of the last move, the 4th-wave is still developing (see illustration 12-4, Diagram A). If the 2-4 trendline is violently broken *before* wave-3's terminus is exceeded, then a 5th-wave failure is indicated (see Diagram B). For the 5th wave Failure to be proven, the entire Impulse pattern must be completely retraced faster than it took to form. Afterward, the market can drift awhile or continue on its course, away from the 5th wave failure. To exceed the end of the 5th-wave Failure the market would need to consume *at least* twice the time taken by the entire Impulse pattern (1-5); usually, much more time is be required before a new high (or low in a downtrend) takes place.

Figure 12-3

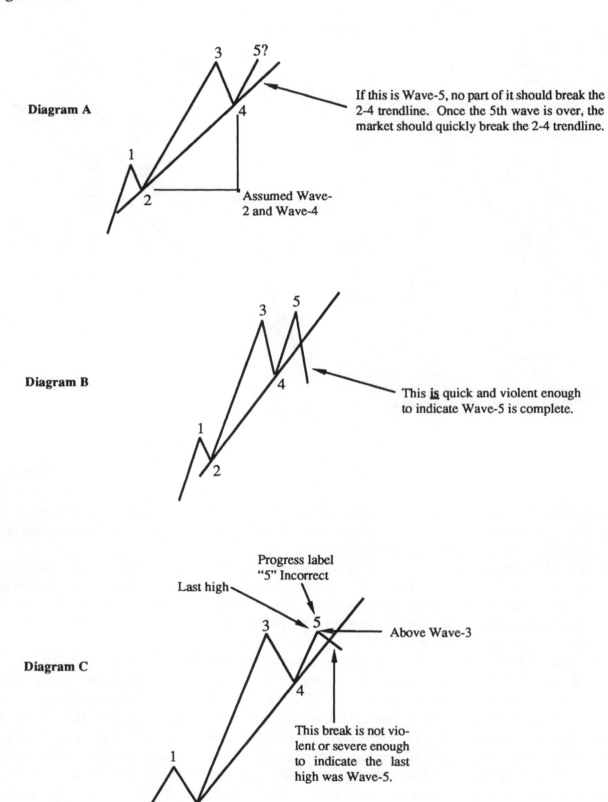

Diagram A

If this is Wave-5, no part of it should break the 2-4 trendline. Once the 5th wave is over, the market should quickly break the 2-4 trendline.

Assumed Wave-2 and Wave-4

Diagram B

This **is** quick and violent enough to indicate Wave-5 is complete.

Diagram C

Progress label "5" Incorrect

Last high

Above Wave-3

This break is not violent or severe enough to indicate the last high was Wave-5.

Figure 12-4a

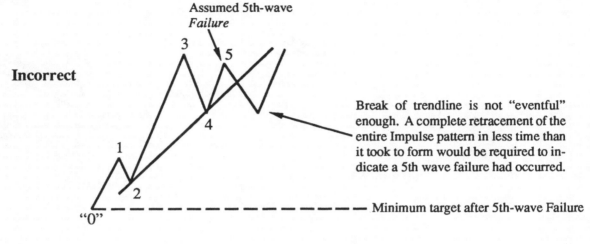

Assumed 5th-wave
Failure

Incorrect

3   5
4
1
2
"0"

Break of trendline is not "eventful" enough. A complete retracement of the entire Impulse pattern in less time than it took to form would be required to indicate a 5th wave failure had occurred.

— — — — — — — — — — Minimum target after 5th-wave Failure

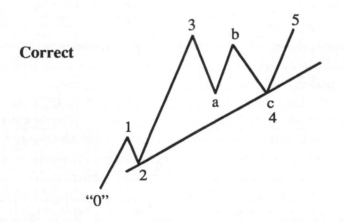

Correct

3   b   5
a   c
1   4
2
"0"

Figure 12-4b

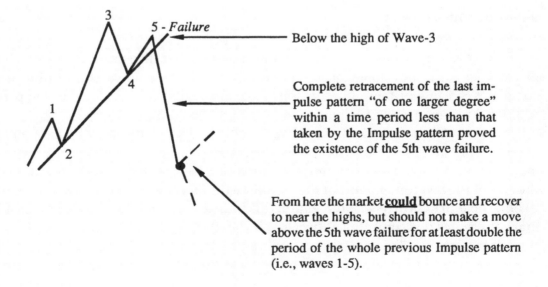

3
5 - *Failure*   Below the high of Wave-3
4
1
2

Complete retracement of the last impulse pattern "of one larger degree" within a time period less than that taken by the Impulse pattern proved the existence of the 5th wave failure.

From here the market **could** bounce and recover to near the highs, but should not make a move above the 5th wave failure for at least double the period of the whole previous Impulse pattern (i.e., waves 1-5).

Figure 12-5

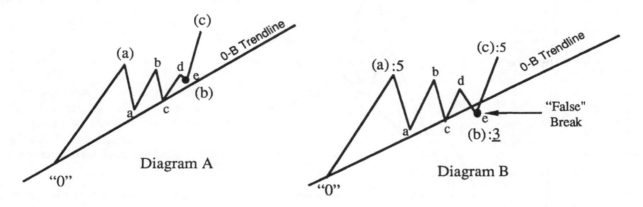

Diagram A

Diagram B

Wave-B

The end of Wave-b is found in a similar fashion to wave-2, with the small difference that the Impulse wave following wave-b (i.e., wave-c) will hardly ever be more than 161.8% of wave-a and the b-wave can sometimes be a Triangle. To better understand this concept, look study Figure 12-5.

If wave-b is a Triangle, there are two ways in which the pattern will channel. In Diagram A, the "0-B" trendline is actually drawn across the c-wave of the Triangle, not the e-wave. To draw the trendline across wave-e would create a break of the trendline by wave-c, an unacceptable situation.

Frequently, if wave-b is a Triangle, after the "0-B' trendline has been drawn across the c-wave of the Triangle, the e-wave will temporarily break that trendline and then reverse (Diagram B). This creates what Elliott called a "false" break. This "false" break should be very short-lived (relatively speaking) and the e-wave <u>must not exceed the end of wave-c</u> of the Triangle. After this "false" break, you may redraw the trendline, but it is not necessary. The larger (c)-wave will not break the 0-B trendline until it is complete (unless the c-wave becomes a Terminal pattern).

Triangular Activity

Through multiple observations, it is often possible to realize that a Triangle in forming even during its early stages of development, sometimes Triangles are evident almost immediately after wave-a completes. A few of these observations have already been discussed. In the paragraphs to follow, additional techniques to anticipate triangular formations will be covered.

Let us assume the market is in a Zigzag rally and you have drawn what you believe to be a 0-B trendline (Figure 12-6). The market starts to advance, but then turns down and breaks the trendline *before* the minimum price and time requirements for wave-c have been met. **If the breakdown does not exceed the end of the last touch point of the trendline (in this case, what you thought was the end of wave-b) and the market turns around again, this indicates triangular development** (Figure 12-7). **If the market creates a second 'false' break, a Triangle is virtually guaranteed** (Figure 12-8).

The patterns in a Triangle frequently act as if relevant, market-created trendlines were non-existent. The corrections within the Triangle break these "trendlines" with the most non-violent, casualness

(Figure. 12-9). This is another important 'early warning signal' of triangular development. A casual, meandering break of an established trendline does not assure a b-wave triangle is forming, but it does virtually guarantee that some type of Triangle of one degree or another is taking place.

Figure 12-6

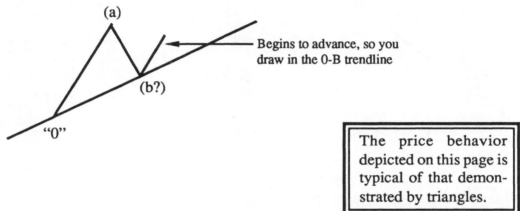

Begins to advance, so you draw in the 0-B trendline

The price behavior depicted on this page is typical of that demonstrated by triangles.

Figure 12-7

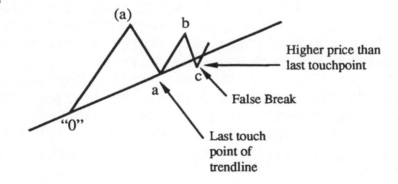

Higher price than last touchpoint

False Break

Last touch point of trendline

Figure 12-8

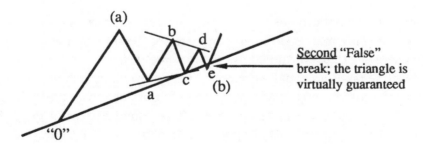

Second "False" break; the triangle is virtually guaranteed

Figure 12-9

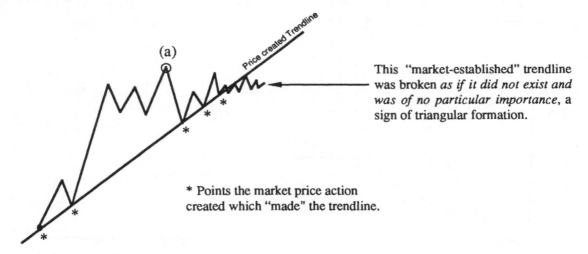

(a)

Price created Trendline

This "market-established" trendline was broken *as if it did not exist and was of no particular importance*, a sign of triangular formation.

* Points the market price action created which "made" the trendline.

Terminal Activity

When using trendlines, the detection of Terminal Impulse activity is very similar to that used in the detection of Triangular behavior. Figure 12-10, Diagram A, shows what usually happens when a Terminal is forming. Observe the break of the 2-4 trendline and notice how the market reacted to it. That is a perfect example of a "non-eventful" trendline break. Notice in this example which wave subdivided, **not** wave-(3), the extension, but wave-(5) which was the Terminal pattern. This is an example of the **Independence** (see Rule page 9-7) of the Subdivision Rule from the Extension Rule. The Extended (3)rd-wave is marked with an "x"; the subdivided (5)th-wave is marked with an "s."

During the formation of a Terminal pattern, part of it will usually break the larger 2-4 trendline. To get only the structure of a Terminal pattern without the early warning of a temporary 2-4 break is rare, but not impossible (see Figure B, 12-10).

Important: The Wave Theory does allow you to speculate when a particular formation (trend) has completed, but the necessary reaction to that formation is quintessential in the verification of your assumptions. In Figure 12-10, the action after wave-(5) must return to the low of wave-(4) to confirm the analysis; if that does not happen, the interpretation is *wrong*. Note: In Figure 12-10, wave-(4) is the minimum target for the drop, 99% of the time it would return to the beginning of wave-(1) or lower.

Real 2-4 Trendline

When the 5th-wave of an Impulse pattern has completed, a correctly drawn 2-4 trendline *must* be broken shortly afterward. "Shortly" is a market defined relative term. To ascertain what relative means to a wave pattern you need to check the time consumed by the 5th wave. If the break from the termination of the 5th wave to the 2-4 trendline takes an equal amount of time or less than that consumed by wave-5, the break adheres to normal behavior thereby confirming wave-5's conclusion. If the break takes more time than wave-5, you must assume wave-5 is developing into a Terminal pattern or the 2-4 trendline is incorrectly drawn and wave-4 is not finished. Maybe your entire interpretation is incorrect. Whatever the circumstances, the market has **not** completed an Impulse sequence (Figure 12-11).

Figure 12-10

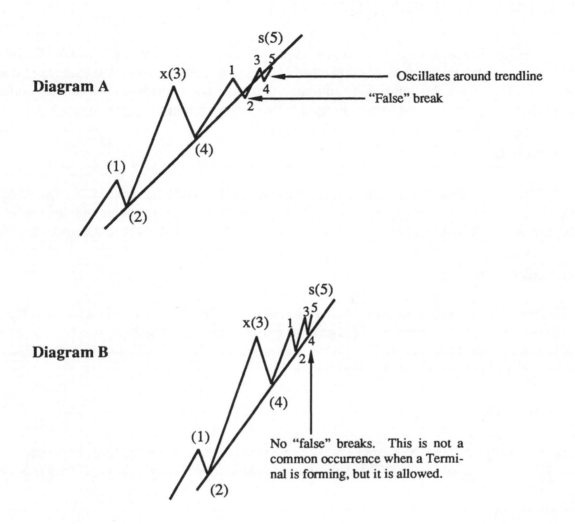

Diagram A

Oscillates around trendline

"False" break

Diagram B

No "false" breaks. This is not a common occurrence when a Terminal is forming, but it is allowed.

Figure 12-11

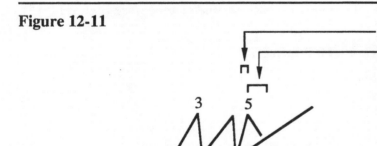

The time consumed by **wave-5** is <u>less</u> than the time taken by the market to return to the 2-4 trendline. If wave-4 & 5 are to be confirmed complete, the move which starts after wave-5 should return to the 2-4 trendline in the same amount of time (or less) as that taken by wave-5. If it takes more time, wave-5 is subdividing **or** wave-4 is not finished.

Impulse Identification with Channeling

With an understanding of the different ways the various Impulse patterns channel, you can frequently use Channeling to decipher which wave of the pattern is going to be the Extension. The earlier you decide which wave is the Extension the better you can trade the trend. The explanations and illustrations to follow will provide you with a clear idea of how each Impulse variation Channels and behaves.

1st Wave Extension

When the first wave Extends in a sequence, the channeling of the pattern should resemble the channeling of a Terminal move (Figure 12-12). The 5th wave should not touch the upper trendline. It will usually stay below it, but on occasion, if wave-2 is very large, it might exceed the upper trendline.

3rd Wave Extension

There are several ways a move with a 3rd wave Extension will channel. No matter which touch points are used to draw the trendlines, they should always be parallel or near parallel (Figure 12-13). Notice, in the diagrams, which of the two corrective phases was more complex (wave-2 or wave-4). Drawn are the typical setups for each 3rd Extension Impulse variation.

5th Wave Extension

There is really only one way to channel a 5th-wave Extension (see Figure 12-14). It is basically the reverse of a 1st wave Extension. The Channeling will tend to expand outward like a "megaphone."

Double Extension

Even though I am uncomfortable with this pattern, due to its extreme infrequency, it is included here for completeness. From experience, this is how a Double Extension should channel (see Figure 12-15).

Correction Identification with Channeling

Flats

To detect which Flat variation is unfolding on you chart, all channel lines need to be *parallel* and *horizontal* with the high and low of wave-a the beginning points. In Figure 12-16a (page 12-13), all Flat patterns have been marked in this fashion. You should study them to understand how the Channeling of the Correction tells you in advance which type of Flat variation is unfolding.

For support and resistance levels in a Flat, the Channel lines should be drawn differently. Draw the Flats' **trendline** across the beginning of wave-a and the end of wave-b, Figure 12-16b (page 12-14). A

Figure 12-12

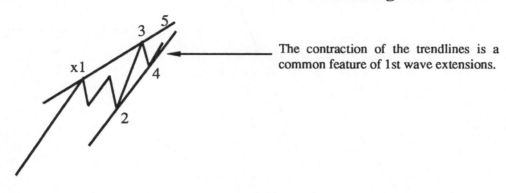

Channeling 1st Extensions

The contraction of the trendlines is a common feature of 1st wave extensions.

Figure 12-13

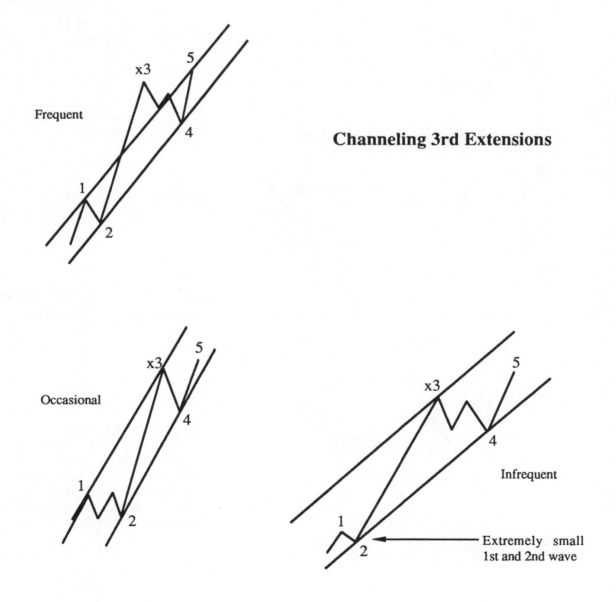

Frequent

Channeling 3rd Extensions

Occasional

Infrequent

Extremely small 1st and 2nd wave

Figure 12-14

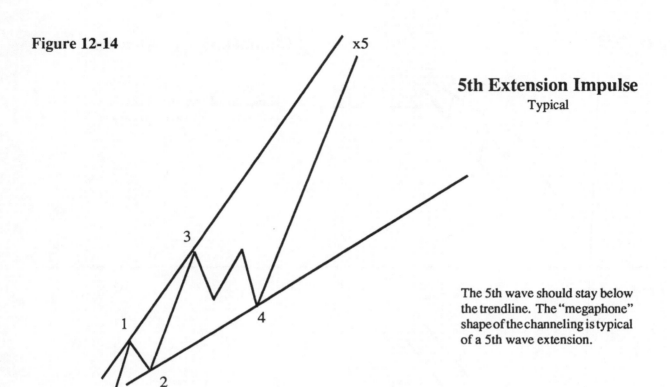

5th Extension Impulse
Typical

The 5th wave should stay below the trendline. The "megaphone" shape of the channeling is typical of a 5th wave extension.

Figure 12-15

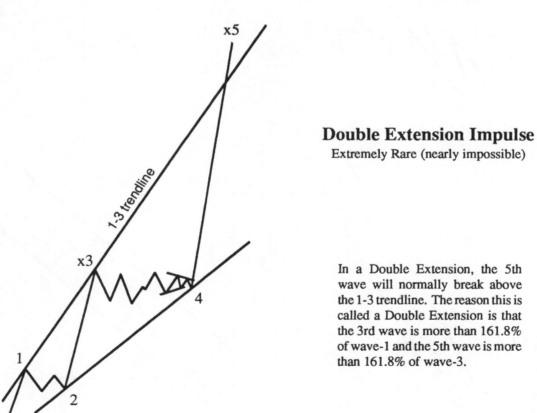

Double Extension Impulse
Extremely Rare (nearly impossible)

In a Double Extension, the 5th wave will normally break above the 1-3 trendline. The reason this is called a Double Extension is that the 3rd wave is more than 161.8% of wave-1 and the 5th wave is more than 161.8% of wave-3.

Figure 12-16a

1.
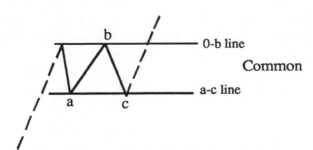
0-b line
a-c line
Common

2.
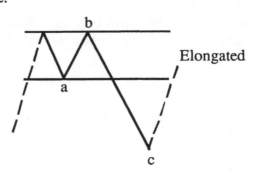
Elongated

3.
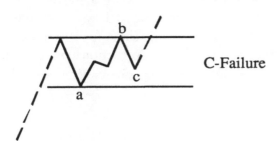
C-Failure

4.
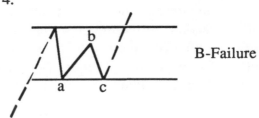
B-Failure

5.
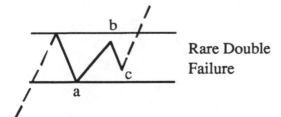
Rare Double Failure

6.
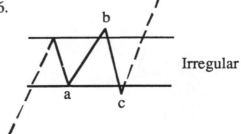
Irregular

7.
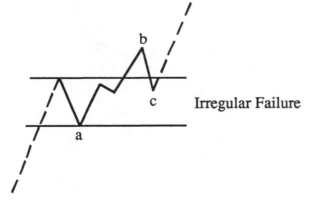
Irregular Failure

8.
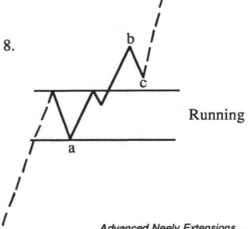
Running

parallel line should be drawn across the end of wave-a to get an idea of the potential support area for wave-c. Even if the b-wave of the Flat makes a substantial move beyond the end of wave-a (see Running Correction in Figure 16-b), you still draw the trendline based on the above rules. Four of the Diagrams from Figure 12-16a have been reproduced below in Figure 12-16b to show how real-time channeling would be administered to the following Flat variations.

Implications of Channeling Flats

The way a Flat channels provides you with subtle clues on the immediate strength or weakness of the market and gives you an idea of how much the market should advance or decline after the correction is over. Most of the clues which can be derived from channeling a correction are dependent on the length of wave-b in relation to wave-a. The larger the b-wave the greater the chances of an explosive move (either up or down) after wave-c completes. The smaller wave-b is in relation to wave-a, the more likely the Flat will either be the first segment of a larger a-b-c **or** the Flat will be followed by an x-wave and another Standard correction. If a Flat channels perfectly (i.e., if wave-c is the same length as wave-a) it will probably be followed by an x-wave and become part of a Complex Correction (see Irregular Failure, Figure 12-16c). Figure 12-16c illustrates the implications of various channeling conditions when working with Flats.

Figure 12-16b

5.   **Double Failure**

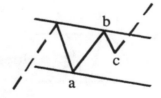

6.   **Irregular**

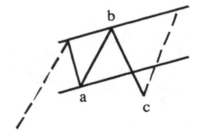

7.   **Irregular Failure**

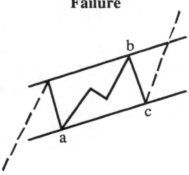

8.   **Running**

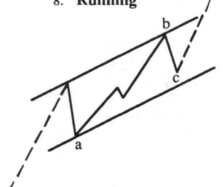

Figure 12-16c

Implications of Channeling

The slightly shorter b-wave (in relation to wave-a) indicates temporary market weakness

5. Double Failure

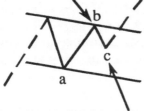

Wave-c's ability to stay comfortably clear of the opposing trendline indicates the weakness initially created by wave-b has been neutralized

The new high for wave-b indicates the power of the uptrend is improving.

6. Irregular

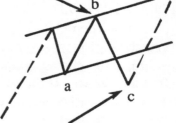

The intial strength implied by the b-waves new high has been negated by the significant break of the opposing trendline

Based on the length of wave-b in relation to wave-a, the market is in a strong position; the next advance should be larger than the advance immediately before wave-a.

7. Irregular Failure

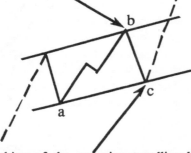

The touching of the opposing trendline by wave-c does not negate any strength indicated by wave-b, but it does indicate this correction may be followed by an x-wave, making the entire correction only part of a larger Complex formation. This concept is explained in further detail under the Zigzag heading (p. 12-16).

This is as powerful as the b-wave in a Standard correction can get. The move to follow this pattern should be much bigger than wave-b and at least 161.8% of the move immediately before wave-a.

8. Running

Wave-c continues to promote the thesis of a powerful trend in effect. It is important the c-wave achieve a price relation to wave-a of at least 61.8%; if it is much less than that, the b-wave is probably not complete.

When wave-b concludes at the same high as wave-a it indicates a relatively neutral market environment. But, if wave-b is also very similar in time to wave-a, an elongated c-wave may occur.

2. Elongated

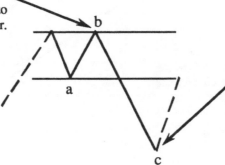

The extraodinarily long c-wave warns that the market is getting weak and that the c-wave will probably not be retraced competely by the next wave of the same degree. An elongated pattern also implies the market is forming a Triangle.

Zigzags

There are not many **Zigzag** varieties, but there are three distinct ways a Zigzag can Channel (see Figure 12-17). Diagrams' A & B show "normal" Zigzag channeling. Either one of the Zigzags in diagram A or B could complete a corrective phase. If the Zigzag channels as in Diagram C, the probabilities highly favor the Zigzag <u>will not</u> conclude the downward correction. It will be part of a complex Double or Triple pattern. If the Zigzag channels as in Diagram C, the wave which immediately follows should **not** retrace all of the Zigzag. If that wave retraces less than 61.8% of the Zigzag, it should be labeled an X-wave. The x-wave will separate two Standard Elliott corrections.

Figure 12-17

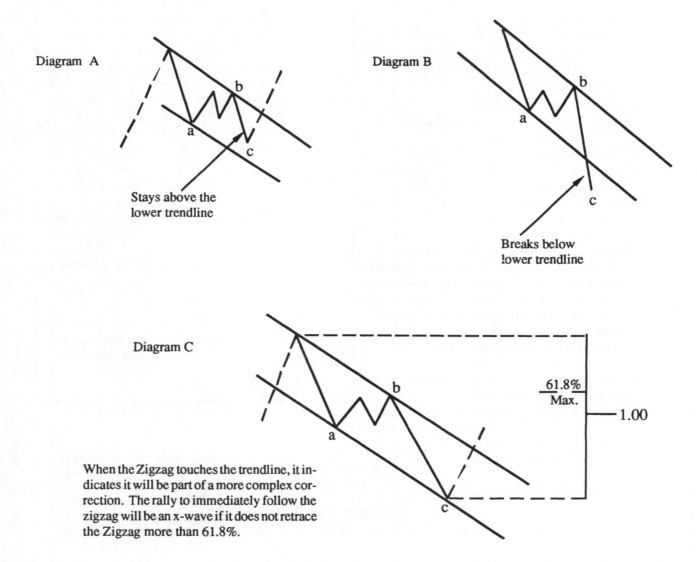

Diagram A

Stays above the lower trendline

Diagram B

Breaks below lower trendline

Diagram C

61.8% Max.

1.00

When the Zigzag touches the trendline, it indicates it will be part of a more complex correction. The rally to immediately follow the zigzag will be an x-wave if it does not retrace the Zigzag more than 61.8%.

Triangles have a base trendline just as Impulse waves do. The base line for a Triangle is the B-D trendline. It should *always* be employed no matter which Triangle variation you are working with. When the trendline gets broken you know the Triangle is over. The trendline on the other side of the Triangle may be drawn three different ways. The most common is the A-C trendline (see Figure 12-18, Diagram A), followed by the C-E trendline (Diagram B). The least common is the A-E trendline (Diagram C).

Figure 12-18

Most Common

Diagram A

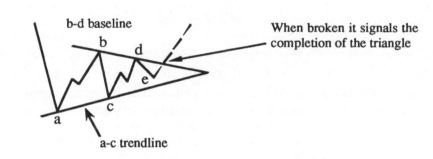

Frequent

Diagram B

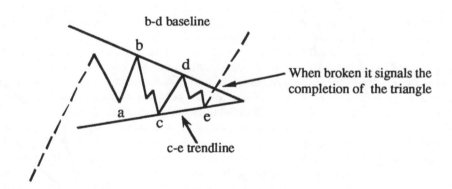

Rare

Diagram C

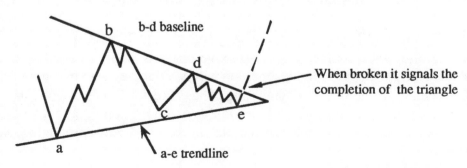

Complex Patterns

All the **Channeling** rules I have developed over the years and the clues which they provide are more reliable in "standard" Elliott patterns. When the market unfolds in a more complex, Non-Standard manner, specific rules are harder to define, but the following I have found to be reliable. Generally, the base trendline will run across only the "0" point and the x-wave(s) *or* the b-waves of the pattern. If the b-waves consume more time than the x-wave(s), the channel will be drawn using the b-waves. If the x-wave(s) is the largest corrective phase, the trendline should be drawn from point "0" across the first x-wave. Below is a list of rules for each category of Complex Correction.

Double & Triple Zigzags

The Double or Triple Zigzag, unlike any other Elliott pattern, can provide an ideal channeling environment. Two parallel lines should contain the entire series of multiple advances and declines (Figure 12-19). This is one of the significant differences between Impulse formations and Double & Triple Zigzags (which are frequently mistaken as Impulse patterns).

Figure 12-19

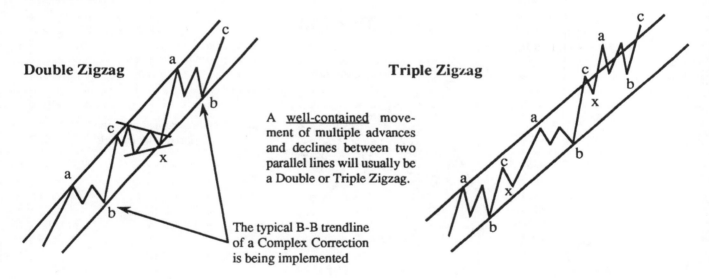

Double Zigzag

Triple Zigzag

A <u>well-contained</u> movement of multiple advances and declines between two parallel lines will usually be a Double or Triple Zigzag.

The typical B-B trendline of a Complex Correction is being implemented

Double & Triple Combinations (which start with Zigzags)

Like Double and Triple Zigzags, usually Double and Triple Combinations will channel within well defined parallel lines, *until* the <u>last</u> corrective phase is nearing completion. Mentioned earlier, most Double and Triple Combinations conclude with a Triangle. As you know from previous discussions, Triangles create problems with Channeling. Therefore, the last few segments of the Triangle, which will conclude the last corrective phase of the Double or Triple Combination, will provide at least one (if not two or more) 'false' breaks of the base parallel line before finally completing. See Figure 12-20 for some examples.

Figure 12-20

Double Combination

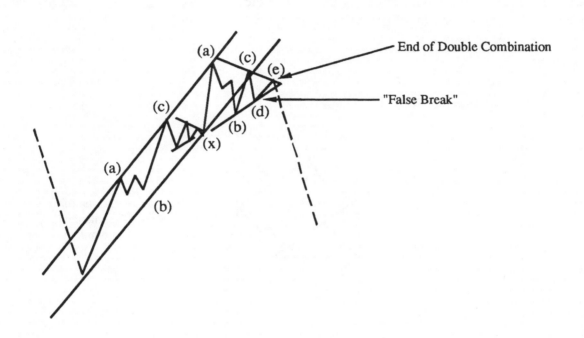

End of Double Combination

"False Break"

Triple Combination

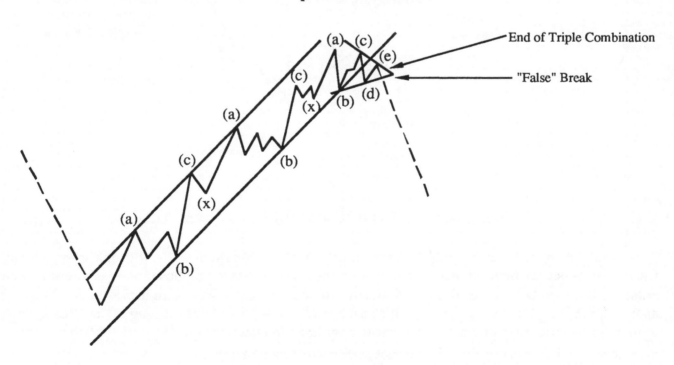

End of Triple Combination

"False" Break

Double &Triple Flats

The x-waves in Double or Triple Flat are almost always *much* smaller than the a-b-c's which they separate, so the Channeling has to be executed using the b-waves of each Flat (see Figure 12-21). When the baseline is broken (if drawn correctly), the pattern should be over. **Note**: A *C-Failure* pattern is a likely candidate for the last Flat of one of these complex formations.

Figure 12-21

Double Flat

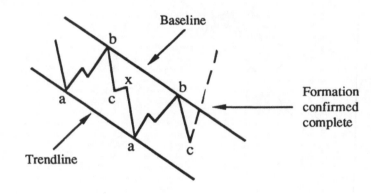

Triple Flat

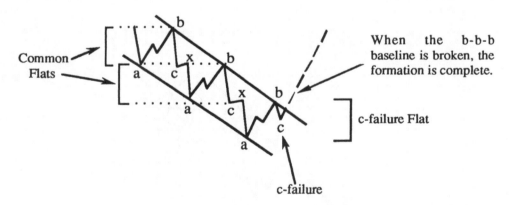

Double & Triple Combinations (which start with Flats)

Because there are many variational possibilities with such patterns, it is difficult to create any Channeling process which will work for all possibilities. The **baseline**, composed of b-wave termination points, will *always* be more reliable than the **trendline** composed of a-wave termination points. The best approach is to continue using the b-wave's baseline as the important channel and realize that a-wave termination points may not create a very clean trendline. In Diagram A of Figure 12-22, the trendline is satisfactory, but in Diagram B it does not perform its function as well.

Figure 12-22

Double Combination

Diagram A

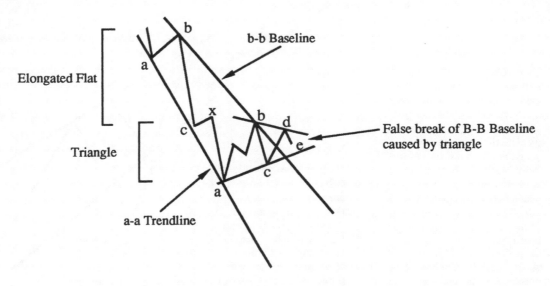

Elongated Flat

Triangle

b-b Baseline

False break of B-B Baseline caused by triangle

a-a Trendline

Triple Combination

Diagram B

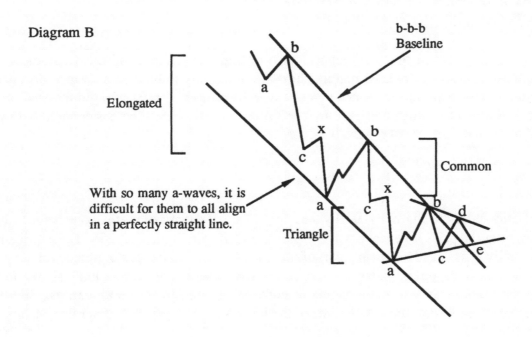

Elongated

With so many a-waves, it is difficult for them to all align in a perfectly straight line.

Triangle

b-b-b Baseline

Common

Identifying the Completion of Waves

When you are having difficulty deciding where each Elliott pattern is beginning or ending, the use of diagonal channel lines will help you identify *exposed points* which can usually be used to formulate reliable Elliot Wave counts. Figure 12-24 gives you an idea of how this technique is implemented.

Figure 12-24

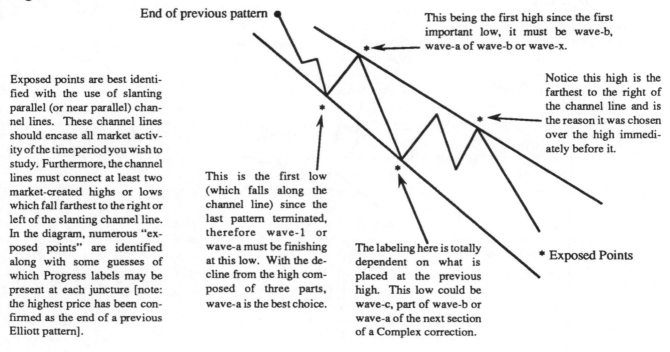

End of previous pattern

This being the first high since the first important low, it must be wave-b, wave-a of wave-b or wave-x.

Notice this high is the farthest to the right of the channel line and is the reason it was chosen over the high immediately before it.

Exposed points are best identified with the use of slanting parallel (or near parallel) channel lines. These channel lines should encase all market activity of the time period you wish to study. Furthermore, the channel lines must connect at least two market-created highs or lows which fall farthest to the right or left of the slanting channel line. In the diagram, numerous "exposed points" are identified along with some guesses of which Progress labels may be present at each juncture [note: the highest price has been confirmed as the end of a previous Elliott pattern].

This is the first low (which falls along the channel line) since the last pattern terminated, therefore wave-1 or wave-a must be finishing at this low. With the decline from the high composed of three parts, wave-a is the best choice.

The labeling here is totally dependent on what is placed at the previous high. This low could be wave-c, part of wave-b or wave-a of the next section of a Complex correction.

* Exposed Points

Advanced Fibonacci Relationships

There are two major categories of Fibonacci relationships. The first, and most common, I have termed **Internal** relationships. Internal relationships are frequently found in standard Elliott patterns. **External**, the other way the market creates Fibonacci relationships, are common in unusual patterns: Non-Standard Complex Corrective and Terminal or Failure patterns. Each type is covered below, with examples, and the places they most regularly occur.

Internal

Internal relationships are characterized by the comparison of the price length of one wave to another with no concern over the price level each wave began or ended (see Figure 12-25). These price ranges will usually partially overlap. This means part of one move will be in the same price level zone as the other. Relationships in Impulse patterns, with a few exceptions, are almost exclusively Internal. Depending on the circumstances, relationships in corrections can fall into both camps. Below is a complete listing of the most common (and some of the less common) ways Internal relationships occur in Impulse patterns.

Figure 12-25

Internal Relationships

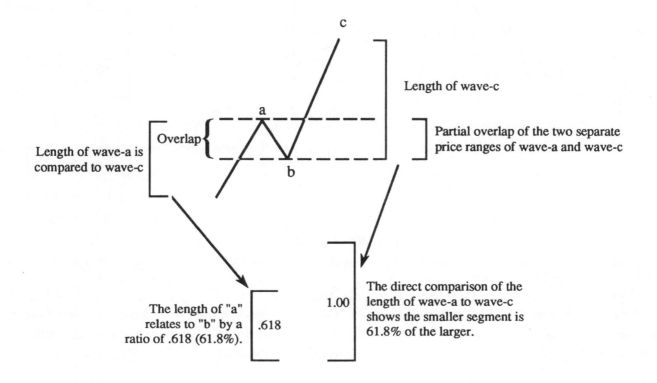

Impulse patterns

1st Wave Extension

When wave-1 is the longest wave in a pattern, wave-3 <u>should not</u> be more than 61.8% of wave-1. When wave-3 is 61.8% of wave-1, wave-5 (applying the rule of Alternation) will usually be 38.2% of wave-3. If wave-3 is 38.2% of wave-1, then wave-5 will usually be 61.8% of wave-3. That is basically all the **Internal** relationships possible in a 1st wave Extension (see Figure 12-26).

3rd Wave Extension

When the 3rd wave extends in a pattern, wave-1 *must not* relate to wave-3 **Internally** by 61.8%. It could be 38.2% of wave-3, but frequently they will not relate Internally at all (study Figure 12-27). When the 3rd wave is the longest, wave-1 and wave-5 should be nearly equal (100%) in price or related by 61.8% *or* 38.2% (ordered from the most to least common relationship). If wave-4 is the complex correction, the 5th-wave should be equal to or greater than wave-1. If wave-2 is the complex correction, wave-5 should be equal to *or* less than wave-1. If wave-1 is extremely small in relation to wave-3, the 5th wave will be 38.2% of the entire distance traveled from the beginning of wave-1 to the top of wave-3 added to the end of wave-4 (refer to Figure 12-27). The 5th wave *must* be less than 61.8% of wave-3 if wave-3 is the Extension.

Figure 12-26

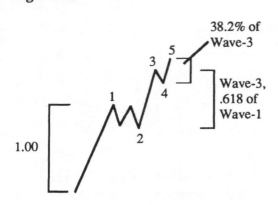

38.2% of Wave-3

Wave-3, .618 of Wave-1

1.00

Figure 12-27

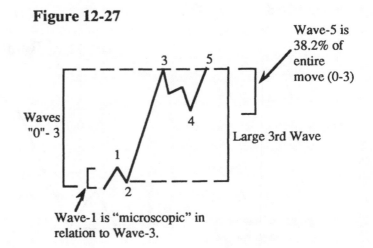

Wave-5 is 38.2% of entire move (0-3)

Waves "0"- 3

Large 3rd Wave

Wave-1 is "microscopic" in relation to Wave-3.

5th Wave Extension

The only possible **Internal** relationship between wave-1 and wave-3, when the 5th wave extends, is that wave-3 will be 161.8% of wave-1. If they do not relate, wave-3 must be *more* than 100%, but *less* than 261.8% of wave-1. If the 5th wave Extension Internally relates to 1 & 3, it will usually be 161.8% of the distance from "the beginning of wave-1 to the end of wave-3" added to the end of wave-4 (see Figure 12-28). Wave-5 *should* be at least 100% of 1+3 *or* 161.8% of wave-3, whichever is shorter. It can be as much as 261.8% of 1+3, but that is about the maximum limit for the length of a 5th-wave extension (refer to Figure 12-28).

Remember, all discussions above are in reference to Internal relationship possibilities. There are other ways that wave-1, 3, and 5 can relate in Impulse patterns, but those relationships fall under the **External** category. It is also possible, under unusual circumstances, for relationships not to occur between certain segments of an Impulsive or Corrective pattern.

Figure 12-28

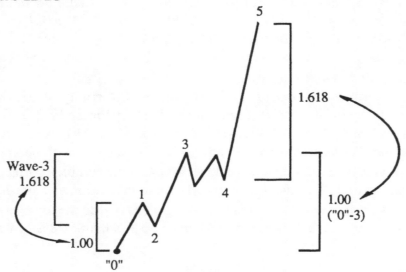

Wave-3 1.618

1.00

"0"

1.618

1.00 ("0"-3)

The 5th-wave Extension could have been as short as 100% of 0-3 or 161.8% of wave-3, whichever is the lessor of the two. The maximum length of wave-5 would be 261.8% of (0-3).

Wave-2 & Wave-4

If wave-2 is the largest price correction of the Impulse pattern, the 4th wave will likely be 61.8% of the price covered by wave-2 (see Figure 12-29, Diagram A). The next choice would be 38.2%. If wave-4 is the larger price correction in the Impulse pattern, wave-2 should be 61.8%, or less likely, 38.2% of wave-4 (refer to Figure 12-29, Diagram B).

Wave-a & Wave-b

Unlike waves 2 &4, waves a & b move in opposite directions. Contrary to the prevailing beliefs of many, Fibonacci relationships are more reliable between patterns traveling in the same direction, **not** opposite directions. As a result, there are no consistently reliable relationships between waves-a & b. The relationship between waves a &b is used mostly in deciding the Structure of the a-wave (see **Retracement Rules** and **Pre-Constructive Rules of Logic**, Chapter 3).

Figure 12-29

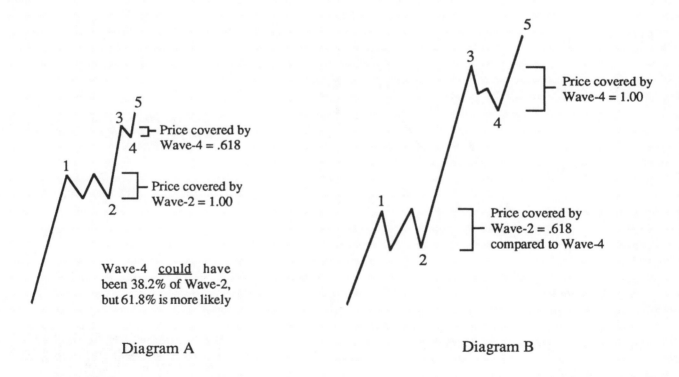

Diagram A

Diagram B

Wave-c (in a Zigzag)

The **Internal** length "limit" of wave-c, in relation to wave-a, is about 161.8% of wave-a (Figure 12-25, page 12-23). If the length of wave-c falls outside of the maximum Internal and External limits (see **External**: c-wave, Zigzags) of its relation to wave-a, then the Zigzag is Elongated and *should* be considered part of a Triangle. It will either be an entire leg of a Triangle or part of a complex pattern which is an entire leg of a Triangle.

The c-wave could also be equal in length to wave-a, which is very common (Figure 12-30, Diagram A). Wave-c should not be less than 61.8% of wave-a, Internally. If so, it is probably due to a very small b-wave Triangle. The Zigzag is probably the entire leg of a Triangle or part of a complex correction which is the leg of a Triangle, Figure 12-30, Diagram B.

Wave-c (in a Flat)

The c-wave in a Flat should not exceed the a-wave by more than 138.2% (see Figure 12-31). If the c-wave is more than 138.2% of wave-a, then it falls into the **Elongated Flat** category. The most common relation between wave-a and wave-c, in Flats, is near equality (refer to Figure 12-32). The next most likely relationship is 61.8%. This can happen in basically two ways, as a c-failure (Figure 12-33, Diagram A) *or* as a b-failure (Diagram B). The minimum **Internal** relationship allowed between wave-a and wave-c is 38.2%. This only happens on rare occasions in what I call a 'severe' failure. This can happen in three general ways. The b-wave retraces most of wave-a (up to, but not exceeding, 100% [see Diagram A of Figure 12-34]). The b-wave retraced wave-a in the area of 81% (Diagram B). The b-wave *must* retrace at least 61.8% of wave-a for this 'severe' failure to still be possible (Diagram C). Of the three, Diagram A is the most common; Diagram B is next in line and Diagram C is very unlikely.

Figure 12-30

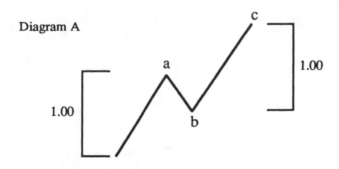

Diagram A

Wave-a and wave-c are equal in this Zigzag; a very common arrangement.

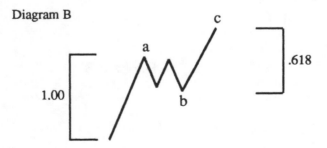

Diagram B

If wave-c were shorter than 61.8% of wave-a, the Zigzag would likely be part of a Triangle. NOTE: You will hardly ever see a Zigzag with a c-wave shorter than 61.8% of wave-a.

Figure 12-31

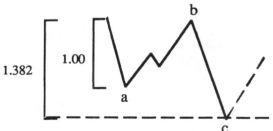

If wave-c is more than 138.2% of wave-a, the market is creating an *Elongated Flat*.

1.382 1.00

a b c

Figure 12-32

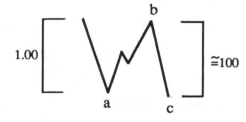

1.00 ≅100

a b c

The most common relationship between wave-a and wave-c of a Flat.

Figure 12-33

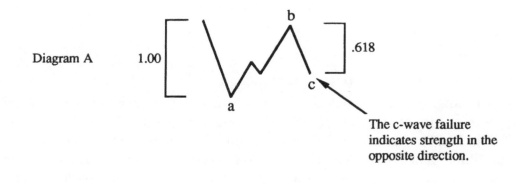

Diagram A   1.00   .618

a b c

The c-wave failure indicates strength in the opposite direction.

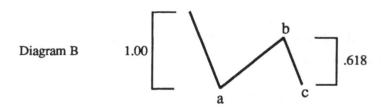

Diagram B   1.00   .618

a b c

Figure 12-34

Diagram A

Diagram B

For wave-c to be only 38.2% of wave-a, wave-b can be no higher than the top of wave-a. If the b-wave is larger than wave-a, the 38.2% of the rally from wave-a's low would all be **part** of wave-b, not the entire correction.

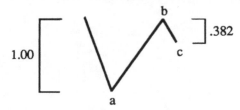

81% of Wave-a

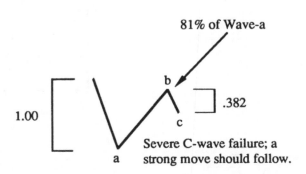

1.00

.382

This would be called a "severe" c-wave Failure. It must produce a rather quick move after wave-c completes.

Severe C-wave failure; a strong move should follow.

Diagram C

The b-wave must retrace at least 61.8% of wave-a.

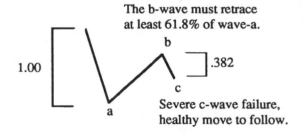

1.00

.382

This pattern is a _Double Failure_.

Severe c-wave failure, healthy move to follow.

Wave-c in Triangles

The c-wave of a **Triangle** usually relates to wave-a by 61.8%, but do not eliminate a Triangle count just because that parameter is not met (Fig. 12-35). If wave-b is bigger than wave-a, wave-c might be 61.8% of wave-b (Fig. 12-36). Remember, Triangles are probably the most common, flexible, varied, and annoying of all Elliott patterns. As long as the market experiences an obvious contraction period composed of five separate segments and there are at least two Fibonacci relationships that occur between the various segments, you have a likely candidate for a Triangle (see Chapter 5 for additional Rules).

Wave-d

Wave-d _only_ occurs in Triangles. It will normally relate to wave-b by 61.8%. Wave-d <u>could</u> relate to virtually any wave in the Triangle by 61.8% or 38.2%.

Wave-e

Wave-e will usually relate to wave-d by 61.8% <u>**or**</u> 38.2%. It could also relate to the larger waves by 38.2%. If it relates to wave-a it would probably be because the b-wave were larger than wave-a.

External

Unlike **Internal** relationships, **External** relationships are based on <u>non-overlapping</u>, but *touching* price ranges. For example, wave-a may be 50 points long and complete at the 500 level. If the c-wave were to relate Externally to wave-a, you would calculate 61.8%, 100%, and 161.8% values of wave-a and add or subtract those amounts from the 500 level (Figure 12-37). Here you <u>are not</u> directly relating the length of wave-a to wave-c; you are measuring wave-a and then subtracting price amounts, based on the Fibonacci ratios, from wave-a's low which relate only to wave-a. In other words, External relations are derived by using one price length and taking Fibonacci ratios in increasing or decreasing scale and adding or subtracting those values directly from a specific price level defined by the termination of one of the waves of the pattern being studied. These levels would be considered potential support or resistance zones. If the market stops and reverses at one of these levels it has confirmed their importance.

Figure 12-35

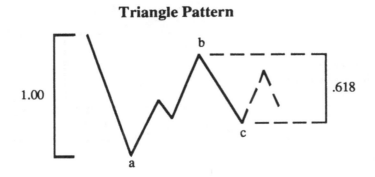

Triangle Pattern

Figure 12-36

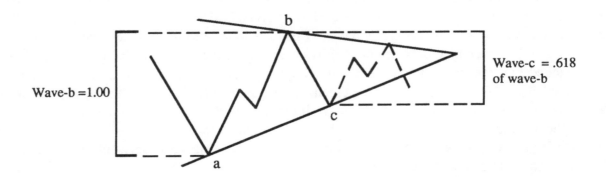

Figure 12-37

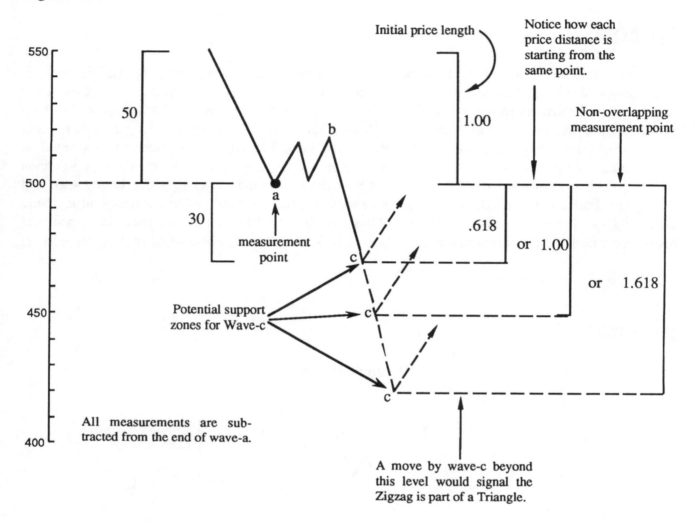

All measurements are subtracted from the end of wave-a.

A move by wave-c beyond this level would signal the Zigzag is part of a Triangle.

Impulsions

1st wave Extension

Since wave-3 should not be much more than 61.8% of extended wave-1, there are not many **External** relationship possibilities for a 1st wave extension. If you take 61.8% of wave-1 in Figure 12-38, and add that amount to the end of wave-1, you have a resistance level for the next advance. If wave-2 retraced any of wave-1 (which is mandatory in Multiwave or higher patterns), the 3rd wave would be too long if it reached that resistance level. The only conclusion is that that is where the entire advance will terminate (Fig. 12-38). If the 1st wave extension is going to form around External relationships, the 61.8% resistance point is the most common. It is possible the entire advance could terminate 38.2% above the 1st wave, but it is not likely and would indicate a fairly high level of market weakness (or strength if the pattern completed downward). A 38.2% External relation would favor that the whole Impulse pattern terminated a larger formation.

3rd Wave Extension

There are no reliable **External** relationships within 3rd wave Extension patterns. Why? The market turning points are *too close* or too *far away* to relate by common Fibonacci ratios (see Figure 12-39).

5th Wave Extension

External relationships are more common than **Internal** relationships when the 5th wave extends. There are two situations where External relations can apply in an Extended 5th wave pattern:

(1) First, the 3rd wave can stop at a point that is 161.8% above the end of wave-1 (Figure 12-40, diagram A). That would be the most likely *if* external relations are involved. The next possibility is 100% above the end of wave-1 (Figure 12-40, Diagram B). If the 3rd wave exceeds the 161.8% level, it will likely be the Extended wave and the 5th wave should be shorter than wave-3.

(2) Second, the 5th wave can stop at a point that is 100%, 161.8% or 261.8% above the top of wave-3 (Figure 12-41). The 161.8% level is the most likely level for the termination of the 5th wave Extension. The 100% level is the next most likely, and the 261.8% level is only likely if the market is terminating a large, long term advance (or decline). Note, in Figure 12-41, at each level there is a comment on the *implications* of each relationship at that point.

Figure 12-38

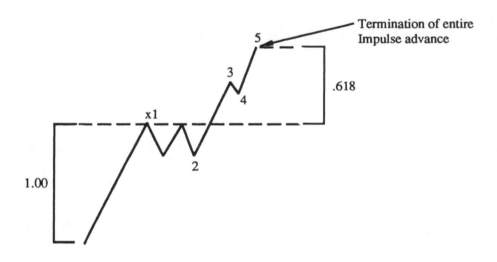

Figure 12-39

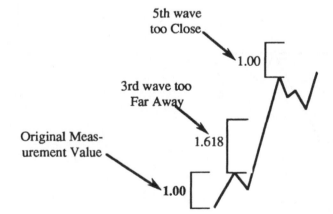

Figure 12-40

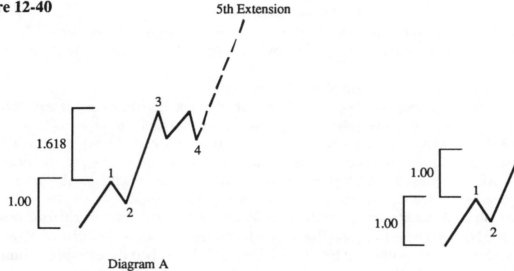

Diagram A

Diagram B

Figure 12-41

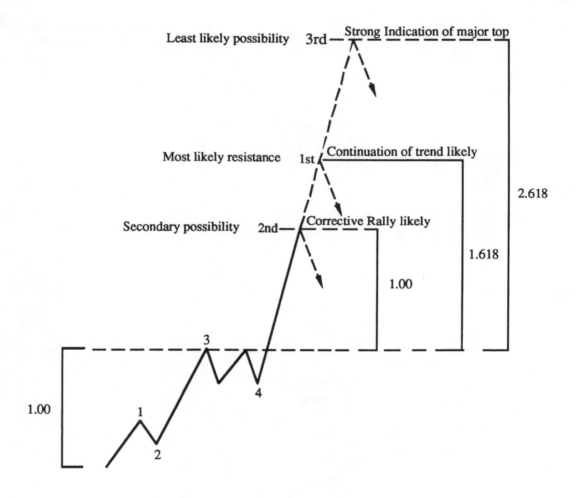

Double & Triple Zigzags and Combinations

Quite often in more complex corrections, after the first **External** support or resistance level has produced a reversal, the market may turn around and break through that level. Almost always, the second push to new price levels will move only 61.8% of the previous break. If there is a third break, it should move 38.2% of the original move. This is what I call the "waterfall effect" (Figure 12-42).

Frequently, the first Zigzag of the complex pattern will have its own External relationships between waves-a & c. Afterward, it is necessary to use the *entire* price value of the first Zigzag as a "yardstick" for the rest of the support or resistance levels. This "entire" value is marked with an asterisk (*) in the diagram below and is next to the <u>larger</u> "1.00."

The **Waterfall Effect** is, by far, most often found in Double Zigzag patterns or complex patterns that start with a Zigzag. To apply the concept to Triple Zigzags and Combinations, work with two groups at a time. In other words, work with the first two corrections as if the pattern were only a Double Zigzag. After completing that study, work with the second two Zigzags in the same way (you will be making use of the middle correction twice). All the same rules should apply as the ones discussed for Figure 12-42.

Figure 12-42

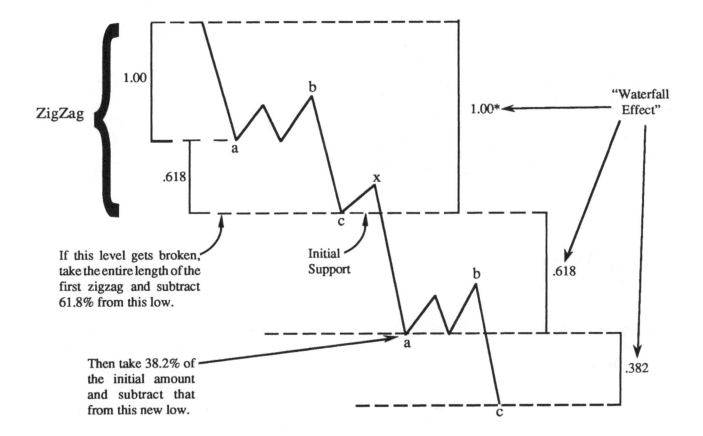

Double & Triple Flats and Combinations

External relationships for Double Flat and Combination patterns will probably not create the **Waterfall Effect** mentioned in the previous paragraph. Usually, one External support or resistance level is all that is needed to stop the market's rise or fall (Figure 12-43). A Triple Flat or Combination pattern will likely exhibit the Waterfall Effect (Figure 12-44).

Figure 12-43

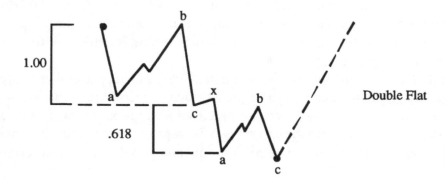

Double Flat

Figure 12-44

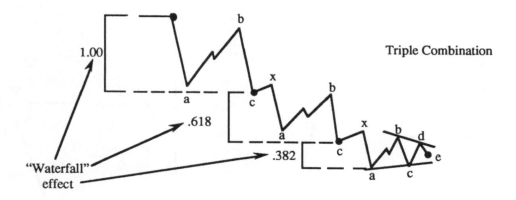

Triple Combination

Missing Waves

When available data creates price action which falls somewhere between the monowave and polywave stage of development, an event <u>never before</u> discussed in Elliott literature comes into play, *"Missing Waves."* This phenomenon, when not understood, can wreak havoc with your short-term assessment of a market's possibilities. When a wave is "missing" it is always a monowave, it is never a polywave or higher pattern. Fortunately, even when a wave is missing from your data, it can be detected indirectly based on the illogical integration of pattern implications and based on the strange development of price behavior inherent with "missing" waves.

Where and when do they occur?

Missing waves occur <u>only</u> on a polywave scale, not multiwave or higher. They are possible in Impulsive polywaves, but are more common in Non-Standard *complex*, corrective polywaves. Standard Elliott corrections **cannot** contain "Missing Waves." Waves will occasionally be *missing* in a formation when the market is preparing to make a <u>significant</u> change in trend.

How do they occur?

All Elliott patterns require a minimum number of monowaves to be correctly formed. An Impulse wave **cannot** be formed with only four monowaves, it *must* contain at least five (5). A correction requires at least three (3) monowaves. As should be clear, to make a specific number of monowaves requires a certain number of data points. When such things as Alternation, Time relationships, Rule of Equality, etc., are considered, it is possible to calculate the number of data points necessary to create a minimally acceptable Elliott *polywave* pattern.

Below is a listing of the minimum number of data points needed for a pattern to have the *chance* of fully developing without a *missing wave*. Possessing the minimum number of data points **does not** eliminate the possibility of a *missing wave* in the pattern, it only decreases the likelihood. Having less than the needed number of data points guarantees there exists a *missing wave* in the pattern (as long as your progress label interpretations are correct). A polywave with *twice* the minimum number of data points required (or more) *should not* be considered a candidate for a *missing wave*. Each one of the listings below includes the starting data point as part of the minimum required to form the pattern listed (Figure 12-45).

POLYWAVE PATTERNS (only):

1. Impulse wave - 8
2. Zigzag - 5
3. Flat - 5
4. Triangle - 8
5. Double Flats & Zigzags - 10
6. Doubles ending with Triangles - 13
7. Triple Flats & Zigzags - 15
8. Triples ending with Triangles - 18

Figure 12-45

Starting Data Point

There are four data points in this diagram. The starting data point is included in the listings above. When you are counting the number of data points which make up a pattern, make sure to count the starting point.

For a pattern to contain a *missing wave*, it must take just the right number of data points to form *most*, but not *all*, of the monowaves needed. What is "*just the right number*"? It is not a specific amount, but a range. *Fifty percent* of the minimum amount required for a formation (or less) basically eliminates the possibility of a *missing wave* and so does a <u>doubling</u> of the minimum amount. Why? At half the minimum, the pattern would be **so** simple it would on register as a monowave <u>or</u> a simpler corrective pattern. *[For example, Double Zigzags will appear as single Zigzags when severe time restraints are placed on their formation; see Figure 12-46]* At twice the minimum level, there would be *too* much detail to allow for a missing wave.

From this it becomes clear that the range of data points which allows for a *missing wave* is half of the minimum *plus one* (+1) or twice the minimum *minus one* (-1). Any number of data points below the minimum (but higher than halfway) indicates a missing wave is a certainty or that your wave count is incorrect. For each additional data point above the minimum, the chances increase geometrically that a *missing wave* will not be present. Once the number of data points reaches twice the minimum level a *missing wave* should be considered impossible.

Figure 12-46

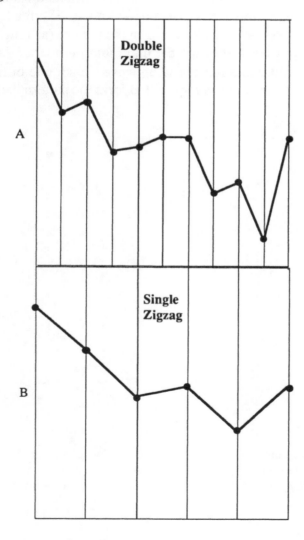

In Diagram A, there are 9 data points which make up a Double Zigzag (the 10th data point confirms the conclusion of the Double Zigzag). If the time element available for the formation of this move were only **half** the currently allotted 10 time units, the Double Zigzag would look like a single Zigzag (see Diagram B).

Working with the same data as in Diagram A, but only plotting every other point, you can clearly see how the detail of a pattern decreases and simplifies. This reduction in the number of data points distorts what actually took place in the market, making the action look like a simple Zigzag instead of a Complex Correction. Fortunately, such a severe reduction in the number of data points generally produces a <u>different</u>, but still identifiable, Elliott pattern. It is when the available data falls somewhere between the two diagrams that a missing wave becomes reality. Even when a wave is missing, it still can be detected due to the strange behavior it creates.

Now that you know *how* they occur, the question comes up, *"Why would a market not have enough data points to create an Elliott pattern?"* All analytical studies must be done within certain limitations. When you want to do some analysis, you have to decide which market to observe, you have to decide when to start you chart, what period of time the chart will cover and what type of data will be used. Will it be yearly data (which may be appropriate for real estate, farming or interest rates). Maybe the study requires shorter-term information such as weekly or daily (which is more appropriate for stock and commodity analysis). No matter what your decision, it will create limits; either on the amount of market action visible for analysis or on your time to analyze.

Whatever period of time you choose to monitor, there is usually shorter-term price data available which would reveal further intricacies of price action. When a pattern on a shorter-term chart completes, there may or may not be enough data points on the next larger chart for the same (but less detailed) pattern to fully develop. When the complexity of a short-term move exceeds the number of data points (time units) available to form the same pattern on a longer term chart, a simplified version of the same move, a monowave or a "missing wave" is unavoidable. An example of a simplified version is a complex pattern turning into a simple pattern (a Double Zigzag into a single Zigzag, see Figure 12-46). Or, if there is too much time disparity between one chart and the next, then the complex pattern on the short-term chart will appear as a monowave on the longer term chart. If the time difference between charts is small, a "missing wave" is virtually inevitable.

For our theoretical example, we have chosen the London Gold am/pm fixes for our "longer-term" data and the continuously traded International Gold Market as our shorter-term data. The London Fixes (the best way to follow Gold, in my opinion) will provide us with two average Gold prices per day. The international Cash market will allow us to monitor the formation of super short-term Elliott patterns.

In Figure 12-47, there are two plots. One is of the am/pm Gold Fix, the other of intraday cash Gold market data. The light line is the **IGM** over several days. The **dot** in the middle of each day and at the end of each day represents the time when the London Fix was released. The **dark line** is the London Gold fixes plotted separately and each connected with a straight line. As you can plainly see, the **IGM** reveals an Impulse pattern over the three-day period. The London data displays a pattern which resembles a near perfect Zigzag; that is how a "missing wave" can occur.

The London Gold plot would force you to assume the market is ready to decline since it appears to be correcting upward. The market action which follows the "Zigzag" will prove that interpretation wrong. If you study the **"Emulation"** section (page 12-38), you should be able, at least in hindsight, to correctly reconstruct the "Zigzag" into a bullish pattern.

Which patterns are susceptible?

The most susceptible patterns to missing waves are *Complex Corrective Polywaves* (corrections that involve x-waves). Virtually without exceptions, it is the x-wave which is "missing" in one of these patterns (see "emulation" for a more detailed description). This is due to the fact that x-waves are almost always the smallest corrective waves of such patterns, so they are the first to disappear as the time element of the Correction is reduced. Much less frequently, Impulsive polywave patterns which have a 1st or 5th wave Extension can "miss" their smaller corrective phase. In a 5th wave Extension that would be wave-2 (as in Figure 12-47). In a 1st wave Extension it would be wave-4.

Figure 12-47

The dark line represents the London Gold Fixes. The lighter line is the continuously traded Cash Gold market. The Cash market reveals an Impulse pattern while the London AM-PM fixes indicate a Zigzag pattern.

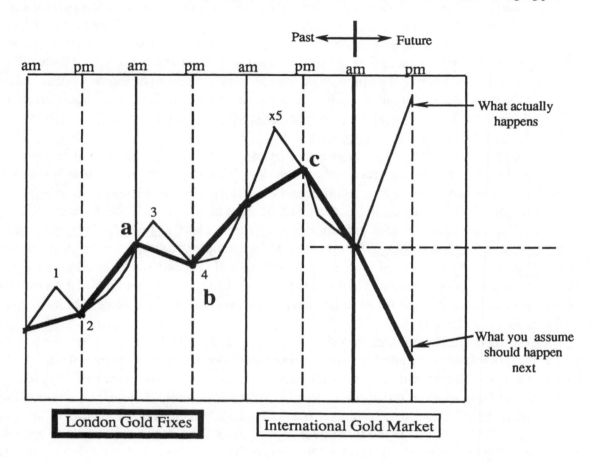

Emulation

Emulation is a tough area. This phenomenon occurs when one pattern (Corrective or Impulsive) is *mimicking* the behavior of a pattern from the opposite **Class**. This obviously can create problems with the analysis process. The two redeeming aspects of this transgression is that it occurs *only* on short-term price action of *polywave* development and almost always provides the astute analyst forewarning clues.

In the early stages of market development, the transition from monowaves to polywaves is, on occasion, unstable. It is one of the few, truly tricky periods you *must* learn to deal with. When one pattern Emulates another, it generally indicates the pattern is "missing a wave." You will have to ascertain when the market is *or* is not revealing all of its behavior to you. The phrase, *"missing wave,"* is used to describe *unseen* market action. It is imperative you understand the missing wave concept before continuing with this section (see "Missing Waves," page 12-34).

Below are numerous patterns that, to the casual observer, would appear to be another Elliott pattern. Despite the *missing waves* which might occur in these patterns, they still have their own recognizable shape and behavior.

Double Failure

Unlike most other "emulations," the Double Failure can only fool you during its formation and for a short time period of time after it is complete. Many of the other "emulations" can keep you off track for some time <u>unless</u> you are closely watching market action. As you can see in Figure 12-48, the Double Failure can temporarily mislead you into the assumption that a Triangle is forming. When the supposed d-wave exceeds the top of wave-a, it indicates the correction is probably over and the market **did not** forming a Triangle. A <u>very</u> minor clue that this pattern is taking place during it formation would come from the application of Fibonacci relationships. In a Triangle, the a & c wave usually relate by 61.8%, **Internally.** In a Double Failure, the c-wave will usually fall at a specific price level important to wave-a from an **External** standpoint (Figure 12-49).

Figure 12-48

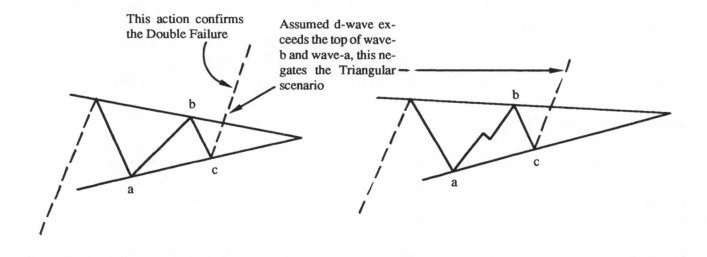

Figure 12-48

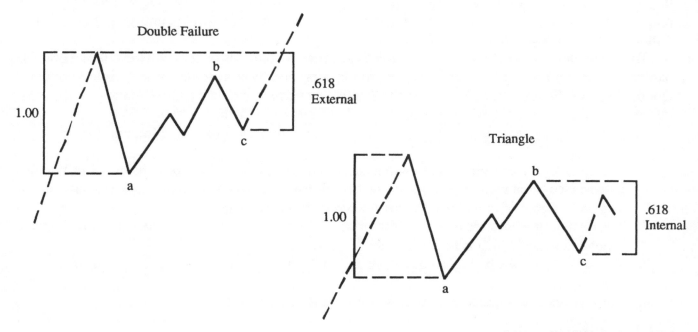

Double Flats

This pattern, if missing the x-wave, could easily be mistaken for an Impulse wave with a 3rd wave extension (see Figure 12-50, Diagram A). On short-term patterns, it is very possible the x-wave will be "missed" since x-waves are usually the smallest corrections in a pattern. There are three major clues the pattern is actually corrective.

First, the supposed 2nd wave (Figure 12-50A) retraced more than 61.8% of wave-1, which indicates wave-1 is more likely corrective (:3) than Impulsive (:5). *[Caveat: it is acceptable for **part** of wave-2 to retrace more than 61.8% of wave-1, but it is not acceptable for wave-2 to terminate beyond the 61.8% level, which it did in the diagram.]*

Second, there is no alternation in time, price or structure between waves 2 & 4 in Figure 12-50A. If there is not alternation of any sort between waves 2 & 4, it does not matter what the pattern looks like; the move is not Impulsive.

Third, attention to Fibonacci relationships could indicate the presumed "3rd extension" is only 161.8% or less of the presumed 1st wave. This, of course, tells you "wave-3" is not the extension and the assumed 5th wave should have been much larger.

If you witness a situation like the one shown in Figure 12-50A (and described above), divide the longest wave (the one that looks like wave-3) in half and assume the halfway point is where a "missing" x-wave occurred (Figure 12-50B). After the pattern terminates, it should be **compacted** to a ":3."

Double & Triple Zigzags

The major difference between the Double Flat and the Double Zigzag is wave-b in the first Zigzag will not retrace more than 61.8% of wave-a. From an **Emulation** standpoint, this makes the Double Flat (missing the x-wave) easier to detect since true 2nd-waves should not terminate at a point that is more than a 61.8% retracement of wave-1. Similar to the Double Flat, the lack of alternation between the assumed waves 2 & 4 would still be present and may be the only early warning the move is actually corrective. Attention to Fibonacci relationships may provide a clue if the presumed "3rd extension" is only 161.8% or less of the presumed 1st wave. This should not happen in a real Impulse pattern unless the 5th wave extends; from the diagram, that is clearly not the case. The next clue will not come until much later when the market retraces the Double Zigzag by more than 61.8% (implying the presumed "Impulse" pattern is actually Corrective).

To the inexperienced Elliottician, the Triple Zigzag or Combination pattern will look like an Impulse pattern. Triple Zigzags, with *missing* x-waves, would not generally be a problem since Triples "emulate" Double Zigzags (Figure 12-52C). Double and Triple Zigzags have nearly the same implications, so mis-interpreting one as the other would not make much difference to the larger trend.

The following are reasons why Figure 12-52A does not make a dependable Impulse pattern:

1. A Triple Zigzag or Combination will usually channel far too well to be considered an Impulse pattern (see "Channeling" - Impulse waves). When a move channels like 12-52A, that is about as good a signal of complex corrective behavior as you can get.
2. The declining waves are too similar in price and time coverage leaving the market without an Extended wave, which is mandatory for Impulsive activity.
3. The action after the presumed 5th wave did not break the 2-4 trendline soon enough (see 2-4 Trendline).
4. Both waves 2 & 4 indicate weakness, breaking the rule of alternation.

It would truly be a dereliction of an analyst's duties to misinterpret a Triple Zigzag which **is not** missing any x-waves (Fig. 12-53). The problem with the proposed count should be obvious; waves-3 & 5 are clearly Zigzag patterns, not Impulsions. That type of development for the odd numbered sections of an Impulse wave is impossible unless the market is formin a Terminal pattern. Due to the Channeling of the move (see Channeling - Terminal patterns), this is without a doubt, not a Terminal pattern.

Figure 12-50

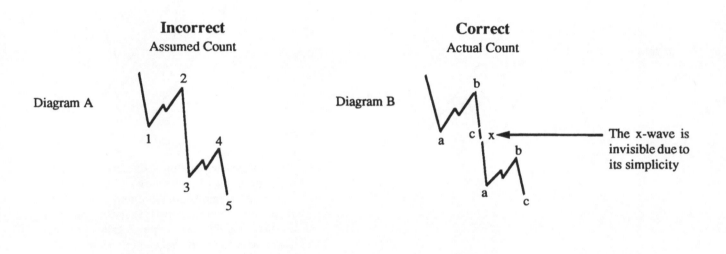

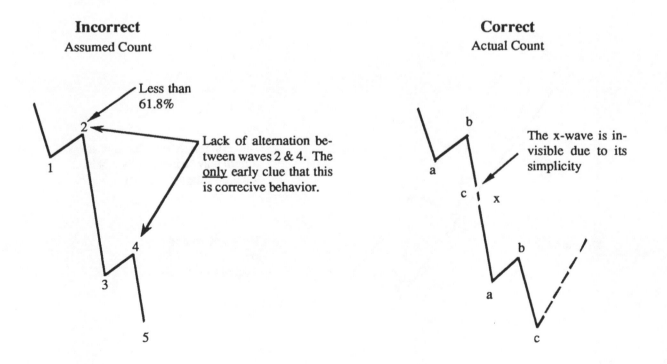

Figure 12-52

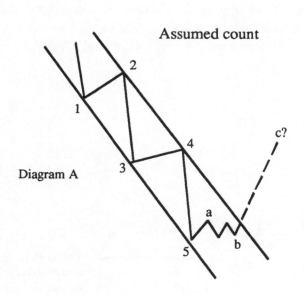

Assumed count

Diagram A

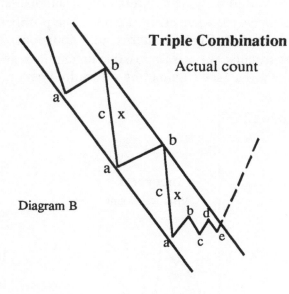

Triple Combination

Actual count

Diagram B

The lack of necessary Impulsive criteria indicates the pattern <u>must</u> be corrective. Each of the X-waves in diagram B should be termed a "missing wave" since there is no visible subdivision for the x-wave. Double and Triple Zigzags and their Combinations are the only corrective formations wqhich closely mimic Impulsive behavior. So, if a pattern "looks" Impulsive, but does not meet important Impulsive parameters, the pattern must be a Complex correction involving X-waves.

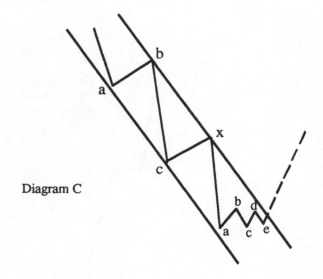

Diagram C

The Triangle is the clue that this pattern may be incorrectly labeled. It has an <u>excessively</u> long a-wave

This same pattern might be mistaken for a Double Combination

Figure 12-53

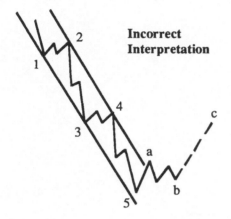

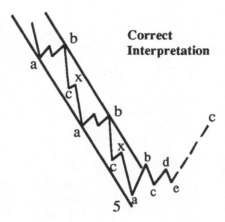

1st Wave Extension

When a wave is *missing* from an Impulse pattern, the resultant price action will almost always look like a Zigzag correction. This makes Impulsions the hardest to detect, especially before they are complete.

When the first wave extends, the 2-wave should almost always be the most complex and time-consuming pattern. Over short periods of time, when the number of data points comprising the polywave is in the <u>range</u> of "missing wave" susceptibility, the 4th wave may not be visible, creating what looks like a Zigzag with a c-wave equal to <u>or</u> shorter than wave-a (Figure 12-54). It cannot be proven this pattern is *missing a wave* until after it is completed.

5th Wave Extension

When the 5th wave extends, the 2nd wave, being the smallest correction, will be the first pattern to disappear during a contraction of the time element. A 5th wave extension with a missing 2nd wave will look like a Zigzag with a c-wave which is equal to <u>or longer</u> than wave-a (Figure 12-55).

Expansion of Possibilities

This is a *completely* new conceptual discovery by the author which, for the first time, helps you overcome the problem of alternate counts. When you are working with price action, there are time periods when the Elliott pattern forming is clear and no <u>reliable</u> alternatives to the interpretation exist. At other times, <u>numerous</u> possibilities exist which can leave many analysts in a quandary.

As remarkable as it may sound, this state of confusion can be exploited to aid you in deciphering the market's position; if not specifically, at least generally. This involves a process I call *"Reverse Logic."*

Figure 12-54

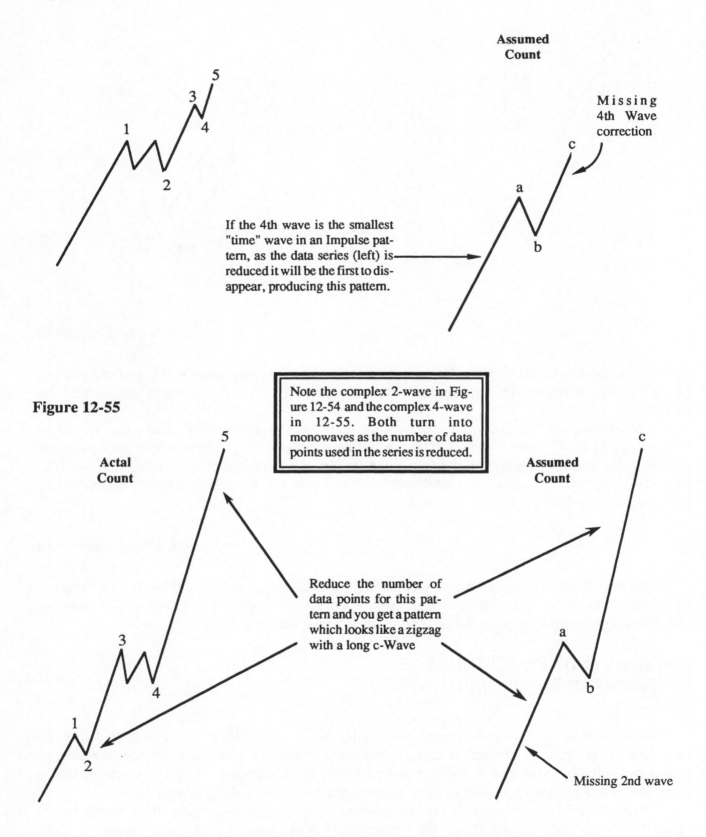

Assumed
Count

Missing
4th Wave
correction

c

a

b

If the 4th wave is the smallest "time" wave in an Impulse pattern, as the data series (left) is reduced it will be the first to disappear, producing this pattern.

1

2

3

4

5

Figure 12-55

Note the complex 2-wave in Figure 12-54 and the complex 4-wave in 12-55. Both turn into monowaves as the number of data points used in the series is reduced.

Actal
Count

Assumed
Count

Reduce the number of data points for this pattern and you get a pattern which looks like a zigzag with a long c-Wave

Missing 2nd wave

5

3

4

1

2

c

a

b

Plainly, the rule states: **When more than one, perfectly acceptable wave interpretation is possible from the same data series, the market *must* be near the center of a Corrective or Impulsive pattern. The more possibilities that exist the closer the market is to the center of a large Elliott formation (wave-b of wave-b; wave-3 of wave-3 or wave-x of a Non-Standard Complex Correction).**

How can the *Reverse Logic Rule* be used to advantage? Whenever the market reaches a point in which careful study reveals numerous possibilities, you should automatically assume the market is toward the middle of a pattern. This knowledge will allow you to eliminate all *alternate* interpretations which indicate a pattern is about to complete. If you keep only the interpretations which indicate the market is in wave-b, wave-3 or wave-x (the middle section of each type of Elliott pattern), usually only one interpretation will remain.

This rule has additional significance and use in trading. If you are waiting to enter a market, but there are numerous interpretations possible, do not trade until the possibilities decrease to only one. Obviously, if there are too many possibilities the market is halfway through its move. Entering at that time would mean more risk and less potential. The only way to take advantage of such a situation is to work with a trend following technique until the count starts to become clear and the possibilities diminish to only one. Another beneficial aspect of this *Rule* occurs when you are already positioned in a market. If you entered the market when only one logical interpretation of the wave pattern is possible and you are currently in profit, do not be scared out when the number of interpretation possibilities increases. That is just a sign that the market has a lot further to go before it can top or bottom out.

Localized Progress Label Changes

If you have been following a market and successfully calling its turning points for awhile, then suddenly something completely unexpected occurs, that is not reason to completely discard your interpretation. A count which has been accurately revealing the path of future price action must have some factors which are correct even if those factors are based only on **Structure**. The most important consideration when making changes to a count is to make them as minor as possible. If minor changes do not work, more significant ones will need to be implemented.

How do you make minor count changes? The vast majority of the time the need for a count adjustment is due to the expansion of a pattern (refer to **Flexibility of Progress Labels**). What you thought was the end of a Correction or Impulse pattern was only wave-a or wave-1 (respectively) of a more complex Correction or Impulse move. This fact provides you with a very simple process of count adjustment. Starting with the very last monowave on your chart, whatever label that monowave has been assigned, drop down to the next smaller Degree title. Using that Title, assume the monowave is only wave-a or wave-1 (whichever is appropriate for the situation) of a larger Correction or Impulse pattern. The same principle would work for segments larger than monowave development.

For example, let us say you observe a large Impulse advance which you assume is the end of a bull market. After correcting, the market makes a new high. The new high indicates the bull market is not over. What do you do about your previous labeling? The high you had marked as the end of wave-5 (of any degree) at the top of the bull market, becomes wave-1 of the same Degree. The degree Title of wave-5 drops down one level. If you had been working with a corrective pattern, what you marked as the end of the Correction would become a lower degree a-wave of a larger corrective pattern.

Appendix

Stock Market Forecast to the Year 2060

by

Glenn Neely

(this forecast was compiled during the late summer of 1988)

This following is a reprint of an article written for the "Foundation for the Study of Cycles" in their September/October 1988 "Elliott Wave" issue. Thank you to Richard Mogey for giving us permission to reproduce the article and Diane Epperson for preparing the camera-ready art work of the original. For more information on the Foundation, call 714-261-7261.

The Future Course of the U.S. Stock Market
An Elliott Wave Viewpoint

by Glenn Neely

THE GROUND WORK: What Does the Theory Furnish the Technician?

The Elliott Wave Theory furnishes a framework within which to organize a market's price action into specific formations over any time period. All market action under the Wave Theory breaks down into two major categories:

1. Action **with** the larger trend.
2. Action **against** the larger trend.

1. Simplistically, price action moving with the trend (of the next larger magnitude) will be constructed differently from that going against the trend. The majority of price movement in the direction of the trend will be constructed of **five** smaller phases or segments (see Figure 1a). Broadly speaking, this type of price action is defined as *Impulsive* (in nature). If a pattern on the largest scale possible is Impulsive, it cannot be completely retraced until at least one more comparable Impulse wave (in the same direction as the first) is completed.

2. Price action that moves in the opposite direction of the next larger trend is usually constructed of **three** smaller segments (see Figure 1b). This type of action is classified as *Corrective* (in nature). When price action is correctively constructed, future price action will usually retrace the Correction completely.

DYNAMIC CONCEPTS

For anyone who has tried, it is usually futile to linearly extrapolate current market or economic action into the future. Ask any corporation how well they predicted their

Glenn Neely established the Elliott Wave Institute in 1983, and teaches the only real-time Elliott Wave course in the world. He is author of ELLIOTT WAVES IN MOTION.

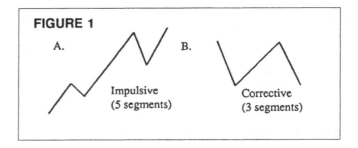

FIGURE 1

A. Impulsive (5 segments)

B. Corrective (3 segments)

own growth and future demand for their products based on previous years' data. Usually their forecasts are significantly different from the final results.

History demonstrates that man's progress is **dynamic and logarithmic**, not static or linear. Look at the historical, logarithmic chart on the U.S. stock market dating back to 1789 (see Figure 2, courtesy Foundation for the Study of Cycles). Sometimes advances occur in spurts, followed by consolidation phases that last for long periods of time. Then again, occasionally the reverse happens. This reveals the market's dynamic behavior. The relatively consistent advance on a log scale for the last 200-plus years demonstrates the logarithmic nature of economic progress.

The Wave Theory supplies the analyst with a forecasting tool that is best implemented on a logarithmic basis. Furthermore, it allows for dynamic development contained within a fractal process. Fractals are part of a relatively new and burgeoning area of science called "Chaos." Chaos deals with the multifarious character of turbulence and the complex geometry of propagation in nature. The association with fractals comes from the Wave Theory's flexibility within a structured format. Repetition of Elliott patterns occurs over and over, each time with a slight variation or new twist. These repetitive patterns can then be combined to create larger patterns of appearance or design similar to the smaller structures. For example, if you examined Figure 1a under a microscope, the long center segment would break down into a pattern which might closely represent the appearance of the entire diagram.

With the introduction of Chaos, science is now able to explain what has, heretofore, been considered random behavior. The numerous similarities between the Wave Theory and Chaos science further validates the applicability of the Wave Theory to the once-thought "random" action of stock and commodity prices.

PRICE ACTION LIMITATIONS

Elliott Wave price patterns force the analyst into specific conclusions. The conclusions are not based on emotion or opinion, but are forced upon the analyst through objective and detailed study. Predictions are derived from the highest probability outcome founded on historical precedent. When applied correctly, the Theory can help the analyst produce short- and long-term forecasts which are, occasionally, pinpoint accurate.

THE ANALYSIS: Implications of the Long-Term Data

A quick overview of the long-term price activity (Figure 2) immediately brings to bear one important fact. The start of the advance, which has been in progress for at least 200 years, cannot be coincident with the beginning of the cur-

rently available long-term data series. Remember, the Wave Theory is a natural law of progression. Its reflections of societal development are present whether someone is around to register them or not. It is only logical to assume that the recording of data would not necessarily coincide with the advent of a multi-century advance.

DETERMINING AN HISTORICAL LOW FOR THE LONG-TERM DATA SERIES

The data we are working with starts in 1789. Obviously, this country was inhabited and growing before that time, so there was economic activity taking place, albeit unrecorded. A quick glance at the start of the data series reveals that the price action was initially drifting sideways for several decades. This is not the way a trend (Impulsive action) begins. The commencement of a trend (under the Wave Theory) must begin with Impulsive action – action that is only minimally retraced by later activity. You can see that the market drifted back and forth many times before finally advancing in the early 1800s. This implies the 20-plus-year consolidation must have been a Corrective phase following an Impulse (trending) pattern.

Methodically administering a host of subtle Elliott Wave techniques, **I deduced that the best point of inception for the last 200-plus year advance was 0.30 (i.e., 30 cents).** The market was most likely at that level around the year 1765 ± 10 years. The following observations helped me to arrive at the above conclusions:

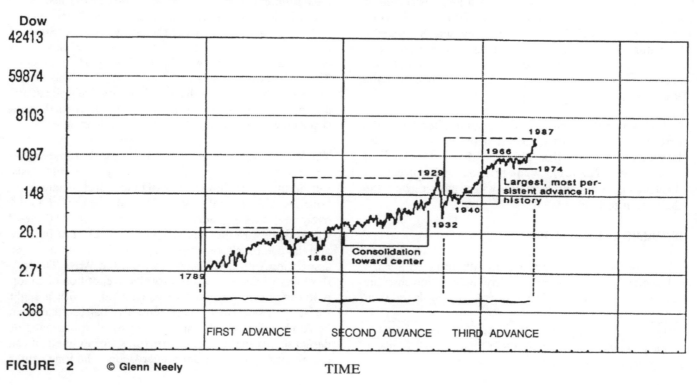

FIGURE 2 © Glenn Neely

TIME

1. For the last 200-plus years, the U.S. economy (based on stock market averages) has been advancing at a great clip (approx. 100,000% since the official projected low around 2.51 in 1789). Based on this evidence, it is safe to assume that the U.S. stock market has been in a trending pattern (Impulsion) since that time *or before*. According to R.N. Elliott, an Impulse wave should contain an Extension (one advance that is *significantly* longer than the other advances). **Since 1789, there has not been an Extended wave** (see Figure 2); the second and third advance are about equal, and the first advance is much smaller. This is a pivotal observation. If the nature of the advance since the 1700s is to be considered Impulsive, there must *eventually* be one advancing wave that is substantially longer than the others. Resting solely on that fact, it is imperative to assume the multi-century economic expansion, probably beginning with the colonization of North America, has not ended.

2. **The large middle section** (second advance) **of the last 200-year advance (1860 to 1929) is Corrective in nature.** Impulse patterns are known for their acceleration phases. These usually occur toward the center of the longest wave. It is immediately evident that the advance from 1860 to the high in 1929 does not exhibit any such acceleration phase toward its center. Actually, the reverse is true; the second advance consolidates toward its center (see Figure 2). This is basically impossible in an Impulse pattern.

3. **From 1940 to 1960, the U.S. experienced the largest, most persistent, least retraced stock market advance in recorded history** (see Figure 2). If a market starts to increase its rate of acceleration after prolonged periods of slower action (relatively speaking), it is an excellent indication that an Extended wave is getting underway. The strongest argument for an extension beginning in 1949 is the gradually increasing market volume over the last forty years as the market continues to make new highs. Volume, as a rule, increases toward the center of an Extended wave, especially if that extension is a third wave.

4. A base trend line can aid in the identification of Corrective patterns of the same degree (i.e., patterns within the same Elliott formation). When employing this technique on the available data (see Figure 4), the trend line runs across the low in 1860 and 1932. **There is no doubt that a Flat pattern (with a C-wave failure), starting in 1835, concluded at the low in 1860.** That Flat pattern contained the typical three segments common to most corrections. The complexity of the Flat correction is far too great to be directly relatable to the virtually vertical decline which started in 1929 and finished in 1932. If the two points identified by the trend line are to have any relation, the Corrective period after 1929 must have been more complex and time consuming than is immediately apparent. **During Elliott's lifetime, he interpreted the price action from 1929 to about 1949 as a 21-year contraction phase** (a Triangle, using Elliott's terminology; see Figure 4). Once a 21-year Triangle is included in the price structure, there is better time corroboration between the two corrective phases

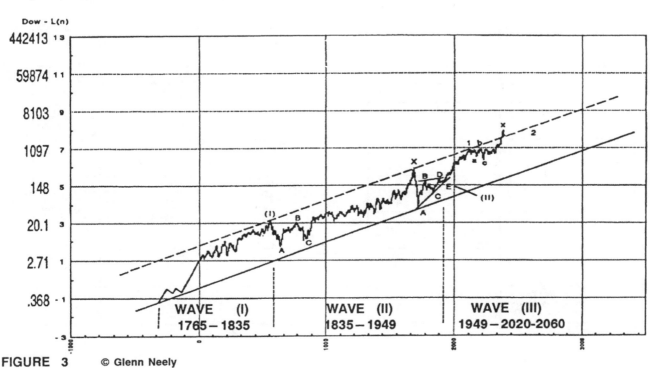

FIGURE 3 © Glenn Neely

along the base trendline (i.e., a 25-year Flat Correction from 1835 to 1860, and a 21-year Triangle).

THE LONG-TERM WAVE COUNT

The previous analysis enabled me to determine that the historical starting point for the U.S. stock market was most likely 30 cents in 1765 (±10 years)! From this it was possible to place a long-term wave count on the historical data series. The results displayed in Figure 3 (p. 219) are:

Wave (I)	1765 to 1835
Wave (II)	1835 to 1949
Wave (III)	1949 to 2020-2060

The following analysis details the placement of the above wave count:

1. **With a starting level of 0.30 (around 1765) established, the advance to 1835 now becomes a very clear Impulse pattern** with a long (extended) first wave (see Figure 4). The 1835 high would be labeled wave (I) of SuperCycle degree.

2. It has already been decided that there was a Flat correction from 1835 to 1860, which was part of SuperCycle wave (II); and that the advance from 1860 to 1929, based on the highest probabilities, was corrective in nature. If that rally had been Impulsive, the low in 1860 would have been the end of wave (II) of SuperCycle degree. Since the advance is Corrective, **the rally from 1860 to 1929 must also be part of SuperCycle wave (II)** (see Figure 4).

When a second wave contains a Corrective advance of that magnitude (1860-1929), the second advance must be some variation of Running Correction. The most common form of Running Correction is the Double Three. A Double Three typically concludes with a Non-Limiting Triangle (*Elliott Waves in Motion,* pp. 5-30, 10-8). This provides a perfect explanation for the 21-year Triangle between 1929 and 1949 that Elliott described. That **Triangle terminated a 115-year corrective phase which finalized SuperCycle wave (II)** (beginning in 1835; see Figure 4).

3. **Wave-3 started in 1949 and is still unfolding.** The powerful implications of a second-wave Running Correction guarantee that **the third wave will be the longest wave of this multi-century advance.** When the third wave is the longest wave of a pattern, it will generally subdivide more than wave-1 or wave-5. As this subdivision occurs, a smaller five-wave move (Impulse pattern) will become evident within the third wave. Inside this larger third wave, wave-1 and 2 will usually mimic the price action of the larger first and second wave. (In this case, the larger first wave begins approximately 1765, and the second wave ends in 1949.) Applying this concept, we can anticipate that the advance from 1949 to some future date will look similar to the price action from 1765 to 1949 (see Figure 4).

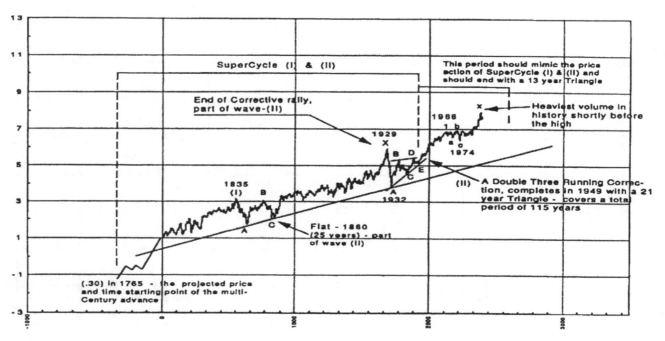

FIGURE 4   © Glenn Neely

From 1949, an Impulse wave is present that concludes at the high in 1966. That high is wave-1 of Cycle degree. The sideward action from 1966 until 1974 would complete a three-phase Flat correction. It was from the 1974 low that the latest "bull market" began in stocks. Once again, **just like the period from 1860 to 1929, the 1980s "bull" market was Corrective in nature** (see Figure 5). In addition, all of the action since the high in 1987 is unfolding into Corrective patterns.

The remarkable parallelism between Supercycle wave (I) and (II) and the latest Cycle wave-1 and wave-2 (Cycle 2 is still unfolding) cannot be discarded. These additional conclusions can be derived from the growing evidence:

1. When measured from the low in 1932 (approximately 55.00), **Cycle wave-1 and SuperCycle wave (I) (ending 1835) relate to each other by 61.8% in price and time.** This development is very interesting when it is considered that the historical prediction for the time and price beginning of the 200-plus year advance was concluded independently of this fact.

2. **The corrections that follow the larger and the smaller first waves are both Flat corrections.**

3. **The two large corrective advances, one from 1860 to 1929 and the other from 1974 to 1987(?), are both Double Zigzag patterns** (see Figure 5). Since the larger advance

from 1860 was a Cycle degree X-wave, it is only logical to assume that the pattern from 1974 to 1987 would be an X-wave of Primary degree (one degree lower than Cycle). As in 1929, **the 1987 high is only part of a larger second wave.**

4. Implementing the 61.8% ratio again, this time to the 21-year Triangle, we can assume that **a 13-year contraction started October 1987** (21 x .618).

5. The maximum extent of the correction, which began in October 1987, is limited to the 1932 percentage retracement of the previous Cycle X-wave (1860-1929). That retracement value was about 50%. Taking 50% of the advance from 1974 to 1987, a level of approximately 1640 is indicated on the Dow. Since the low the day after the 1987 crash was 1706 (very close to 1640), **there is good reason to believe that the 1987 crash low is the corollary to the 1932 bottom.** It cannot be ruled out that the crash low may be exceeded. If so, it should not be broken by more than about 100 Dow points.

6. **Measuring the advance for SuperCycle wave (I), calculating 61.8%, and adding that amount to the end of Super-Cycle wave (I), you arrive at precisely the 1929 top** (see Figure 5). This is a very typical stopping point for an X-wave in a Double-Three Running Correction. Checking the length of Cycle wave-1 (1949 to 1966), multiplying by

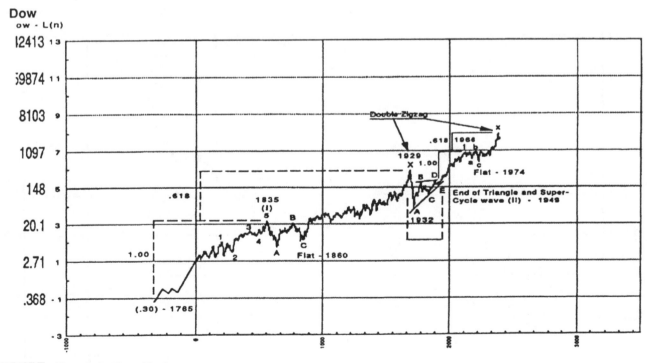

FIGURE 5   © Glenn Neely

the fraction .618, and adding that amount to the top of Cycle 1 puts you exactly at the 1987 high.

The fact that multiple measurements, when applied to wave segments of one lower degree, produce the same results is beyond coincidence. There is no doubt that Cycle wave-1 and wave-2 are, so far, mimicking the action of SuperCycle waves (I) and (II).

CONCLUSIONS

The inherent power of a 115-year "Running Double-Three Correction" (SuperCycle wave (II), 1949) has astounding implications for the future. This allows us to arrive at a resolution on the most probable market and economic activity for the next 70 years:

1. **The economic contraction which started in October 1987 will be of ONE LOWER magnitude in severity than the Depression of 1932.** Why? The Cycle 2-wave currently in progress is of one lower degree than the correction in 1932. As the market moves sideward for the next 13 years, economic conditions should gradually improve as they did from 1932 to 1949.

2. SuperCycle wave (III) is going to be the Extended wave of the Impulse pattern (see Figure 6). In other words, **SuperCycle wave (III) will be much longer than Super-**

Cycle wave (I). This means the U.S. market, right after the turn of this century, should begin to advance again. This advance should last for decades, creating the biggest bull market of all time (see Figure 6). The minimum expectation for an Extended wave is 161.8% of the previous Impulse wave of the same degree. Measuring the length of SuperCycle wave (I), calculating the above ratio, and adding the product to the end of SuperCycle wave (II) (1949) produces the incredible, minimum target of (you'd better sit down) **over 100,000 basis the Dow.** By applying time techniques to the Wave structure, that price level should not be achieved any earlier than the year 2020, and **no later than 2060!**

FUNDAMENTAL JUSTIFICATION AND COMMENTS

The seeds are currently being sown for an international boom of unprecedented proportions. Over the next few decades, the majority of third world countries should have become industrialized. This will bring dramatic improvements in the standard of living for all citizens of such countries. International competition should invoke jealousy and a "keep up with the Joneses" mentality, forcing such communist countries as Russia to move toward a more productive, capitalistic economy.

After the year 2000, the world should begin moving into the most optimistic period ever experienced. This psychology

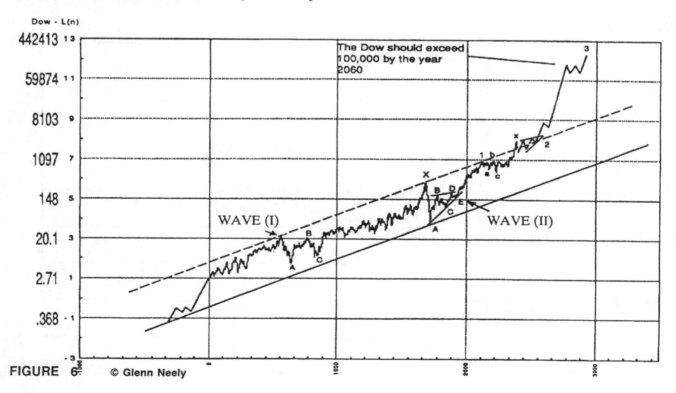

FIGURE 6 © Glenn Neely

should be in full swing by the year 2050. Toward the middle of the next century, there should be a virtual lack of wars, famines, depression, etc.—a sustained period of prosperity and peace.

For those of you who find a 100,000 Dow incomprehensible, consider that the market, from my predicted low of 0.30 up to the 1987 high of approximately 2700, has increased in value almost one million percent in a little over 200 years. A move from 2700 to 100,000 is *only* 4900%. The analysis allows 70 years for this to take place. Thus, a move to 100,000 in 70 years is highly plausible in comparison to the historical record.

I realize the conclusions outlined in this article characterize a near utopian society; but remember, all conclusions are based on what the Wave Theory virtually "guarantees" will be the most incredible economic period of all time. The latter (conclusions) is logically derived from the former (analysis).

Elliott Wave Institute

The Elliott Wave Institute, established in 1983 and the pioneering force in this area of analysis, will continue to pursue its research into pattern formation and the development of new techniques useful to students of the Elliott Wave Theory. The products and services listed below are the most up-to-date results of this endeavor.

Products and Services

WaveWatch - the only source of innovative techniques to assist the Elliott Wave analyst. A virtual gold mine of ideas, WaveWatch takes you through the application of these ideas while describing every thought process necessary to arrive at an appropriate and logical conclusion. All rules are applied to actual market action.

Elliott Wave Telephone Course - The oldest and longest running, REAL-TIME training course available. This unique and exclusive course is conducted **one-on-one** over the telephone to each student. Contact with Mr. Neely is made three times per week for four months allowing the student to experience a multitude of market conditions and trading strategies. Participation in Mr. Neely's class is <u>very</u> limited. Contact the Institute for current availability or future, expected vacancies. Openings are infrequent and may require you be put on a waiting list.

Mastering Elliott Wave - The new, definitive guide to the Elliott Wave Theory. This is the most pioneering work since R.N. Elliott's original discoveries. Not like any other book, it is the **only one** to present a scientific, objective approach to Elliott Wave analysis allowing for confident conclusions again and again even through the most treacherous market environments.

The Art of Trading - our acclaimed two-day Elliott/Gann seminar. The first seminar of this series was conducted at the Embassy Suites Hotel in Los Angeles. It featured detailed discussions of standard Elliott and Gann approaches along with a multitude of new techniques discovered by Mr. Neely. There are plans for a New York and London seminar and another one in Los Angeles. If you have wish to be informed of future engagements, please make sure your current address is on file with the Institute.

Consultation - For those of you who need a little assistance now and then with your wave counts, phone consultation is available on a personal basis with Mr. Neely. Financial arrangements must be made before beginning this service.

To receive more information on the above products and services, call or write the:

Elliott Wave Institute
1278 Glenneyre, Suite 283
Laguna Beach, CA 92651
Office: (714) 497-0949   TeleFax: (714) 497-0983